The Concepts of
Human Evolution

Photo by F. W. Schmidt, Manchester

Grafton Elliot Smith

This Symposium was held to mark the centenary of the
birth of Sir Grafton Elliot Smith.

SYMPOSIA OF THE ZOOLOGICAL SOCIETY OF LONDON

NUMBER 33

The Concepts of Human Evolution

(*The Proceedings of a Symposium organized jointly
by The Anatomical Society of Great Britain and Ireland and
The Zoological Society of London, held at The Zoological Society of London
on 9 and 10 November, 1972*)

Edited by

PROFESSOR LORD ZUCKERMAN

*The Zoological Society of London
University of East Anglia, Norwich, England*

Published for

THE ZOOLOGICAL SOCIETY OF LONDON

BY

ACADEMIC PRESS

1973

ACADEMIC PRESS INC. (LONDON) LTD.

24/28 Oval Road

London NW1

U.S. Edition published by

ACADEMIC PRESS INC.

111 Fifth Avenue,

New York, New York 10003

Library of Congress Catalog Card Number: 73-7036

ISBN: 0-12-613333-6

PRINTED IN GREAT BRITAIN BY

J. W. ARROWSMITH LTD., BRISTOL

CONTRIBUTORS

ASHTON, E. H., *Department of Anatomy, University of Birmingham, Birmingham, England* (p. 71)

DANIEL, G., *St. John's College, Cambridge, England* (p. 407)

DAY, M. H., *Department of Anatomy, St. Thomas's Hospital Medical School, London, England* (p. 29)

DIAMOND, I. T., *Departments of Psychology and Physiology, Duke University, Durham, North Carolina, U.S.A.* (p. 205)

FLINN, R. M., *Department of Anatomy, University of Birmingham, Birmingham, England* (p. 71)

GARNHAM, P. C. C., *Department of Zoology, Imperial College of Science and Technology, London, England* (p. 377)

GOODMAN, M., *Department of Anatomy, Wayne State University, Medical Research Building, Detroit, Michigan, U.S.A.* (p. 339)

LEAKEY, R. E. F., *National Museums of Kenya, P.O. Box 40658, Nairobi, Kenya* (p. 53)

MARTIN, R. D., *Department of Anthropology, University College, London, England* (p. 301)

OXNARD, C. E., *Department of Anatomy, The University of Chicago, Chicago, Illinois, U.S.A.* (p. 255)

POWELL, T. P. S., *Department of Human Anatomy, University of Oxford, Oxford, England* (p. 235)

SPENCE, T. F., *Department of Anatomy, University of Birmingham, Birmingham, England* (p. 71)

WEBSTER, K. E., *Department of Anatomy, University College London, Gower Street, London, England* (p. 169)

WEINER, J. S., *MRC Environmental Physiology Unit, London School of Hygiene and Tropical Medicine, University of London, London, England* (p. 23)

ZUCKERMAN, S., *University of East Anglia, Norwich, England* (pp. 3, 71, 449)

PARTICIPANTS IN THE DISCUSSION

ALLBROOK, D., *Department of Anatomy, University of Western Australia, Perth, Australia, and Department of Anatomy, University of Cambridge, Cambridge, England*

BISHOP, W. W., *Department of Geology, Bedford College, London, England*

FORDE, D.,† *International African Institute, London, England*

FORTES, M., *Department of Social Anthropology, University of Cambridge, Cambridge, England*

GOLDBY, F., *7 Blomfield Road, London W9 1AH, England*

GRIFFITHS, R. K., *Department of Anatomy, University of Birmingham, Edgbaston, Birmingham, England*

HEWETT-EMMETT, D., *Department of Anthropology, University College, London, England*

JAIRAZBHOY, R. A., *37 Hillside Road, Northwood, Middlesex, England*

JOEL, C. E., *39 West Street, Great Gransden, Sandy, Bedfordshire, England*

JOLLY, A., *University of Sussex, Falmer, Brighton, Sussex, England*

JOSEPH, J., *Department of Anatomy, Guy's Hospital Medical School, London, England*

KRAUS, G., *The New Diffusionist Press, 39 West Street, Great Gransden, Sandy, Bedfordshire, England*

LEACH, E., *Provost's Lodge, King's College, Cambridge, England*

MEGAW, J. V. S., *Department of Archaeology, University of Leicester, Leicester, England*

MOLLESON, T. I., *Sub-Department of Anthropology, British Museum (Natural History), London, England*

NAPIER, J., *Department of Anatomy, Birkbeck College, London, England*

PIGGOTT, S., *Department of Archaeology, The University, Edinburgh, Scotland*

RENFREW, A. C., *Department of Archaeology, University of Southampton, Southampton, England*

STRINGER, C. B., *Department of Anatomy, University of Bristol, Bristol, England*

VONDRA, C. F., *Department of Earth Science, Iowa State University, Ames, Iowa, U.S.A.*

WOOD, B. A., *Department of Anatomy, Middlesex Hospital Medical School, London, England*

WORTHINGTON, E. B., *International Biological Programme, 7 Marylebone Road, London, England*

† Deceased.

ORGANIZER AND CHAIRMEN

ORGANIZER

PROFESSOR LORD ZUCKERMAN, *The Zoological Society of London and University of East Anglia, Norwich, England*

CHAIRMEN OF SESSIONS

N. A. BARNICOT, *Department of Anthropology, University College London, London, England*

A. J. E. CAVE, *The Zoological Society of London, London, England*

GLYN DANIEL, *St. John's College, Cambridge, England*

J. Z. YOUNG, *Department of Anatomy, University College London, London, England*

CONTENTS

GRAFTON ELLIOT SMITH: THE SCIENTIST AND THE MAN

Sir Grafton Elliot Smith (1871–1937)

S. ZUCKERMAN

Grafton Elliot Smith and Piltdown

J. S. WEINER

GRAFTON ELLIOT SMITH AND THE FIRST AUSTRALOPITHECINE FOSSILS

Locomotor Features of the Lower Limb in Hominids

M. H. DAY

Australopithecines and Hominines: A Summary on the Evidence from the Early Pleistocene of Eastern Africa

R. E. F. LEAKEY

Some Locomotor Features of the Pelvic Girdle in Primates

S. ZUCKERMAN, E. H. ASHTON, R. M. FLINN, C. E. OXNARD and T. F. SPENCE

COMPARATIVE ANATOMY AND EVOLUTION OF THE FOREBRAIN

Thalamus and Basal Ganglia in Reptiles and Birds

K. E. WEBSTER

The Evolution of the Tectal-Pulvinar System in Mammals: Structural and Behavioural Studies of the Visual System

IRVING T. DIAMOND

PRIMATE SYSTEMATICS AND THE TARSIUS PROBLEM

The Chronicle of Primate Phylogeny Contained in Proteins

MORRIS GOODMAN

Distribution of Malaria Parasites in Primates, Insectivores and Bats

P. C. C. GARNHAM

GRAFTON ELLIOT SMITH: EGYPT AND DIFFUSIONISM

Elliot Smith, Egypt and Diffusionism

Closing Remarks to Symposium

GRAFTON ELLIOT SMITH: THE SCIENTIST AND THE MAN

Chairman: A. J. E. Cave

Symp. zool. Soc. Lond. (1973) No. 33, 3–21

SIR GRAFTON ELLIOT SMITH
1871–1937

S. ZUCKERMAN

University of East Anglia, England

Sir Grafton Elliot Smith died nearly 36 years ago. Not many among those who are here today to commemorate the centenary of his birth knew him, and even fewer worked with him. I share with our Chairman, and with two or three others, the distinction of belonging to a small band of survivors which did both, and which vividly remembers the enormous influence that Elliot Smith wielded in his day. At a time when the experimental method in biology had been all but taken over by physiologists and biochemists, he dominated the world of anatomy, in the same way that Rutherford, his close friend, dominated the world of experimental physics. By a strange coincidence, both men had been born in the same year, both had come to this country from the Antipodes, and both died in the same year.

Whilst this Symposium has been organized jointly with the Anatomical Society of Great Britain and Ireland, it is appropriate that it should be held in the rooms of the Zoological Society of London. As I speak, I can see Elliot Smith in our Old Meeting Room next door, either on his feet addressing the meeting, or alert with interest as someone else discoursed. He was also a member of the venerable dining club associated with this Society, and with which he last dined in April, 1935, on an occasion when Smith Woodward and Tate Regan of the British Museum, R. H. Burne of the Royal College of Surgeons, J. P. Hill of University College, Julian Huxley, and Le Gros Clark were also among the company. I mention this because when, at the end of 1932, Elliot Smith suffered the stroke from which he never fully recovered, he was living nearby in Albert Road. Some two years after his seizure, when he had partly recovered, he sent the Secretary of the Club a note which began with the sentence:

> "Two years ago my chief reason for choosing this house to live in was its proximity to the Zoo and the Zoological Club, but immediately after moving in I was taken ill and that defeated the very purpose of the move."

His letter ended with the request that he be reinstated as a member. These lines tell only too plainly the pleasure he gained from his association with this Society, to which I in turn had the good fortune of being introduced by him.

Elliot Smith was born at Grafton in New South Wales, the son of a schoolmaster, and his interest in the biological sciences began to reveal itself when he was still a schoolboy. He attended the Medical School of the University of Sydney in its early days, when it counted amongst its staff a galaxy of talent which at that time would have been difficult to match almost anywhere else. His undergraduate career was brilliant, and after qualifying in Medicine, he became a demonstrator in the anatomy department. There he began those studies of the brain, and particularly of those of its parts concerned with smell, which immediately made his reputation as a first-class scientist.

After two highly fruitful years of original research in Australia, he moved to England to continue his studies in Cambridge, where he found himself in the intellectual society of men like Gaskell, Langley, Gadow and Shipley, and where his friendship with Rutherford, who had come from New Zealand a year earlier, began. By 1900 he had established a worldwide reputation for his studies of the cerebrum, and in particular of its commissural systems. Soon after he had been elected a Fellow of St John's, he accepted an invitation to the Chair of Anatomy in Cairo.

It had been his intention not to become involved in Egyptology, but soon after he had taken up his new post he was pressed to examine and pronounce upon an enormous collection of Egyptian skeletal material dating from the predynastic era to the XVIIIth dynasty, which had been uncovered in preparation for the raising of the original Aswan dam. It was in this work that he formed his earliest association with Wood Jones, who had recently qualified at the London Hospital, and who had already made his mark as a man of striking originality.

This new anthropological interest, the first development from which was a study of the techniques of mummification, combined with the extension of his comparative neurological studies to the Egyptian brain, and in particular to the visual cortex, turned his attention to the problem of human evolution, on all aspects of which his curiosity had become aroused. About this I shall say more later.

In 1909 he accepted an invitation to the Chair of Anatomy in Manchester, where he remained until 1919. From then until a year before his death in 1937, he was head of the new Institute of Anatomy and Embryology at University College in London, which had been established with the help of the Rockefeller Foundation.

Elliot Smith stood out in the British world of science, and particularly in that of anatomy, from the moment he arrived in Cambridge. Sir Arthur Keith (1950) tells in his autobiography of a letter Lady Keith received in 1906 from Professor J. Symington, the head of the Department of Anatomy of the University of Belfast, during the course of a meeting of the Anatomical Society which was then taking place in that city. In it he said:

> "I think the event of the Meeting has been the rise of Elliot Smith in the estimation of all the anatomists present to the very first place among the men who work at our subject."

Towards the end of their working lives, Keith and Elliot Smith were hardly the best of friends, and it is interesting that the former nonetheless went out of his way to refer to this letter when writing of his own life.

Anatomy in those days was not the subject which we know today. It was far more stereotyped, more descriptive, and less experimental than now. But all the same, there were many distinguished personalities and famous names in the environment which Elliot Smith was to dominate from then on. To name only a few, there were J. P. Hill and J. T. Wilson, old colleagues from Australia; D. J. Cunningham, the author of the voluminous textbook which is still a standard work in the English-speaking world; F. G. Parsons of St Thomas's Hospital, who was to set W. E. Le Gros Clark on the successful course he was to pursue in anatomy; and J. E. S. Fraser, the embryologist at St Mary's, best known for his textbook on bones, which all candidates for the Primary FRCS examination believed they had to master. And there was Arthur Keith himself, whose scientific qualities I usually felt were almost in inverse proportion to his widespread influence and charm. He was a distinguished looking figure who laboured assiduously in physical anthropology, and who had an outstanding ability to charm non-specialist audiences. He was immensely friendly, and the door of his Conservator's room on the ground floor of the Royal College of Surgeons was always open. But I must confess that I found that the diagnostic procedures he employed in matters osteological savoured more of divine inspiration than of normally accepted scientific method.

No doubt Elliot Smith's fame would have been ensured had he devoted himself entirely to comparative neurology, and never turned his mind to man's physical and social evolution. But enduring though his contributions to our knowledge of the brain have proved to be, he is better remembered today for his excursions into the anthropological field, and in particular for his powerful and combative espousal of the diffusionist school of social anthropology, than for his other researches.

Elliot Smith was always a vigorous crusader for his own views, and he never shrank from controversy. J. T. Wilson, who had left Sydney for the Chair of Anatomy in Cambridge, and who remained one of Elliot Smith's close friends and admirers to the end, noted this fact with sadness in the official obituary which he wrote for the Royal Society (Wilson, 1938):

> "One cannot but regret that at times his lucid thought was apt to express itself in a somewhat overforceful and pungent style, so that those to whom he was personally unknown could hardly be expected to discern the thoroughly genial and friendly personality concealed by the trenchant language of the acute and unsparing critic."

I do not quite share J. T. Wilson's regret. Controversy and criticism are necessary parts of the scientific process, and in the context of the present Symposium my own view is that much which has been published as contributions to our knowledge of human ancestry, but which later proved worthless, would have enjoyed a shorter life had there been more Elliot Smiths in the world. Knowledge which is neither challenged nor changed tends to become sterile dogma, and those who sustain it usually encourage and breed mediocrity. Equally, change without critical challenge can become unintended conspiracy, certainly in the anthropological field, with certain anatomists turning their science inside out trying to uphold some false hypothesis—as in the celebrated case of the Piltdown skull. Challenge and scepticism are always called for, as Elliot Smith well knew.

In dealing with the opposition which his diffusionist views inspired, he was wont to say that people are by nature conservative in their attitudes, and resistant to any change of opinion. His Huxley Memorial Lecture of 1928 deals with this theme under the title *Conversion in science* (Elliot Smith, 1928). Pages are also devoted to it in the long introductory chapter to his *Human history* (Elliot Smith, 1930).

> "The inertia of tradition" he wrote, "and the lack of courage to defy it when new evidence fails to conform to it seem to be potent to blind all, except the ablest and most fearless of men, to the most patent facts."

And as he also said:

> "One of the strange ironies of the attainment of human rank is that the acquisition of speech, which opens to Man almost unlimited opportunities for extending the range of his knowledge, at the same time provides him with the means for evading the effort of independent thought by offering him ready-made conventions of speech as well as of customs and ideas. The vast majority of mankind thus accepts without question the guidance of tradition, and by sheer inertia loses the ability to observe or interpret evidence in any sense other than the conventional one that has been instilled into them by custom."

Elliot Smith castigated his intellectual opponents with homilies of this kind. But in labouring his own views he usually failed to see that his message applied as much to himself as to others.

I must, however, immediately add that it would be wrong to say—as many have said—that he shut his eyes to new evidence, or that his mind was closed to new argument. I could illustrate the point with several examples, but let me mention one small story recently told me by Professor A. J. E. Cave. Elliot Smith held the view, based on his comparative studies, that the parieto-occipital fissure of the brain is a consequence of the development of the corpus callosum. The young Cave presented him with two brains from the Leeds Anatomy Department in which, despite the congenital absence of the corpus callosum, the parieto-occipital tissue was conspicuously developed. When shown these specimens Elliot Smith—to quote Professor Cave's own words— "grinned and grunted, 'Well I boobed there all right'." But I also remember another occasion when, at a meeting of the Anatomical Society, he reacted sharply to the same young anatomist when he had the temerity to question Elliot Smith's peculiar views about the cause of thinning of the parietal bones in ancient Egyptian crania.

Elliot Smith's anthropological interests I would classify under four main heads: first, his general interest in the theory of evolution; second, his concern to understand the unique selective conditions which determined the emergence of man within the Order Primates; third, his interest in what he called the human pedigree, as determined by the study of fossil remains; and finally, and perhaps most important, his interest in the growth and diffusion of culture. All four of these interests are likely to be touched on in this Symposium.

There is no need for me to say much about those of Elliot Smith's writings in which he discussed the general theory of evolution, or in which he spoke about the importance of selection, although it should be said that in some passages in which he dealt with these matters, his writing was, by modern standards, frequently loose. For example, it would not be difficult to pick holes in the following general proposition:

"Studying a series of different mammals belonging to any given natural Order, it is found that those which wander away from their original home and become subjected to new environments and new conditions of life, new food to search for, and new dangers to overcome, are more rapidly transformed than those that stay at home."

This kind of writing (Elliot Smith, 1930) dates well before the emergence of knowledge about population genetics, theories of gene drift, and so on.

On the other hand, no one could complain about the precision of his exposition when he speculated about the factors which may have been responsible for the cerebral development which marked the evolution of mammals in general, and of primates in particular. Here his argument is spelt out in several writings, but in greatest detail in *The evolution of man,* first published in 1924 (Elliot Smith, 1924).

In the first of these essays Elliot Smith expounds the basic proposition that man's structural and other characteristics indicate a close kinship with the great apes, and that the latter in turn are more closely allied to the catarrhine primates of the Old World, than to the platyrrhine primates of the New World. As a further necessary proposition, he also accepted that all primates derive from some primitive stock, which also gave origin to the Spectral tarsier of Borneo and Java.

Elliot Smith then turned to the great development of the brain, which distinguishes mammals from all other vertebrates. This was due mainly to the growth of a neopallium, a term which Elliot Smith had himself coined as far back as 1901. This structure constitutes a "unifying organ" which receives information through visual, acoustic and tactile pathways. It is also characterized by a motor area, for controlling the voluntary responses of the body to the sensory inputs which the brain receives.

The "new breed of creatures" which the early mammals constituted rapidly exploited every mode of life, ranging from the flight of the bat to the aquatic life of the seal. Most mammalian groups became committed to one particular kind of life, and by so doing sacrificed both their primitive simplicity and an ability to adapt to new conditions. Only the few which retained their primitive mammalian characteristics—the Order Primates in short—kept open an evolutionary path which tens and tens of millions of years later was to lead to the emergence of Man. These primitive mammals were the small creatures which had adapted to an arboreal form of life, and from which the tree shrews and tarsiers also evolved.

Animals tied to the ground, Elliot Smith argued, are ruled by the sense of smell. On the other hand in arboreal mammals the visual, tactile and acoustic sensory pathways become dominant, leading to a considerable growth of the neopallium, and to a reduction of the more primitive olfactory structures of the brain. Binocular and stereoscopic vision, he contended, are also particularly valuable in an arboreal environment, and in the course of primate evolution were therefore "selected", as was also a capacity for skilled movements of the hands and fingers. Hence, Elliot Smith argued, the variety of primitive features one finds in the human hand, as opposed to its specialization in a hooved or aquatic mammal. The two other changes which he emphasized as being important in the course of man's evolution were the adoption of the erect attitude by some protohuman or animal precursor, and the development of speech, a development which I do not believe he ever spelt out in the detail which it deserved.

Much of Elliot Smith's analysis constitutes speculation. But it is speculation which still provides as useful a framework as any other by which to explain the steps whereby man attained his pre-eminent position in the animal world. It was in the development of this general theme, which Elliot Smith had first sketched in an address on the evolution of man which he delivered 60 years ago to the 1912 meeting of the British Association for the Advancement of Science (Elliot Smith, 1913), that he and F. Wood Jones, the prodigy who was his one-time protegé, parted company.

In 1912 Elliot Smith had been given some specimens of *Tarsius* by his friend, Dr Charles Hose, who was then working in Borneo. Elliot Smith then asked Wood Jones to study the animal's reproductive system. Six years later Wood Jones published his classic work, *Arboreal man* (Wood Jones, 1918a), in which, as Elliot Smith put it, his former assistant not only amplified the basic thesis of which he, Elliot Smith, was the author, but had introduced certain modifications of his own. Later that year—I am again quoting Elliot Smith (1924)—Wood Jones

"emancipated himself more definitely from any influence my views may previously have had, and issued his remarkable speculations in the form of a brochure called *The problem of man's ancestry* (Wood Jones, 1918b), in which he attempted to exclude all the Apes and Monkeys from Man's ancestry and to derive the Human Family directly from a *Tarsius*-like animal".

Not long afterwards, in 1919, there took place in the old Meeting Room of the Zoological Society, now part of our Library, a discussion on the "Zoological position and affinities of *Tarsius*" (Woodward *et al.*, 1919). This discussion was a milestone in the history of primate studies and was instigated by Wood Jones himself. Many of the "heavy-weights" of the period participated.

Smith Woodward, the British Museum's leading palaeontologist, spoke first, and was followed by Elliot Smith, who was perhaps unnecessarily scathing in his remarks, although the printed record does not indicate that he ever referred to Wood Jones by name. Then came J. P. Hill, whose criticisms were levelled on embryological grounds. Wood Jones "batted" fourth, and no doubt out of respect to those of his elders and betters who had already spoken, put his case with great moderation, but without qualifying his view that *Tarsius* was more closely related to man than the latter was to the anthropoid apes.

Mr R. I. Pocock, the distinguished zoologist, then the Society's Curator of Mammals, spoke next, but again in opposition to Wood Jones. He was followed by Chalmers Mitchell, Secretary of the Zoological Society, who thanked Wood Jones for initiating the debate, at the same time remarking that Wood Jones had been "more reticent" in the discussion than he usually was in addresses to more popular audiences, where "he attacked Darwinian evolutionists and Huxley in particular". The discussion was summed up by Professor E. W. MacBride who again added to injury by remarking that Wood Jones had "rather shrunk during the discussion" from a view "to which he had liberally committed himself in recent books published by him", namely that *Tarsius* and man agreed in retaining some primitive characteristics which monkeys and apes had lost.

One or two of us taking part in this Symposium knew every participant in that symposium of 50 years ago—although I doubt if any of us was old enough to have been present. It must have been a remarkable occasion because of the polemics. And it is equally remarkable because, in retrospect, Wood Jones was insisting that *Tarsius* was what all to-day accept it to be—a true primate—and not just some aberrant lemur.

Elliot Smith's third preoccupation with human evolution was the elucidation of man's pedigree as revealed by the fossil record. Here he hardly displayed any greater critical faculty than did other contemporary anatomists, zoologists and palaeontologists who were concerned with the problem. He had, however, learnt from his studies of Egyptian skeletal material how unreliable are some of the osteological features which are used to diagnose the sex of human remains, and to that extent he was wary of hasty judgements based upon anatomical characters. He took part in the celebrated debate about the Piltdown skull—a debate in which, as we now know, no one distinguished himself. From the point of view of anatomical history, it can only be seen as a protracted charade in which famous scientists pitted opposing views about an assemblage of bony fragments which someone had apparently brought together in order to deceive (Weiner, 1955). Arthur Keith changed sides in this debate, and also fatally crossed swords with Elliot Smith. Keith got it into his head that Smith Woodward, who had been responsible for the original reconstruction of the skull, had fitted the bony fragments together incorrectly. Having failed in his attempt to persuade his anatomical and zoological colleagues that this was so, he carried the dispute into the open, and finally attacked Elliot Smith at a meeting of the Royal Society when the latter was discussing the significance of the endocranial cast of the skull as reconstructed by Smith Woodward.

"It was a crowded meeting," Keith (1950) tells us in his autobiography, "and it so happened that he [Elliot Smith] and I filed out side by side. I shall never forget the angry look he gave me. Such was the end of a long friendship". Keith then goes on to say that even though he knew himself to be right, "the Royal Society looked on me as a brawler, and continued to frown on me for my outspoken criticism".

I cannot visualize Keith as a brawler. In person he was altogether too mild and kind. But equally, as I have already said, I never was able to regard him as more than a superficial scientist—whatever his influence in the anatomical world of his period. He would hardly have survived in the critical environment of today! In retrospect I even find it difficult to see how he sometimes got by in the more mellow days of his own period. For example, in his autobiography he also tells us about a "brush" which he had with Karl Pearson, then the country's foremost biometrician. Keith had published a paper in which, to quote his own words, he had "made the usual assumption—namely, that men and women with big brains are usually of greater ability than those with smaller heads and brains". Where such an assumption could have come from it is difficult to see, and Pearson set out to show that it was groundless. As Keith puts it: "he applied his biometrical engine to Cambridge students, and found that those who just scraped through their examinations had as big heads as the men of Tripos I". To which Keith adds, "there was a flaw, I thought and still think, in Pearson's reasoning. The brain is much more than an organ for passing examinations: it is the organ of feeling and of doing—functions which are difficult to measure". It is difficult to imagine what this has got to do with the size of the brain.

But Keith was always one for sweeping generalizations. Some may remember his hormone theory of the differentiation of racial types, a theory which he put forward in all seriousness, in a period when the nature of no single hormone was understood. He never did understand why Elliot Smith refused to hold with this quite empty generalization! Some will also remember Keith's description of war as "Nature's pruning fork". Here again, he was bitterly disappointed when Elliot Smith, like many others, criticized him publicly for his trivial simplification. But Keith was only one of many well-known biologists of those days who were always searching for some sweeping generalization—too often generalizations which could not be tested scientifically, and which were, in consequence, intellectually sterile.

I have digressed somewhat in talking about the differences Keith had with Elliot Smith, beginning with their clash over the Piltdown skull. Fossil skulls, of course, were Keith's stock in what I shall call scientific trade—by which I mean no offence—and whilst Elliot Smith had many interests, he was also fascinated by any fossil that might have some bearing on the line of human descent. For example, he wrote a great deal about the fossils from Chou Kou Tien which were first described by Zdansky and Davidson Black, under the name *Sinanthropus pekinensis*, and about the australopithecines, of which the first specimen was described by Professor Raymond Dart in 1925. I shall refer to Elliot Smith's views about *Sinanthropus* in a moment, when I speak of him as I knew him. About *Australopithecus* he was much more cautious than some of his contemporaries, in spite of his loyalty to his former research student Dart, and to his old friend Broom, both of whom (with the aid of Le Gros Clark) succeeded finally in elevating the australopithecines to a position of importance which I and my Birmingham colleagues alone seem to reject as unjustifiable on any anatomical evidence.

In his contribution to a volume called *Early man*, a series of lectures delivered for the Royal Anthropological Institute and published in 1931 (Elliot Smith, Keith *et al.*, 1931), Elliot Smith refers to the australopithecines in the same category as anthropoid apes, as he also does in his little book *The search for man's ancestors* published in the same year (Elliot Smith, 1931). I cannot ever remember him discussing Dart's find in any other terms. In his "autobiography", *Adventures with the missing link*, Dart refers in several passages to Elliot Smith, but without any indication that he had ever persuaded the latter to believe the views he (Dart) was promoting about the Taungs fossil (Dart, 1959).

Dart talks of a visit he paid to London early in 1931, and of his failure to convince a meeting at the Zoological Society of his opinions. And he also refers to a dinner after the meeting, at which he mentions my name as one of those present. He also tells of a monograph on *Australopithecus* which he had hoped Elliot Smith would arrange to be published by the Royal Society, but which, judging by what Dart says in his book, was "rejected". What I do remember is Elliot Smith's dismay when he read the manuscript, which he showed me, and which he regarded as a document falling well below the standards of a scientific contribution.

Most of Elliot Smith's writings on man's social evolution focus on the question of the spread of human culture, a subject in which he became more and more interested after he had begun to study the skeletal remains of the early Egyptians. The essential thesis which he was concerned to propagate was that the major steps in man's cultural evolution were unique occurrences, and that having occurred, they spread or diffused from their points of origin. To give updated examples —not those which led Elliot Smith into such vigorous controversy over a large part of his life—we know that neither the internal combustion engine nor the jet engine was invented time and time again in different parts of the globe. Correspondingly, we know that different people depend on the same basic information for the development of some practical idea—for example, radar. The second part of Elliot Smith's thesis was that the home of civilization was the Nile Valley, and that the critical steps in man's social evolution first occurred there—the cultivation of crops, the domestication of animals, the working of metals, the storage of harvested food, the invention of the wheel and so on.

This concept about the diffusion of culture, which will be the subject of the fifth section of this Symposium, was opposed to what Elliot Smith regarded as the prevailing belief of ethnologists and cultural anthropologists that different groups of men passed independently through the same historical phases during the course of their social evolution. Elliot Smith's own conversion from this view he attributed— as I understand it—to the influence of Dr W. H. R. Rivers, and in particular to an address given by the latter in 1911 to the Anthropological Section of the British Association (Rivers, 1911). In this lecture Rivers argued that where similarities in customs and institutions are found in different parts of the world, they are not the consequences of the human mind always working the same way, but the result of the inter-mixture of peoples, and of the merging of cultures.

It was this basic proposition which Elliot Smith transformed into his concept of the diffusion of culture. This started him off on a search, which continued to the end of his working life, for proofs to support his general theory, and for arguments to confound those by whom it was denied. Today it is difficult to understand why the debate was so heated. Even when one reads the recorded symposia in which Elliot Smith pitted his own against opposing views, it is difficult to see why the discussion generated so much passion. In one such debate (Elliot Smith, Malinowski *et al.*, 1928) Elliot Smith illustrated his thesis by referring to the simple wooden match, about which he said:

> "If some European traveller who was unaware ... was roaming in a part of the world where no white man had ever been before, and found there a wooden match, he would inevitably conclude that the match afforded certain evidence of contact, direct or indirect, with someone who had benefited by the English invention. If, however, he were not a

mere man-in-the-street, but an ethnologist faithful to the orthodox theory of his creed, he would have to assume that so obvious a mechanism must have been invented independently by the uncultured people of the country where he had picked up the match."

He was followed in this symposium by Professor Malinowski, then a leader of the functionalist school of anthropology. Malinowski's view was that "every cultural achievement is due to a process of growth in which diffusion and invention have equal shares". Inventions are made and re-made time after time in different places by different men along slightly different routes, and all are based upon an infinity of small steps in the course of the development of knowledge. Malinowski then referred to Elliot Smith's reference to the match as

"a puerile example—sometimes used by those who believe that culture can be contracted only by contagion and that man is merely an imitative monkey".

Where Malinowski had found the wooden match in use among the Aborigines of New Guinea and Australia it was not a part of their culture. It had been mechanically imported and supplied to them by the trader . . . it had been put into their hands by another society which has never succeeded in diffusing into the Melanesian culture the chemistry, physics and engineering which underlay the development of the match.

Gordon Childe, one of the most distinguished of the names which grace the history of British social anthropology, and whom I met through Elliot Smith, wrote—in a book which he published in 1951 (Childe, 1951)—that he was certain that Elliot Smith, while

"the founder of the English Diffusionist school, had no intention of reviving theological dogmas in his polemic against Tylor and his concept of evolution. Yet that in effect is what Diffusionism has led to".

But as Childe went on to say:

"the 'conflict' between Evolution and Diffusion is entirely fictitious. Diffusion is a fact. The transfer of materials from one territory to another is archaeologically demonstrated from the Old Stone Age onwards. But if material objects can thus be diffused, so can ideas—inventions, myths, artistic designs, institutions. Evolutionists have never denied this. For 'evolution' does not purport to describe the mechanism of cultural change. It is not an account of why cultures change—that is the subject-matter of history—but of how they change."

I am not qualified to enter into the argument, but I do believe that, for all his exaggeration, Elliot Smith recognized all this. In his little book, *The diffusion of culture*, one of the last works which he published on the subject (Elliot Smith, 1933), he wrote,

"my colleagues and I have repeatedly been accused of claiming that all
the arts and crafts, as well as the customs and beliefs, of the whole world
came from Egypt! There is no warrant for so stupid a statement. What
we do claim is that when after hundreds of thousands of years of inertia
men at last began to build up civilisation, the process was begun by the
inhabitants of Egypt, who, by inventing the practice of agriculture
started the process of making a civilised state. They created the needs
and the ideas which provoked men to go on devising new inventions".

And he went on to say:

"the charge is repeatedly made that those who accept the validity of the
principle of diffusion deny the possibility of a custom or belief being
invented twice independently. What we do in fact maintain is not the
impossibility of such a coincidence, but our inability to find any evidence
to show that it has happened".

Today we know, of course, that many of Elliot Smith's archaeological facts were not as firm as he supposed them to be. We all, I suppose,
also accept Gordon Childe's proposition that culture spreads or can
spread by a process of diffusion. For example, it is no argument against
the thesis of diffusion that we now know that metallurgy is just as old in
Hither Asia as on the Nile; or that "there is no unique cradle of civilisation"—such as Elliot Smith supposed the Nile Valley to be. As a
recent correspondent in *The Listener* puts it, the diffusionist hypothesis
only needs to be deemed an invalid principle "when it closes the mind
to alternative explanations" (Crossland, 1971).

If Elliot Smith were alive today, I feel sure, however, that he would
still be debating the validity and significance of his diffusion theory, and
with the same vigour by which he impelled many to accept his arguments that diffusion was the major factor in the spread of human
culture. But however contentious and however given to exaggeration, I
also feel that he would have been quick to renounce views which he
might have been advancing given that they were shown to be wrong.
The pity is that he lived and worked before the days of radio-active
methods of dating and, so far as physical anthropology went, before
modern statistical methods of analysis became available.

I first met Elliot Smith in January of 1926, a few days after I had
arrived from South Africa with two letters of introduction in my
pocket, one addressed to him, and the other to Arthur Keith. They were
given to me by Professor M. R. Drennan, the man under whom I had
first studied anatomy, and who had sharpened a natural interest—
which someone born in South Africa could hardly avoid having—in
anthropological matters. Incidentally, I find it interesting that Drennan's name is not even mentioned in Dart's autobiography, for the

two knew each other well, and Drennan was the first Professor of Anatomy to be appointed in South Africa—and I suppose, one of its first professional physical anthropologists. But that is another story.

Both Elliot Smith and Keith received me kindly, and in spite of the fact that I had come to England to study clinical medicine, neither discouraged me from pursuing certain enquiries I had begun on the comparative anatomy of the primates. I had arrived with a report of a study I had made into the growth changes in the skull of the baboon. I did not know that my typescript could be treated as a finished paper, and was surprised when Elliot Smith immediately suggested that he submit it for publication in the *Proceedings of the Zoological Society of London*, where it duly appeared (Zuckerman, 1926). Having discovered that watching baboons in the wild had been a part of my upbringing, he then encouraged me to speak about their social behaviour to an undergraduate Anthropological Society in University College. This paper was published in 1929 in *The Realist*—a journal which had a short life under the editorship of Archibald Church and Gerald Heard, the latter of whom had attended the meeting which I had addressed, and who had asked me to write up my lecture, which I had delivered from notes (Zuckerman, 1929). This paper subsequently became the basis of a book which Elliot Smith and W. J. Perry, the social anthropologist who was his co-worker, and also Professor B. Malinowski—another of my audience—encouraged me to write (Zuckerman, 1932).

In retrospect I do not find it at all surprising that the interests of the people around Elliot Smith in his Institute in Gower Street appealed to me much more than did those of my clinical teachers.

Whatever the research interests of some of the professors of anatomy who were then in Chairs, the teaching of the subject was very much a routine and dreary affair when Elliot Smith took charge of the new Institute of Anatomy and Embryology at University College, London. But it did not take him long to revivify the Department and extend its intellectual interests. Charles Singer, the medical historian, was there. Herbert Woollard, who was one of the first British anatomists to reintroduce the experimental approach to anatomy, was there. His *Recent advances in anatomy* was a shot in the arm to what had become a dead subject (Woollard, 1927). W. J. Perry, Elliot Smith's collaborator in expounding diffusion theory, was there. Popa and Una Fielding, who rediscovered the pituitary portal vessels, were there. Even Sir Bernard Spilsbury, the renowned forensic expert of the day, had a room. The speed of the changes introduced by Elliot Smith must have been remarkable, for in Abraham Flexner's book on medical education (Flexner, 1925), published only five years after Elliot Smith had moved from Manchester, and in which he castigated anatomists for the dead hand they had imposed on their subject, one finds the sentence: "In England, notable progress has been made [in breaking with the sterile Edinburgh tradition] at Cambridge and above all at University College". University College meant Elliot Smith.

One day, towards the end of my second year in University College Hospital, I was asked to go and see Elliot Smith. I remember him sitting at his desk, and pausing a moment before asking me rather solemnly what it was I intended to do with myself—whether I proposed becoming a surgeon or a scientist. I replied, "I don't know, sir. What would you advise?". There was a slight pause, and then came the judgement—"not a surgeon"—which at the time I took as an indication of reports he may have had about my shortcomings as a medical student. Soon after, he asked whether I would care to be a candidate for the Prosectorship of the Zoological Society of London, an office which dated from Victorian days, and which had been started at the instigation of the great T. H. Huxley. It was vacant after the death of Dr. John Beattie, who had followed Dr. Sonntag, a man who had devoted himself to a study of the comparative anatomy of monkeys and apes. Largely owing to Elliot Smith's influence, I was appointed, and through that appointment, I was able to carry on with a variety of studies, including the experimental investigation of the primate menstrual cycle and the influence of its phases on the social life and family behaviour of monkeys and apes. In those days the Prosectorship, which has now lapsed, also automatically carried with it a part-time demonstratorship at University College.

I now come to the part I have left out—Elliot Smith's interest in the Chinese fossils popularly known as Pekin man. In the late twenties, Davidson Black, Professor of Anatomy at the Peking Medical College, had enthused Elliot Smith with his descriptions of the fossil remains of early man in China, and Elliot Smith had written a number of papers and monographs in which he had followed and defended Black's thesis that *Sinanthropus*, the new genus to which Pekin man was assigned, represented a form of man which was generically distinct from all other known types, including the early ape-man of Java, *Pithecanthropus*. One day Elliot Smith was asked to give a lecture on the subject to the Royal Anthropological Institute, but as he had written so much about it, he asked whether I would undertake the task, and in order to ease it, suggested that I used whatever statements on the subject he had made and whatever other materials he had—including Davidson Black's own papers, as well as the casts of the remains which were in University College.

B

Setting about the task, I soon concluded that Elliot Smith's belief that the Peking remains were generically distinct from the *Pithecanthropus* fossils of Java was untenable. It seemed to me that unequal significance had been attached to the peculiarities not only of *Sinanthropus* and *Pithecanthropus* but also of Neanderthal man of Europe, and that consequently there were difficulties in the classification of all these fossils. I was inevitably driven to one of two conclusions—either *Sinanthropus* should be included in the genus *Pithecanthropus*, or alternatively the accepted classification of all extinct Hominidae should be revised. As the inclusion of *Sinanthropus* in the genus *Pithecanthropus* would in no way straighten out difficulties in connexion with Neanderthal man, I felt that a complete revision of the classification was plainly required. A rational classification seemed to imply the subdivision of the Family Hominidae into two sub-families, the Palaeanthropidae and the Neanthropidae—both terms having been coined by Professor Elliot Smith himself to emphasize cultural contrasts in the industries associated with these various types. These terms are not used today, but the same idea is implicit in the presently accepted classification of the genus *Homo* into the two groups, *Homo erectus*, my Palaeanthropidae, and *Homo sapiens*, my Neanthropidae.

I took a draft of the paper I had prepared to Elliot Smith, and instead of a defence of the views which I had questioned, I received from him a letter (7 November, 1932) which said:

> "I think the manuscript is excellent and raises a most important issue in a way which points the satisfying solution of it. On the enclosed piece of paper I have jotted down what it might mean for those who have to name the various creatures. This leads to the somewhat surprising result that after all some justification may be provided for Keith's use of the term 'Palaeanthropus palestinus'!"

A week later he wrote again, saying:

> "In your talk at the Anthropological Institute on the 23rd I hope you will include the discussion of your new proposals for the classification of the Hominidae on the lines of the document you have already written. If you were to do so that report [could] get into the January number of *Man* and so secure the publicity which would arouse an interest in the subject".

As it turned out, the paper was not published in *Man* but in the *Eugenics Review* (Zuckerman, 1933). I do not recall that it attracted much attention at the time, although some ten years later Le Gros Clark (1940) did give me the credit of having been the first to suggest that *Sinanthropus* and *Pithecanthropus* should be classified together, and approved the scheme I had proposed. But if that is almost the last I have heard of my paper, the meeting at which it was delivered, and over which Elliot Smith himself presided, stands out in my memory for a much more impelling reason than any of the arguments I was putting forward.

When we left the Anthropological Institute, just off Gower Street, Elliot Smith asked whether I would accompany him home. This was in the days before many people had cars, and we had to change buses at Camden Town in order to get to his house in Albert Road. While we were waiting for our second bus, he took out his handkerchief, and as he wiped the side of his nose, he remarked strangely,

> "I am wondering which of my lenticulo-striate vessels are leaking. All day I have felt as though I've a cold on the right side of my nose, and there's been a slight tingling in my fingers. I wonder how serious a stroke it's going to be".

I was more than a little shaken, but he seemed as calm as could be. I took him home, told his wife that he wasn't well, and called next morning—I did not live far away—to see how he was. Two or three days later, he was in University College Hospital following a stroke, from the effects of which he never fully recovered.

Professor H. A. Harris, who had been a member of the staff of the University College Department of Anatomy before Elliot Smith arrived from Manchester, was in charge while Elliot Smith was away. He had the title of Professor of Clinical Anatomy, and later succeeded J. T. Wilson as Professor of Anatomy in Cambridge. He was vigorous, ambitious and possessive, and not the most popular member of the staff of the Department. He refused to allow anyone to visit Elliot Smith in hospital. One day, however, when I was about to leave to take up an appointment at Yale, I insisted on seeing him to say goodbye. Elliot Smith had been prepared for my visit, and was sitting up in bed with a cheroot stuck between his twisted lips. His opening words, in a voice which I hardly recognized, were "Why have you kept away?". I didn't dare give him the reason.

For all the controversies in which he engaged, Elliot Smith, as I have said, was immensely kind. He undoubtedly helped create the intellectual climate of his day—both within and outside his subject—to an extent which I believe no other anatomist of this century has done. If he did not shrink from controversy, equally he did not retreat in the face of adverse comment, in the hopes that by not acknowledging it, it would go away. He did not pull his punches. I recall the scathing review he

wrote of Arthur Keith's *New discoveries relating to the antiquity of man*.
But controversy, as I have said, is implicit in the method of science,
and had nothing to do with the man, a man whom Cave described, in a
commemorative volume published shortly after Elliot Smith's death
(Dawson, 1938), as being marked with the simplicity of greatness, and
by an innate modesty. In this judgement he was doing no more than
echo what Sir John Stopford, who became the distinguished Vice-
Chancellor of the University after succeeding Elliot Smith in the Chair
of Anatomy, had said. To him Elliot Smith was characterized by a
restraint and modesty which at times almost approached humility.

Let me conclude, therefore, by referring to an extraordinary book
which has recently been published with the title *The Piltdown Men*, by a
Mr Ronald Millar (Millar, 1972). The author declares himself to be a
layman, and his painstaking work is clearly the result of assiduous
reading. But it is equally the reading of a man whose training has not
provided him with the capacity necessary to assess matters about which
he has no professional knowledge, nor the patience to find out what the
Piltdown men were like from those who knew the people concerned. I
mention this here because in his book Mr Millar paints an extraordinary
picture of Elliot Smith, and concludes—from whom he got the idea, I
cannot conceive—with the suggestion that it was Elliot Smith who had
been responsible for the Piltdown hoax. *The Times Literary Supplement*
has already published my rebuttal of Mr Millar's observations (Zuckerman, 1972). But I would like to end by repeating what I said in its
columns—Elliot Smith would never have falsified in order to refute
those who may have differed from him. His own opinions, right or
wrong, were founded on facts which were open to all. Equally, and to
the best of my judgement, he would always have abandoned views
which he may have held when persuaded by evidence that they were
wrong. The more he fought for his views, and however heated the
argument, the brighter the light that was shed and the wider the
horizon became. What more can be asked of any scientist?

REFERENCES

Childe, V. Gordon (1951). *Social evolution*. London: Watts & Co.
Crossland, R. A. (1971). Revolution in prehistory. *The Listener* **1971**: 211.
Dart, Raymond A. (1959). *Adventures with the missing link*. London: Hamish
    Hamilton.
Dawson, W. R. (ed.) (1938). *Sir Grafton Elliot Smith. A biographical record by his
    colleagues*. London: Jonathan Cape.
Elliot Smith, G. (1913). President's address. *Rep. Br. Ass. Advmt Sci.* **1912**: 575–
    598.

Elliot Smith, G. (1924). *The evolution of man*. London: Humphrey Milford, O.U.P.
Elliot Smith, G. (1928). *Conversion in science* (Huxley Memorial Lecture). London: Macmillan & Co.
Elliot Smith, G. (1930). *Human history*. London: Jonathan Cape.
Elliot Smith, G. (1931). *The search for man's ancestors*. London: Watts & Co.
Elliot Smith, G. (1933). *The diffusion of culture*. London: Watts & Co.
Elliot Smith, G., Malinowski, B., Spender, H. J. & Goldenweiser, A. (1928). *Culture. The diffusion controversy*. London: Kegan Paul.
Elliot Smith, G., Keith, A., Parsons, F. G., Burkett, M. C., Peake, H. E. & Myres, J. L. (1931). *Early man, his origin, development and culture*. London: Ernest Benn.
Flexner, Abraham (1925). *Medical education*. New York: The Macmillan Company.
Keith, Sir Arthur (1950). *An autobiography*. London: Watts & Co.
Le Gros Clark, W. E. (1940). The relationship between *Pithecanthropus* and *Sinanthropus*. *Nature, Lond*. **145**: 70.
Millar, Ronald (1972). *The Piltdown men*. London: Gollancz.
Rivers, W. H. R. (1911). The ethnological analysis of culture (Presidential Address). *Rep. Br. Ass. Advmt Sci*. **1911**: 490–499.
Weiner, J. S. (1955). *The Piltdown forgery*. London: Oxford University Press.
Wilson, J. T. (1938). *Obituary Notices of Fellows of The Royal Society*. No. 6: 323–333. London: The Royal Society.
Wood Jones, F. (1918a). *Arboreal man*. London: Edward Arnold.
Wood Jones, F. (1918b). *The problem of man's ancestry*. London: S.P.C.K.
Woodward, A. Smith, Hill, J. P., Wood Jones, F., Pocock, R. I., Cunningham, J. T. & Chalmers Mitchell, P. (1919). Discussion on the zoological position and affinities of *Tarsius*. *Proc. zool. Soc. Lond*. **1919**: 465–498.
Woollard, H. H. (1927). *Recent advances in anatomy*. London: J. & A. Churchill.
Zuckerman, S. (1926). Growth-changes in the skull of the baboon, *Papio porcarius*. *Proc. zool. Soc. Lond*. **1926**: 843–873.
Zuckerman, S. (1929). The social life of the Primates. *The Realist* **1929** (July): 72–88.
Zuckerman, S. (1932). *The social life of monkeys and apes*. London: Kegan Paul.
Zuckerman, S. (1933). *Sinanthropus* and other fossil types. *Eugen. Rev*. **24**: 273–284.
Zuckerman, S. (1972). The Piltdown men. Letter in *Times Literary Supplement* 27 October 1972: 1287.

Symp. zool. Soc. Lond. (1973) No. 33, 23–26

GRAFTON ELLIOT SMITH AND PILTDOWN

J. S. WEINER

MRC Environmental Physiology Unit, London School of Hygiene and Tropical Medicine, University of London, London, England

Lord Zuckerman (1972) has firmly and clearly denounced the slur which has been cast on Sir Grafton Elliot Smith in the book on the Piltdown forgery just published by Mr Ronald Millar (1972). I entirely support Lord Zuckerman in dismissing Mr Millar's case against Elliot Smith and indeed it is a most weakly argued case. It rests more on an attempt to exonerate Charles Dawson than on any positive evidence incriminating Elliot Smith. As practically all the relevant evidence both on the technical and human aspects of the forgery as well as the arguments leading to Charles Dawson as the likely perpetrator are to be found in Weiner (1955) and as Mr Millar would like the reader to think that he has demolished my arguments, I feel obliged to draw attention to a number of fallacies and misunderstandings in his book. A full account of these would take a long time but I shall confine myself to a few major and, I think, sufficient points.

There are five issues which stand out. First there is the matter of the timing and sequence of the many different phases of the Piltdown discoveries which stretched from about 1898 to 1915. The second concerns the issue of the artificial staining of the fragments. The third concerns the question of the anatomical knowledge required for the forgery, and the fourth concerns what Mr Millar calls the "hindsight" allegations against Dawson by a number of my informants. The fifth concerns the motive of the forger.

Mr Millar is not prepared to look at all closely at the sequence of events for on this account alone, it is impossible, indeed ridiculous, to think of Elliot Smith as the perpetrator. I need only quote what Mr Millar himself says on page 234: "At the time of the 'planting' of the fossils, Smith was in what might be considered a backwater appointment in Cairo". Where are the proofs that Elliot Smith made the many repeated visits to the Barkham Manor site or the Sheffield Park site that would have been needed? How did he manage to arrange that Dawson followed him at the appropriate times to find the planted material and to make the announcements to the British Museum or the public? In fact, the events of the 1915 finds alone are so complicated that for Elliot Smith to have been tied up in these let alone with events before he even got to England, I find completely incredible. Since Mr Millar has used so much of my material I feel he should have examined the sequence of events much more carefully. He does, I am glad to say, even on his cursory analysis, realize that the timing is a major reason for exonerating Teilhard de Chardin.

The question of the anatomical knowledge required by the perpetrator I have, of course, discussed in my book, but Millar thinks that only an anatomist of Elliot Smith's competence could have perpetrated the forgery. I must emphasize that for someone like Dawson, with 15 or 20 years of interest in archaeology and evolution, a man of undoubtedly high intelligence, frequently in contact with palaeontologists, who knew the Hunterian collections at the Royal College of Surgeons well, and who had in his possession a cast of the Heidelberg jaw amongst other things, it would be by no means difficult to acquire the necessary knowledge. For years the Diploma in Oxford put non-biologists such as social anthropologists and archaeologists through a course of skull measurement and comparative palaeontology and primatology, and this is still done in several places. In Dawson's day the material was so sparse that an intelligent and dedicated person with Dawson's background would certainly have realized what to do when he came to articulate an ape jaw with a human cranium especially if he had by his side a human jaw with its flat molar wear and articular condyle obviously different from that of the ape mandible.

Just as with the timing of events and the anatomical aspects, Mr Millar has not grasped the full significance of the artificial staining. He thinks that Dawson's use of potassium bichromate (regretted by Smith Woodward) was quite innocent, ignoring the fact that the fossils did not need any hardening. But he should know that there were not one but three staining techniques employed. About the staining bichromate he makes a glaring error by saying on page 229: "Potassium bichromate is not detectable at all". He seems to have overlooked the many tests in the scientific reports which show chromate in many of the fragments. There was also staining by iron ammonium sulphate in some fragments in addition to the bichromate, and thirdly there was the staining of the canine by Vandyke Brown. It really is no use saying that Dawson was using potassium bichromate in an innocent or conventional way. In fact the jaw was apparently already chromate stained when retrieved from the gravel in the presence of Dawson and Smith Woodward and taken by the latter to London.

Then Mr Millar thinks he has demolished the veracity of those of my informants who told me they had strong suspicions of Dawson from the early days. Here Mr Millar may not have grasped a most important point from his reading of my book. Let me therefore say quite un-equivocally that before the public announcement of the hoax in November 1953 (and after the technical work had been finished) I had, during the preceding weeks, carried out my main enquiries in Sussex and I did this, contrary to the unwarranted assertion by Mr Millar, without any prejudice whatever against Dawson. In fact until all the scientific work had been done I had not concerned myself with a possible perpetrator at all. Dawson, Smith Woodward, de Chardin, were merely names to me. When I first spoke to Mr Saltzman in Lewes, he had no inkling that the forgery had been unmasked. I asked him about Mr Dawson in the most general way and discovered the strong suspicions which he harboured of him. Nor did I reveal the unmasking of the hoax to Mr Pollard when he gave me the most astounding information regarding the accusations that his friend, Mr Harry Morris, had levelled at Dawson before 1916. Morris, who had not only told Mr Pollard but others including Captain St. Barbe, had taken the trouble to write down that he regarded Dawson as a faker of the implements. Morris had been dead many years when I came on to the scene and I am afraid was denied the benefit of the hindsight which Mr Millar would like to attribute to him.

Finally, as to motivation, it is ridiculous to insinuate that Elliot Smith needed advancement through malpractice. Elliot Smith was elected FRS in 1907 and appointed Professor of Anatomy in Manchester in 1909. The Piltdown discovery was announced in 1912. Charles Dawson's ambitions are documented in my book.

There are many other points which are incorrectly argued or simply misunderstood by Mr Millar in his book. He gives an inaccurate rendering of some of the results reported by Dr Oakley as Dr Oakley has pointed out to me.

In conclusion I think it is worth asking whether there is any point to Mr Millar's book apart from the totally unconvincing attempt to incriminate Grafton Elliot Smith His book purports to be a history of anthropological discovery giving the background that made the Piltdown affair possible and, in his estimation, the opportunities and motivation for Elliot Smith rather than Dawson, but the real significance of the history of palaeontological controversy eludes Mr Millar. Briefly it is that this branch of science has remained extremely refractory to exact methods of analysis and interpretation and this has allowed a good deal of emotionalism, crankiness and ill-informed journalism to flourish in a field which commands great public interest. The Piltdown episode highlighted the need for a rigorous basis for this discipline. Like Lord Zuckerman and his colleagues in Birmingham as well as other professional anthropologists, I have strongly urged (as in my study of

Swanscombe Man) that it is feasible and indeed essential that objective, statistical techniques be applied in the comparisons of fossils and particularly fossil fragments, before any far reaching claims are made as regrettably continues to be the case. Piltdown has demonstrated that dating by both direct and indirect methods can be put on an objective basis. Thirdly, the Piltdown investigation indicated that it is quite possible to put forward in human palaeontology, testable hypotheses so that new fossil finds are assessed for their affinities, not to arbitrarily chosen comparative material but to the whole corpus of well attested and quantified remains.

Unless these things are done human palaeontology will remain a low-level science which may often be fun to pursue in an easy-going way but, as you may have gathered, will involve a great waste of time in order to clear up ill-considered and sensational claims.

REFERENCES

Millar, Ronald (1972). *The Piltdown men.* London: Gollancz.
Weiner, J. S. (1955). *The Piltdown forgery.* Oxford: Univ. Press.
Zuckerman, Lord. (1972). The Piltdown men. Letter in the *Times Literary Supplement* 27 October 1972: 1287.

CHAIRMAN'S SUMMING-UP

We have greatly enjoyed Professor Zuckerman's reminiscences, curtailed though they were, of Professor Elliot Smith in whose honour this Symposium is held—he could have gone on for 24 hours quite easily. He has given us from the heart an account of a man who, as he truly says, and I fully agree, revolutionized British anatomical science, and was a personality the like of which we have not seen either in our own subject or in allied subjects since. He was a great fire at which disciples warmed their minds as well as their hearts.

I should also like to thank Professor Weiner for his remarks on a remarkable and memorable hoax. If motivation is wanted, in my posthumous memoirs I will leave my views as to who the nigger in the woodpile was.

GRAFTON ELLIOT SMITH AND THE FIRST AUSTRALOPITHECINE FOSSILS

CHAIRMEN: A. J. E. CAVE AND J. Z. YOUNG

The first two papers in this session, by Professor M. H. Day and by Mr. R. E. F. Leakey, were given under the chairmanship of Professor A. J. E. Cave. Professor J. Z. Young, FRS, took the Chair for the final paper, by Professor Lord Zuckerman, Professor E. H. Ashton. Dr. R. M. Flinn, Professor C. E. Oxnard and Mr T. F. Spence.

Symp. zool. Soc. Lond. (1973) No. 33, 29-51.

LOCOMOTOR FEATURES OF THE LOWER LIMB
IN HOMINIDS

M. H. DAY

Department of Anatomy, St. Thomas's Hospital Medical School,
London, England

SYNOPSIS

Evaluation of the stance and gait of Lower and Middle Pleistocene hominids has in the past been hampered by lack of both suitable fossil materials and appropriate methods of analysis. In recent years multivariate statistical techniques have been employed in order to try and extract locomotor information from some of the available fossils, information that is at best complementary to anatomical analysis. However, the advent of new and more complete specimens from both East and South Africa has confirmed some of the earlier inferences. Anatomical examination of the original fossils has revealed the dangers of working from casts alone.

Over 50 specimens that relate to the lower limb and pelvis are now known from Lower Pleistocene deposits and a considerable number from Middle Pleistocene deposits. Almost every bone of the lower limb is represented, some several times. Of particular importance is KNM ER 803, since all of these remains belong to one skeleton. All of the new material has contributed greatly to our knowledge of early hominid locomotor anatomy and discloses evidence for two co-existent morphological patterns that are distinct in a number of respects. The lower limb material attributed to fossil Hominidae is reviewed and the functional implications of these observations are discussed.

INTRODUCTION

The study of the evolution of the stance and gait of fossil members of the Hominidae has been hampered in the past by a lack of both appropriate methods of analysis and sufficient suitable fossil materials. In recent years, multivariate statistical techniques have been employed in attempts to extract locomotor information from some of the available fossils, information that is at best complementary to anatomical analysis (Day, 1967; Day & Wood, 1968). However, the advent of new and more complete specimens from both East and South Africa, coupled with anatomical reappraisal of some of the older specimens, has confirmed some of the earlier inferences drawn from the recognition of functionally significant constellations of anatomical characters and revealed the dangers of trigonometric and statistical analyses by those who have not examined the original materials.

Over 50 specimens that relate to the lower limb and pelvis are now known from Lower Pleistocene deposits and a considerable number from Middle Pleistocene deposits. Almost every bone of the lower limb is represented in the fossil collections, and some several times.

Through the courtesy of Museum Directors and Heads of Departments in Nairobi, Pretoria, Johannesburg and Leiden, I have been permitted to examine all of the extant hominid lower limb material—it has been rewarding and at times revealing.

In reviewing the principal specimens relating to the lower limb, it may be appropriate to consider them in anatomical sequence and to consider those that are usually regarded as being of Lower Pleistocene origin before those from the Middle Pleistocene.

LOWER PLEISTOCENE REMAINS FROM EAST AND SOUTH AFRICA

Sts 14

The most complete early hominid pelvic remains known are those from Sterkfontein, a pelvis upon which has been based a good deal of locomotor speculation. Broom, Robinson & Schepers (1950) believed that its features were consistent with upright posture and bipedal gait. This viewpoint was widely accepted and later elaborated by Le Gros Clark (1967).

However, reservations have been expressed and as late as 1967 it was stated that this australopithecine pelvis had some quadrupedal features (Zuckerman, Ashton, Oxnard & Spence, 1967). It was concluded, particularly in relation to the orientation of the gluteal muscles, that " . . . even if these hominid forms could have been bipedal, their gait could never have been like that of man". Here the categorical denial of the possibility of australopithecine gait being "like that of man" is preceded by only a conditional acceptance of bipedality in these forms. It seems fair to conclude that Zuckerman *et al.* (1967) still regarded the question of quadrupedality versus bipedality in the Australopithecinae as open.

Recently I examined the Sts 14 pelvis for the first time and I was considerably disturbed by the extent of its restoration and by the way in which it had been reconstructed. There is little doubt that the current reconstruction contains major anatomical errors, particularly on the left side.

In the reconstruction and orientation of a primate pelvis one of the rules that surely must be observed is that the symphysis pubis lies in the midline and runs anteroposteriorly. In this specimen the symphyseal surface is preserved on the left pubis but not on the right (Fig. 1a). In this reconstruction, the symphyseal surface is placed well to the right of the midline (Fig. 1b). The posterior view (Fig. 1c) shows how the pubis has been rotated inferiorly and medially so that

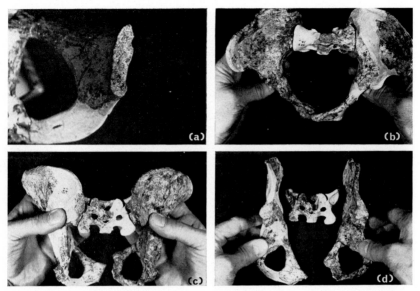

Fig. 1. Sterkfontein Pelvis (Sts 14). (a) The symphyseal surface of the left pubis. (b) The pelvic inlet showing the left symphyseal surface placed well to the right of the midline. (c) Posterior view showing how rotation of the left pubis has resulted in the symphyseal surface being obliquely placed. (d) Comparison of the two hip bones orientated in the plane of the ilium to show the incorrect reconstruction on the left side.

the symphyseal surface now lies obliquely. Comparison of the two halves, both orientated in the plane of the iliac blades, once again displays the misplacement of the left pubic bone (Fig. 1d).

Examination of the left side discloses where the error lies since the left acetabulum is not round and the pubis, ischium and ilium on that side are all separated by plaster infill (Fig. 2a). The same view displays the broken and distorted condition of the pubis on the right. It is clear that the profile of neither half is accurate. Indeed, while it is true that this specimen shows many important anatomical features very few measurements can be taken safely upon the fossil, or upon the available casts, while it is in its present condition.

SK 50

The pelvic fragment from Swartkrans is very badly crushed and shows numerous cracks all over its surface, many of which are matrix-filled. Step faults occur at many of these cracks, in particular on the iliac pillar which remains prominent and raked forward. Posteriorly the iliac bone consists of a thin skin of heavily cracked bone buttressed

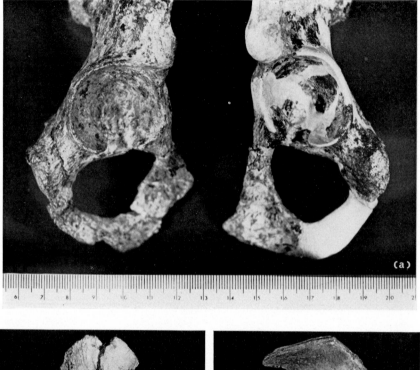

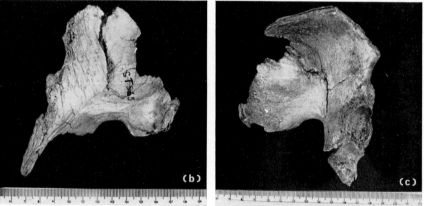

FIG. 2. (a) Sterkfontein pelvis (Sts 14). Comparison of the two hip bones to show that the left acetabulum is not round and that the three components of the hip bone are all separated by plaster infill. In addition the broken and distorted condition of the right pubis can be seen. (b) Sterkfontein pelvis (Sts 65). Internal view. A right ilio-pubic fragment from Sterkfontein showing a sciatic notch, a shallow iliac fossa, a small auricular surface and a pre-auricular groove. (c) Swartkrans pelvis (SK 3155 (b)). A new pelvic specimen from Swartkrans that includes much of an adolescent hip bone. The acetabulum is complete, there is a sciatic notch and an S-shaped iliac blade.

internally by a mass of plaster, in fact the major crack in the area shows an angulation for which there is no anatomical justification. The ischium seems to have been sheared off exposing the cancellous bone and the inferior ramus of the pubis is badly distorted and angulated laterally with a step fault at the junction.

In summary, it would be unwise to say more than that there is anatomical evidence here of a small acetabulum, of some backward extension of the ilium, of a pronounced rectus femoris attachment with evidence of a reflected head, and of a marked ilio-psoas groove medial to the anterior inferior iliac spine. There are no anatomical measurements that can be safely taken on this fossil.

Sts 65

This specimen (Fig. 2b) consists of several fragments all under the same museum number and not all of which are hominid. The major specimen is a right ilio-pubic fragment known to Broom (C. K. Brain pers. comm.) but never published by him. It consists of the upper half of the acetabulum, the iliac blade, the anterior superior iliac spine, much of the iliac crest and the superior pubic ramus. Behind the acetabulum the sciatic notch is well shown leading back to a prominent and rather unusual posterior inferior iliac spine which projects laterally as a stout process. Internally the fragment displays a shallow iliac fossa and a small auricular surface below which lies a well defined pre-auricular groove. Further details of this specimen include a well defined pectineal line, a marked anterior inferior iliac spine with an impression for the reflected head of rectus femoris as well as an impression for the upper attachment of the iliofemoral ligament; medial to the anterior inferior iliac spine there is a clear ilio-psoas groove. The iliac pillar is raked forward in front of a marked hollowing of the lateral aspect of the bone which gives rise to a clear S-shaped curvature of the iliac blade when viewed from above. Careful inspection of the external surface discloses the upper third or more of the middle gluteal line separating the attachment of gluteus medius from the underlying gluteus minimus. Clearly the anterior fibres of gluteus medius reached well forward, at least as far as the forwardly raked iliac pillar.

SK 3155 (b)

A new pelvic specimen (Fig. 2c) from Swartkrans consists of a large portion of an adolescent hip bone. The ilium is almost complete save for the loss of the anterior superior iliac spine and the posterior portion of the iliac crest. Only the acetabular portions of the ischium and pubis remain. The acetabular morphology is of particular interest since it is complete and shows the non-articular area in the floor of the acetabulum. The widest, or principal weight bearing portion of the acetabulum is arranged so that a line drawn through it at right angles would cut the iliac crest just behind the anterior superior iliac spine as in upright bipeds. The presence of a sciatic notch is related to the backward extension of the ilium while from above the S-shape of the iliac blade is well shown. The anterior inferior iliac spine is well developed for the straight head of rectus femoris while further marks disclose the attachment of the reflected head of this muscle and the upper attachment of the iliofemoral ligament.

In summary, the early hominid pelvic material described (in addition to the well known material from Makapansgat and Kromdraai) displays a group of pelvic features that can be recognized as typical of these early hominids; these characters include among others a small acetabulum, a backward extension of the ilium enclosing a sciatic notch, a pronounced anterior inferior iliac spine, an ilio-psoas groove, an S-shaped iliac blade and a distinct iliac pillar. This constellation of characters appears to be clear anatomical evidence of upright posture and bipedality in these early forms but, as has been formerly suggested (Zuckerman et al., 1967) not necessarily of precisely modern human character. Nonetheless, there would seem to be no anatomical evidence of quadrupedalism in any of these pelves.

FEMORAL REMAINS FROM EAST AND SOUTH AFRICA

Sts 14

This femur (Fig. 3a, b, c), only briefly mentioned by Broom et al., (1950), was found with the pelvic remains bearing the same number and may be associated with these remains. Recently its "head size" and its "neck angle" have been used in conjunction with measurements taken from photographs and new estimates of the femoro-condylar angle, in an attempt to calculate the reconstructed length of this femur by trigonometric means. From this insecure basis, an estimate of australopithecine stature has been made (Lovejoy & Heiple, 1970). Careful examination of the original fossil reveals the specimen to be a useless jumble of glued fragments surmounted by a crude plaster head and neck. There are no measurements that can be taken safely from this specimen, therefore any calculations relating to it should be viewed with extreme caution. Morphologically its only features worthy of note are that the bone was probably gracile in overall character and that the neck may have been somewhat flattened and deep.

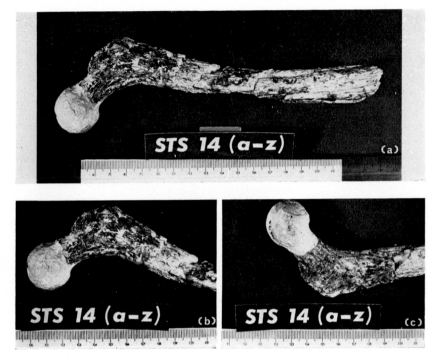

Fig. 3. Sterkfontein femur (Sts 14). (a) Anterior view of complete specimen showing the poor condition of this femur. The "head" is entirely plaster reconstruction. (b) Anterior view of the "head", neck and upper end of the shaft to show the lack of surface detail. (c) The "head" and "neck" in posterior view to show the lack of surface detail and lack of trochanters.

SK 82 and SK 97

These two fragments are the upper ends of two right femora from Swartkrans. It appears that they were never described by Broom, nor accepted by Le Gros Clark as hominid since they do not figure anywhere in his extensive writings in this field. Later, however, they were diagnosed as hominid by Napier (1964) and later still he was fully vindicated by the discovery of O.H. 20 from Olduvai Gorge (Day, 1969), a femoral neck with very similar morphological features. In the study of O.H. 20 new evidence of habitual upright posture came to light. In all three specimens, clear evidence was found of the groove (well known in man) for the obturator externus tendon as it winds round the back of the femoral neck in extension and hyperextension of the hip joint—a groove not present in the femora of quadrupedal and knuckle walking primates. More recently a vertical groove has been identified

on the anterior aspect of these femoral necks, in line with the lesser
trochanter; a groove clearly made by the tendon of ilio-psoas muscle
in full extension and hyperextension of the hip joint. The presence
of the groove has been confirmed on the originals. It would appear
that the recognition of this significant morphological feature reaffirms
that extension and hyperextension of these hip joints was habitual
and not occasional.

A canonical analysis using 10 parameters from the femoral neck
alone produced a discrimination between samples of modern man,
Gorilla and *Pan* (Fig. 4). Interpretation of the analysis confirms the

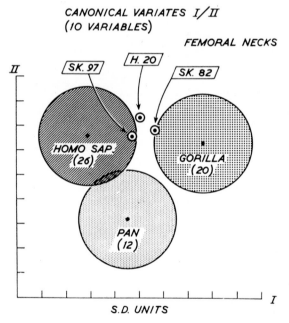

FIG. 4. Canonical variates I and II plotted against each other to show the maximum
discrimination between the major groups. Three fossil femora have been interpolated.

similarity of the fossil remains to each other and their non-identity
with the other groups tested. This is of particular interest since it
has been suggested recently that most of the distinctive features of
these fossils fall individually within the range of normal for a population
of North American Indians (Lovejoy & Heiple, 1972). It would seem
that such a conclusion is premature since it overlooks the effect of the
combination of small head, long neck, flat neck, non-flared greater
trochanter and posteriorly directed lesser trochanter. This constellation

of anatomical features has recently been observed, wholly or partly, in such specimens as KNM ER 815, KNM ER 738, KNM ER 1505 all specimens from East Rudolf, as well as a femoral neck from the Omo region (unpublished). So far it has not been possible to trace a modern human femur with this precise combination of features.

It seems clear that one distinctive anatomical pattern is beginning to emerge, in relation to the upper end of the femur in early hominids, from four separate sites in both East and South Africa.

Sts 34 and TM 1513

The two well known distal ends of femora from Sterkfontein (TM 1513 is also labelled Sts 1513) are very little reconstructed although a good deal of surface bone is missing from the lateral condyle of TM 1513 and the medial condyle of Sts 34.

Recently a previous minimum estimate of the femoro-condylar angle (Le Gros Clark, 1947) has been criticized as being too low and a new estimate has been given. The prominence of the lateral condyle in the region of the patellar groove has been re-emphasized as part of the mechanism that counters lateral dislocation of the patella in full extension of the knee in forms with a high femoro-condylar angle (Heiple & Lovejoy, 1971). I agree with these findings and I can add some observations suggesting the presence of a knee-locking mechanism in the form of asymmetrical grooves for the menisci on the flattened condylar surfaces. The presence of grooving for the tendon of popliteus in full flexion could represent evidence of an unlocking mechanism similar to that known in man.

TM 3601

This is a distal femoral specimen of unknown provenance and consists of broken femoral condyles with well-exposed cancellous bone. The importance of this specimen rests in the fact that the medial and lateral aspects of the condyles are intact. The lateral condyle shows an epicondyle with a good attachment above for the lateral head of gastrocnemius and below a clear attachment for popliteus with a groove running posteriorly and smoothing off the edge of the articular surface when the tendon leaves the groove in full extension of the knee. In the medial condyle there is evidence of the attachment of the medial ligament of the knee and medial head of gastrocnemius muscles. Although it is not known from which of the South African sites this specimen derives, its provenance is of less importance since none of the South African sites is precisely dated. The morphological features that it shows seem to relate closely to those already recognized from the two Sterkfontein specimens and, at the very least, underlines their validity as undeformed specimens.

Below the knee, much of the early hominid locomotor evidence derives at present from Olduvai in the form of the O.H. 35 tibia and fibula (Davis, 1964), the O.H. 8 foot (Napier & Day, 1964; Day & Wood, 1968) the O.H. 10 toe (Napier & Day, 1966; Day, 1967) and the O.H. 43 metatarsals (unpublished).

Of all these remains the O.H. 8 foot is still the most convincing evidence of early hominid bipedality since it is a truly associated functional complex of unmistakable locomotor significance. Similarly the anatomical features of lateral tilt and torsion in a great toe phalanx are telling indicators of bipedality. Recently the statistical evidence advanced in support of the anatomical conclusions reached in relation to O.H. 8 and O.H. 10 has been somewhat harshly criticized (Oxnard, 1972). Suffice it to say that these criticisms will be answered elsewhere, but in any event they can have little impact upon the bulk of Lower Pleistocene locomotor evidence available to anatomists familar with the fossil material.

MIDDLE PLEISTOCENE REMAINS FROM EAST AFRICA AND ASIA

The well known lower limb fossil remains from Java and Peking (Figs 5 & 6) have for many years constituted the sum total of the lower limb material available for study in relation to this period. Indeed since the loss of the original Peking material even casts of these femoral remains have been hard to find. Recently new finds at Olduvai Gorge have augmented the meagre fossil record.

O.H. 28

The recovery of a femoral shaft and a closely associated pelvic fragment from Bed IV, Olduvai Gorge, was an event of the greatest importance since the bones were associated with an Acheulean industry and have been dated on palaeomagnetic and stratigraphic grounds to about 500 000 years B.P. (Leakey, M. D., 1971; Day, 1971). The remarkable similarities between the femur and those known from Peking has led to its attribution to *Homo erectus*; similarities that include marked platymeria, a distal minimum-breadth point, and convexity of the medial border of the shaft. The associated pelvic fragment is unique in that there are no other pelvic remains known of comparable age. Morphologically it is also a notable fossil in that it is quite different from the known Lower Pleistocene pelvic remains in having a very large acetabulum and a massive vertical iliac pillar—

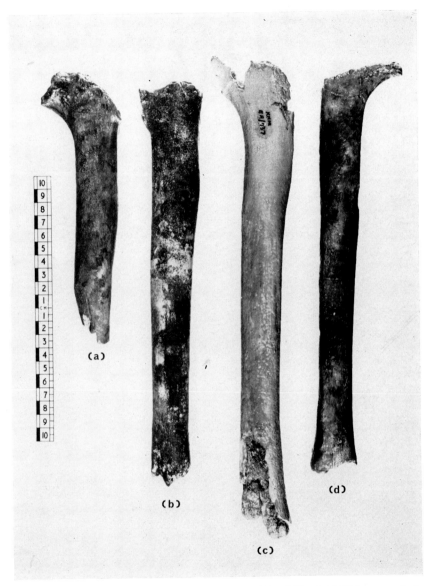

FIG. 5. (a) Pekin femur I, anterior aspect. (b) Olduvai Hominid 28 femur, anterior aspect. (c) KNM ER 737 femur, anterior aspect. (d) Pekin femur IV, anterior aspect.

a pillar unmatched by that of any sapient pelvis so far examined. Further studies are in progress in order to reconstruct the missing areas and to assess the locomotor implications of its unusual anatomy.

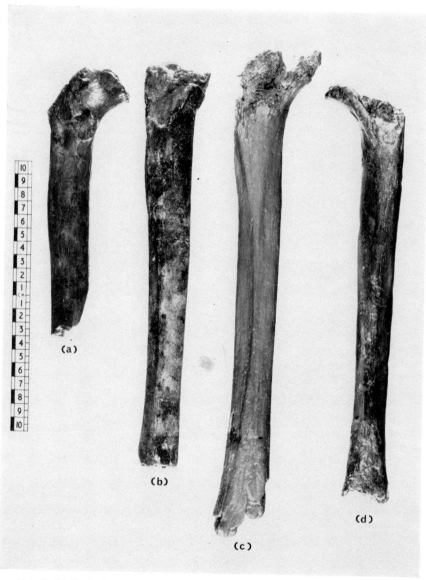

FIG. 6. (a) Pekin femur I, posterior aspect. (b) Olduvai Hominid 28 femur, posterior aspect. (c) KNM ER 737 femur, posterior aspect. (d) Pekin femur IV, posterior aspect.

The Peking femora

These femora were lost with the remainder of the Peking remains but fortunately the descriptions given by F. Weidenreich are detailed and well illustrated. Even so vigorous attempts were made to locate a set of casts of these remains and they were finally found by Miss Molleson in the American Museum of Natural History (Department of Vertebrate Palaeontology), New York. The generous loan of this complete set of casts permitted direct comparisons to be made between them and the O.H. 28 femur and the Trinil femora.

It has been asserted previously in a definition of the species *Homo erectus* that the limb bones of this form " . . . not distinguishable from those of *H. sapiens*" (Le Gros Clark, 1964) and also that " . . . in no characters has it been satisfactorily demonstrated that any of these limb bones (Peking) are distinguishable from those of *H. sapiens*". (Le Gros Clark, 1964). Once again, it appears that the effect of a combination of morphological features has been overlooked. Even though any one feature taken alone may be shown to lie within the enormous range of normal variation documented for such a prolific species as our own, a recognizable group of consistently recurring characters must be acknowledged even in a small sample.

The Trinil femora

The longest known Middle Pleistocene lower limb remains are the six femora from Trinil recovered by Dubois (Dubois, 1926, 1927, 1932, 1934, 1935a, b, 1937). These femora have been a source of controversy ever since their recovery. Recently an opportunity was taken of re-examining them anatomically, radiologically and from the viewpoint of provenance and dating (Day & Molleson, 1973). Anatomically and radiologically we could find no convincing locomotor features, or features of taxonomic significance, that would distinguish Femora I–V from those of modern man, despite Dubois' prolific writings giving contrary and at times confused views. Here we are in agreement with a number of previous authors (Manouvrier, 1895; Hepburn, 1897; Weidenreich, 1941; Le Gros Clark, 1964). Femur VI, however, does not appear to be human and has been taken out of the fossil record of man.

Clearly the anatomical findings demanded a reappraisal of the provenance and dating of the material. Extensive analytical tests were performed in order to try and clarify the situation. The analytical evidence neither confirms nor denies the Middle Pleistocene antiquity of the Trinil femora. Until the dating of the Trinil femora is conclusively proved it would seem wise to retain an open mind concerning the question of including the anatomy of these bones within the range of normal variation for Middle Pleistocene *Homo erectus*.

New hominine material from East Rudolf

That there should be pelvic and femoral differences between Lower and Middle Pleistocene hominid remains is hardly unexpected in view of the possible time interval involved. What was unexpected, however, was the recovery of new hominine material from a Lower Pleistocene layer at East Rudolf.

KNM ER 737

This new femoral shaft (Figs. 5 & 6) is quite unlike other hominid femora from the same horizon and it has been attributed to the genus *Homo* (Leakey, R. E. F., 1971; Day & Leakey, in press). It is remarkable for its similarity to both the O.H. 28 femur and the Peking femora.

KNM ER 803

This specimen (Fig. 7) consists of a single closely associated partial skeleton comprising 20 fragments including a femoral shaft, parts of both tibiae and fibulae, two metatarsals and a number of toe phalanges as well as some upper limb bones and teeth. (Leakey, 1972; Day & Leakey, in prep. The femur is very similar both morphologically and metrically to KNM ER 737 and thus has similarities with the O.H. 28 and Peking femora from more recent horizons. The tibia is straight with a powerful attachment for the soleus muscle. The third left metatarsal is well buttressed dorso-ventrally to withstand propulsive forces and shows a degree of torsion consistent with the formation of a transversal tarsal arch. All of these features are indicative of an upright stance and a bipedal gait.

This new material has strengthened the growing body of evidence (Day & Leakey, in press, in prep.; Leakey & Wood, in press) for the contemporaneity of two hominids in the Lower Pleistocene of East Africa, at least one from the genus *Homo* and at least one from the genus *Australopithecus*.

CONCLUSIONS

From this reappraisal of the older evidence and the examination of new material relating to the lower limb in hominids, it seems that some conclusions can be drawn.

1. The anatomical evidence for upright posture and bipedal gait in both Lower and Middle Pleistocene hominids is convincing. This does not mean that both groups had precisely the same stance and

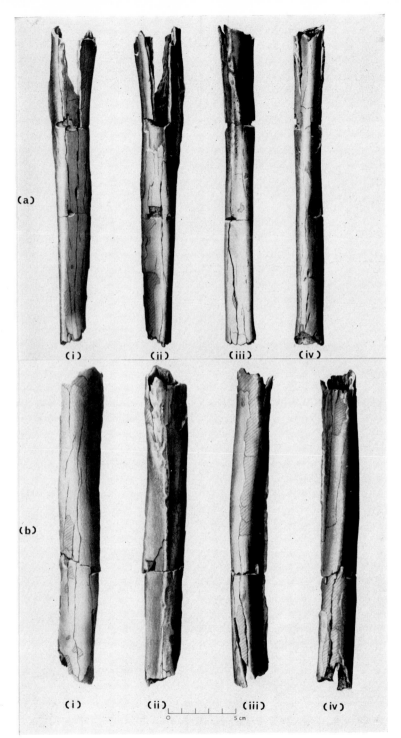

Fɪɢ. 7. (a) KNM ER 803, left tibia. (i) Medial view. (ii) Lateral view. (iii) Posterior view. (iv) Anterior view. (b) KNM ER 803, left femur. (i) Anterior view. (ii) Posterior view. (iii) Lateral view. (iv) Medial view.

gait or that either or both had a stance and gait that was modern sapient in form.

2. There is no indisputable osteological evidence of quadrupedalism in the known australopithecine lower limb material from East or South Africa.

3. That two patterns of lower limb morphology can be distinguished from the Lower Pleistocene period in East Africa, only one of which is known from the Middle Pleistocene although we may anticipate the other in view of the cranial evidence.

4. We may expect these two locomotor patterns to blur as we find more materials from earlier levels.

5. A proper evaluation of the precise functional meaning of the morphological differences that we can recognize will only come from a combination of anatomical, biomechanical and statistical studies making use of all the available fossil evidence.

ACKNOWLEDGEMENTS

It is a pleasure for me to acknowledge the kindness and assistance of many colleagues including Mr R. E. F. Leakey, Director of the National Museums of Kenya; Professor P. V. Tobias, Head of the Anatomy Department, University of Witwatersrand, Dr C. K. Brain, Director of the Transvaal Museum, Pretoria; Dr L. D. Brongersma, lately Director of the Leiden Museum; Dr A. C. Walker, Dr M. D. Leakey and Miss T. I. Molleson.

I am also grateful to Mr R. E. F. Leakey, the Wenner-Gren Foundation for Anthropological Research and the Royal Society of London for financial assistance.

REFERENCES

Broom, R., Robinson, J. T. & Schepers, G. W. H. (1950). Sterkfontein ape-man, *Plesianthropus. Transv. Mus. Mem.* No. 4: 1–117.

Davis, P. R. (1964). Hominid fossils from Bed I, Olduvai Gorge, Tanganyika. A tibia and fibula. *Nature, Lond.* **202**: 967–970.

Day, M. H. (1967). Olduvai Hominid 10, a multivariate analysis. *Nature, Lond.* **215**: 323–324.

Day, M. H. (1969). A robust australopithecine femoral fragment from Olduvai Gorge, Tanzania (Hominid 20). *Nature, Lond.* **221**: 230–233.

Day, M. H. (1971). Postcranial remains of *Homo erectus* from Bed IV, Olduvai Gorge, Tanzania. *Nature, Lond.* **232**: 383–387.

Day, M. H. & Leakey, R. E. F. (in press). New evidence of the genus *Homo* from East Rudolf, Kenya (1). *Am. J. phys. Anthrop.*

Day, M. H. & Leakey, R. E. F. (in prep.). New evidence of the genus *Homo* from East Rudolf, Kenya (3).

Day, M. H. & Molleson, T. I. (1973). The Trinil femora. (In *Human evolution* (ed.) Day, M. H.), *Symp. Soc. Study hum. Biol.* **11**: 127–154.

Day, M. H. & Wood, B. A. (1968). Functional affinities of the Olduvai Hominid 8 talus. *Man* **3**: 440–445.

Dubois, E. (1926). On the principal characters of the femur of *Pithecanthropus erectus*. *Proc. K. ned. Akad. Wet.* **29**: 730–743.

Dubois, E. (1927). Figures of the femur of *Pithecanthropus erectus*. *Proc. K. ned. Akad. Wet.* **29**: 1275–1277.

Dubois, E. (1932). The distinct organisation of *Pithecanthropus* of which the femur bears evidence, now confirmed from other individuals of the described species. *Proc. K. ned. Akad. Wet.* **35**: 716–722.

Dubois, E. (1934). New evidence of the distinct organization of *Pithecanthropus*. *Proc. K. ned. Akad. Wet.* **37**: 139–145.

Dubois, E. (1935a). On the gibbon-like appearance of *Pithecanthropus erectus*. *Proc. K. ned. Akad. Wet.* **38**: 578–585.

Dubois, E. (1935b). The sixth (fifth new) femur of *Pithecanthropus erectus*. *Proc. K. ned. Akad. Wet.* **38**: 850–852.

Dubois, E. (1937). The osteone arrangement of the thigh-bone compacta of Man identical with that, first found, of *Pithecanthropus*. *Proc. K. ned. Akad. Wet.* **40**: 864–870.

Heiple, K. G. & Lovejoy, C. O. (1971). The distal femoral anatomy of *Australopithecus*. *Am. J. phys. Anthrop.* **35**: 75–84.

Hepburn, D. (1897). The Trinil femur (*Pithecanthropus erectus*) contrasted with the femora of various savage and civilised races. *J. Anat. Physiol.* **31**: 1–17.

Leakey, M. D. (1971). Discovery of postcranial remains of *Homo erectus* and associated artefacts in Bed IV at Olduvai Gorge, Tanzania. *Nature, Lond.* **232**: 380–383.

Leakey, R. E. F. (1971). Further evidence of Lower Pleistocene hominids from East Rudolf, North Kenya. *Nature, Lond.* **231**: 241–245.

Leakey, R. E. F. (1972). Further evidence of Lower Pleistocene hominids from East Rudolf, North Kenya. *Nature, Lond.* **237**: 264–269.

Leakey, R. E. F. & Wood, B. A. (in press). New evidence of the genus *Homo* from East Rudolf, Kenya (2). *Am. J. phys. Anthrop.*

Le Gros Clark, W. E. (1947). Observations on the anatomy of the fossil Australopithecinae. *J. Anat.* **81**: 300–333.

Le Gros Clark, W. E. (1964). *Fossil evidence for human evolution.* Chicago: Chicago University Press.

Le Gros Clark, W. E. (1967). *Man apes or ape-men?* New York: Holt, Rinehart and Winston.

Lovejoy, C. O. & Heiple, K. G. (1970). A reconstruction of the femur of *Australopithecus africanus*. *Am. J. phys. Anthrop.* **32**: 33–40.

Lovejoy, C. O. & Heiple, K. G. (1972). Proximal femoral anatomy of *Australopithecus*. *Nature, Lond.* **235**: 175–176.

Manouvrier, L. (1895). Discussion du *"Pithecanthropus erectus"* comme précurseur présumé de l'homme. *Bull. Soc. Anthrop. Paris.* **6**: 12–47.

Napier, J. R. (1964). The evolution of bipedal walking in hominids. *Archs Biol., (Liège)* **75**: 673–708.

Napier, J. R. & Day, M. H. (1964). Hominid fossils from Bed I, Olduvai Gorge, Tanganyika. *Nature, Lond.* **202**: 969–970.

Napier, J. R. & Day, M. H. (1966). A hominid toe bone from Olduvai Gorge. *Nature, Lond.* **211**: 929–930.

Oxnard, C. E. (1972). Some African fossil foot bones: A note on the interpolation of fossils into a matrix of extant species. *Am. J. phys. Anthrop.* **37**: 3–12.

Weidenreich, F. (1941). The extremity bones of *Sinanthropus pekinensis. Palaeont. sin.* New Ser. D. **5**: 1–150.

Zuckerman, S., Ashton, E. H., Oxnard, C. E. & Spence, T. F. (1967). The functional significance of certain features of the innominate bone in living and fossil primates. *Proc. anat. Soc. Gt Br. Ire.* **101**: 608.

Discussion

Zuckerman: Do I understand you to say that we have two distinct groups of hominoids right down to the Lower Pleistocene? Does each group show the same indications of an upright bipedal gait?

Day: I think that I have been assiduous in avoiding taxonomic argument, because frankly I think that more heat than light is generated in this subject if we get ourselves involved in the taxonomic value of individual features. It may be that bipedality is a taxonomically valid character in distinguishing what we sometimes call australopithecines from what we may call early hominines; but it may not be valid since they may both have the same form of locomotion. I think, however, that we are beginning to get evidence of a distinction, in the sense that there is a unity in the locomotor information at the Middle Pleistocene level, and a unity at the Lower Pleistocene level. Now for the first time, in ER 737 and ER 803, we can see features that we can recognize from Middle Pleistocene material beginning to occur in the Lower Pleistocene. Without calling on the cranial evidence for the presence of the genus *Homo* in the Lower Pleistocene, and new evidence which I think we may hear about later today, my locomotor findings seem to bear out this suggestion.

Young: Mr Chairman, might I ask through you whether Professor Day or perhaps Dr Napier would elucidate for those of us who are not so expert in this subject, in what respect the locomotion is not quite fully bipedal.

Day: As I would expect from Professor Young, this is a question of considerable value, but I think that one of the difficulties we have had in the past is that it is easy enough to say "bipedal" as opposed to "quadrupedal", because we know bipedal means two-legged and quadrupedal means four-legged. Immediately you start to qualify these terms, then you are in trouble, because you cannot define precisely what you mean. Dr Napier's definition was to use the terms "heel-toe" or "striding", but again, how do you define a stride precisely? You can possibly define it by a number of characteristics, but you cannot define it metrically or statistically. As soon as you start to try and pin down precisely what you mean—a shuffling gait has been used, a bent-knee gait, a flexed-hip gait—it becomes highly dangerous. I do not think that we have the necessary biomechanical information as yet, and this is the area from which it must come. The statisticians can help, because they can reduce morphology to numbers. The 10 femoral neck measurements that I took mean at most 20 points, and it is the relationships of these 20 points to each other that is all I have discriminated, nothing else. If it helps in interpretation, then well and good, but what it means in purely locomotor terms I would be hard pressed to say. The people who will do it in the end might be people who come from Professor Davis' school; those will thoroughly evaluate in engineering terms what is happening in joints, in relation to ligaments and in relation to muscles.

CAVE (CHAIRMAN): Could these Pleistocene creatures have done a route march (that to me is what bipedalism means) so as to satisfy a sergeant major? That is a rough and ready definition, but is what I mean by true walking.

DAY: The rough and ready answer is yes, but how far is a route march?

WOOD: I want to say, in partial answer to Professor Young's question, that we only have one extant model of hominid bipedalism, and that is the model shown by the population of *Homo sapiens*. This problem is like the question: have you stopped beating your wife? Is this fossil bipedal or not? This has meant in the past that if it is not like *Homo sapiens* it cannot be bipedal. We are now realizing that there have been other models of bipedalism. Increasing evidence indicates that *Australopithecus* had developed a distinct gait, which does not have to be identical to ours to be recognized as bipedal, which it surely was. In addition there are recognizable steps leading to the gait of modern man. Thus, confusion has arisen in the past because the modern model was used as the *sole* example of a bipedal hominid locomotion, and anything unlike the modern state was immediately relegated to sub-bipedal status. The modern model must not be ignored, as it will hold the key to unravelling the biochemical and other implications of the abundant hominid post-cranial material now being found in East Africa.

BISHOP: As a geologist I am most impressed by the care with which Professor Day has studied and presented his material. I appreciate the need for multi-variate and other detailed analyses of fossil bones to facilitate comparison and permit more accurate identification. At the same time I am worried that having gone to such trouble Professor Day is happy to take the fossil material and place it into only two vague divisions, Lower and Middle Pleistocene. I am aware of the difficulties of making clear the age relations of particular fossils in a short presentation. However, it is important to emphasize for Peking, Trinil and all the South African cave sites (Swartkrans, Sterkfontein, Kromdraai and Makapan) that we do not know what age they are. Indeed as these localities lack evidence from the law of superposition we do not even have any data of relative age apart from subjective assessments based on the fossil assemblages. There is as yet no reliable isotopic dating evidence for these localities. The value of the Olduvai, East Rudolf and Omo areas, is that they comprise simple stacks of sediments providing relative age relations and the strata lend themselves to isotopic (potassium–argon) and palaeomagnetic dating. We must differentiate between dated and un-dated specimens just as clearly as between different taxa or locomotor adaptations. I am worried that we are in danger of working on two very different levels of accuracy; "splitting" the fossils following detailed morphological studies but then "lumping" them into over-simplified time divisions.

Professor Day refers to the "time-lapse" between the Middle and Lower Pleistocene, and other speakers have spoken of time in general. If we differentiate between specimens which have a time context and those which do not have any, it is obvious that many of the specimens discussed by Professor Day have none. There is merely an inference that the Middle Pleistocene group of specimens is younger than those assigned to the Lower Pleistocene group.

Much of the new material from East Rudolf and from Olduvai is from neatly documented time contexts. Could Professor Day say for the post-cranial material that he has discussed what is the probable upper age of the youngest material? Also he has used the term "Lower Pleistocene". Although it is probable that some specimens are older than the generally accepted base of the Pleistocene, could he venture a suggestion as to the age of the oldest specimens, that he has referred to the Lower Pleistocene?

DAY: May I say first of all that we in anatomy are delighted that geologists are beginning to react against terms that they have difficulty in defining. We are also delighted that geochronology has grown into a science which begins to depend upon numbers. Fortunately the numbers have an error bar and sometimes the numbers differ according to who has worked them out. I studiously avoided, in my presentation, other than the vaguest and broadest of geochronological statements. I merely stuck to two names that I understand in general terms. I realize that there is a difficulty in defining the point at which the Lower Pleistocene becomes the Middle Pleistocene and so on, and also I realize that the South African material is virtually undated, but there are also quite reasonable faunal associations that have been utilized in the past and dating estimates appear in authoritative books, such as that of Dr Oakley (1964)*.

---

* Oakley, K. P. (1964). *Frameworks for dating fossil man.* London: Weidenfeld & Nicholson.

I only, therefore, put my material into the broadest of geochronological contexts, and I think really I can add little more to that, other than to say that the general bracket for the East Rudolf material is given as somewhere between 1 million and 2·6 million. I think that Mr Leakey, who is presenting the next paper, will explain this in greater detail, because I believe that he has new information to present concerning the sequence about which I have been talking.

MOLLESON: Recent dating, by Dr Garniss Curtis, of volcanic debris in Putjangan and Kabuh beds, at various sites in Java, indicates a dating of about 0·8 million years for the Kabuh beds. The Trinil femora were found by Dubois in these beds. The Olduvai femur comes from Bed IV which has been tentatively dated as younger than 0·7 million years. Whether any of the femora are contemporary with the deposit, derived from older beds, or intrusive from younger deposits, can of course be questioned.

A SPEAKER: Professor Day answered Professor Young saying that as yet we could not go about defining bipedalism for the reasons that he gave. If we cannot define something we mean, can we go about using it as taxonomically significant?

DAY: I think this is a point that I made. It is questionable that this can be done, because I think there is likely to be more than one form of bipedality. Taxonomy, properly practised, is based upon a far wider range of characters than simply one aspect of locomotion. Other aspects of locomotion may come into it too, such as prehensility, and then we have cranial, dental and other morphological features to consider.

GRIFFITHS: I am a little cautious about asking this, but are you saying effectively that trying to make a locomotor classification is not assisting answering the question as to whether these animals are in fact relatives of man, because we have assumed that bipedalism heretofore is the unique feature which distinguishes man from others? Yet we are now saying that we are not able to define bipedalism. If one assumes bipedalism to mean two feet and therefore that the hands are doing something else, is just studying bipedalism significant? What you are saying effectively is that we are gathering more evidence for classifying these fossils, but we are not gathering any evidence to assist us in deciding how they relate to man?

DAY: In the first place I would not agree that bipedalism is the unique feature that distinguishes man from others. I think there are many other features in man that are also unique. Secondly, I am also saying that bipedalism has been used as a taxonomic feature and may well be used as a taxonomic feature in the future, but, at present, I am not entirely sure what I mean by it, and I am not entirely sure what other people mean by it, and

C

until this has been agreed, not only on an anatomical and a biometrical basis, but also on a truly biomechanical basis, it would be unwise to regard bipedality as a unique feature by which you separate man from others.

CAVE (CHAIRMAN): Is it assumed, Professor Day, that all these fossil femora, shafts and so on, are standard for the type? One thinks of the variation of the neck of the modern human femur. A personal study made some 50 years ago revealed a dramatic range of variation in modern man, including almost ursine forms of femur. It is always assumed that new finds are typical of their kind. They may be, but—

DAY: This is an entirely valid point. We can only work within the range of the materials that we have, but when you find two femoral necks which show a constellation of characters which is present in both, and then you find a third, and now I think there are six, which are all—not exactly the same, but in which the same group of features can be discerned—then you have to accept it.

ALLBROOK: I wonder if Professor Day could please tell us why it was he picked on the Trinil femur to compare with those of Peking and Swartkrans? Was the point in analysing them all together because we really have no firm dating for any of them? Is that the reason?

DAY: No, it is not. The point I was trying to make there was that the Trinil femora have been for many years put up as being the model on which Middle Pleistocene hominine locomotion has been based, and the Peking femora have been virtually ignored—ignored to the extent that casts are not available in this country at all. The writings of Le Gros Clark and others have more or less dismissed Weidenreich's assessment of this material and the features that he described have been taken as being simply within the range of modern man. But on examining these again, as I was forced to do because I had a femur on my hands from Bed IV at Olduvai, I found a constellation of characters that I had seen before in Weidenreich's monograph (1941)*. Then, when we eventually got casts (and I have a pair with me that I should be delighted to show you) the similarities were striking. In effect it is like the mandibles; the North African mandibles are extremely similar to the Peking mandibles and people have even talked about genetic continuity between North Africa and the Far East. I think we are now also getting evidence of a similar nature from East Africa. Here was a femur from East Africa very similar in form to those from Peking and yet the classical Middle Pleistocene femur was from Trinil. Many workers have said previously that the Trinil femora are sapient in form. There are only two possibilities; one is that the Trinil femora fall within the normal range for Middle Pleistocene man, the other is that the Trinil femora are wrongly dated.

* See list of references, p. 46.

I will confess, concerning the Trinil femora, that we even considered whether we had another Piltdown situation. I can say that in fact we certainly do not. The Trinil material is all genuine fossil and remarkably preserved. These are the reasons why I looked at Trinil.

JOSEPH: I did not intend to ask this question until you said that the points that you made about one femur were confirmed by another and another and another. What I find surprising—and perhaps I am not the only one—is that as the result of measurements made on one terminal phalanx of one big toe of one animal, you and Professor Napier have made very definite statements about the way in which that animal walked. Why do you not wait until you have found another terminal phalanx and another terminal phalanx and at least another one? This is one of the reasons why a number of the definite statements that are made are found to be false later on.

DAY: A very good point. In fact there are a number of them. One of them has been known for a long time—it is from Kromdraai. It is not written up in Broom's monograph (1946)*, and was never recognized for what it was. There are also other phalanges with 803, and these will be published in due course.

* Broom, R. & Schepers, G. W. H. (1946). The South African fossil ape-men. The Australopithecinae. *Transv. Mus. Mem.* No. 2.

Symp. zool. Soc. Lond. (1973) No. 33, 53–69.

AUSTRALOPITHECINES AND HOMININES: A SUMMARY ON THE EVIDENCE FROM THE EARLY PLEISTOCENE OF EASTERN AFRICA

R. E. F. LEAKEY

National Museums of Kenya, P.O. Box 40658, Nairobi, Kenya

SYNOPSIS

Hominid fossils from the Plio/Pleistocene of eastern Africa have been recovered from seven localities of which the most extensive is East Rudolf in northern Kenya. This site comprises approximately 700 square miles of fluvial and lacustrine sediments representing a broadly continuous sequence of deposition from the Pliocene (circa 5·0 million) to the Early Pleistocene (circa 1·0 million). More than 70 hominid fossils have been recovered from East Rudolf since June 1968 when the area was initially explored.

At East Rudolf, one hominid model, *Australopithecus* sensu lato, can be documented as a chronospecies over a period of two million years. During this period there appears to have been little significant morphological change apparent in the elements preserved, both at East Rudolf and elsewhere in East Africa. This species may have achieved a plateau of specialization during the late Pliocene that precluded further adaptation. The fossils indicate a wide range of variation which might be explained in part on the basis of sexual dimorphism. The relative abundance of specimens, from what appears to have been a savanna habitat, might be significant in an understanding of social organization and behaviour. Post-cranial remains are not common, but there have been major additions to the collection, particularly with respect to the femur in which a long neck, small head and short robust, shaft are consistent characters of functional significance.

Evidence for a second contemporary hominid model is provided by cranial and post-cranial fossils from East Rudolf and other sites in eastern Africa. Dental characters combined with other features such as enlarged cranial capacity imply a model that was distinct from *Australopithecus* in behaviour, with a locomotor adaptation better suited to efficient bipedalism.

A distinctive third hominid model has recently been discovered at East Rudolf from deposits older than 2·6 million years. Both cranial and post-cranial material show features that indicate the need to re-examine our present concepts of hominid evolution, especially the status of *Australopithecus* and other proposed Pleistocene hominid lineages.

INTRODUCTION

Prior to 1960, most of the evidence for the evolution of man during the early Pleistocene was confined to southern Africa where, from 1924 onwards, a series of limestone caves in the Transvaal yielded a number of fossils that included both hominid and other vertebrate forms. While the faunal evidence suggests a lower to middle Pleistocene age for the Transvaal cave breccias, no precise data on the chronology is available. The absence of dating and the nature of the cave sites has unfortunately made it difficult to fully interpret the evolutionary significance of these hominids.

In 1959 the first significant hominid fossil was discovered in East Africa at Olduvai Gorge, Tanzania. Since then a total of seven additional sites have provided fossil evidence of early man in this part of the continent. These sites include: Chemeron, Chesowanja, Kanapoi, Lothagam Hill, and East Rudolf in Kenya; Peninj in Tanzania; and the Omo Valley in Ethiopia. All of these localities represent sedimentary formations where the stratigraphic succession is relatively clear and K/Ar dating is possible.

The greatest body of evidence for early hominid development has been obtained from the large site in northern Kenya known as East Rudolf. This area will be considered in some detail, but first a summary of the more important sites which have recently yielded fossil hominids will provide a useful background before describing the East Rudolf material.

EAST AFRICAN HOMINID FOSSIL LOCALITIES

Olduvai Gorge, Tanzania

This site has been worked intensively since 1931 by the late Dr L. S. B. Leakey and his wife Dr M. D. Leakey. The geological history of the Gorge has been studied in detail and is described by Hay (1963, 1967). This study has also provided evidence on the paleoenvironment of the early hominids in this area. While radiometric dating at Olduvai is hampered by the scarcity of suitable horizons, several dates have been obtained (Curtis & Hay, 1972). The lowest tuff in Bed I, tuff I B has been dated at approximately 1·8 million years (Curtis & Hay, 1972).

Over 40 hominid specimens have now been recovered from horizons throughout the sequence at Olduvai Gorge; the principal specimens include the cranium, Old. Hom. 5 named *Australopithecus boisei* (Leakey, 1959) and some material which has been assigned to *Homo habilis* (Leakey, Tobias & Napier, 1964). The former is widely accepted as an early Pleistocene robust form of *Australopithecus*, probably specifically distinct from *A. robustus*. The *Homo habilis* material is more complex in so far as the type specimen, Old. Hom. 7 is held by some to represent a gracile example of *Australopithecus* while the paratype from just above the Lemuta Member is clearly referrable to *Homo*. The presence of two hominid forms in the lower horizons at Olduvai is thus generally accepted and disagreement tends to focus on nomenclature. In addition a calvaria, Old. Hom. 9 (Leakey, 1961), which appears to be

attributable to *Homo erectus*, has also been recovered from Olduvai introducing a third form of hominid.

In addition to the hominid fossils, Olduvai is renowned for the evolutionary sequence of lithic material provided by assemblages of stone implements, often associated in living floor contexts, from the earliest levels up to the uppermost quite recent Bed IV deposits.

Peninj, Tanzania

This small locality on the west side of Lake Natron was first discovered in 1963. Preliminary investigations located a well preserved hominid mandible (Leakey & Leakey, 1964) from deposits dated by K/Ar at between 1·4 and 1·6 million years (Curtis, 1967), the mandible has been provisionally attributed to *Australopithecus* cf. *boisei* (Tobias, 1965).

Chemeron, Kenya

This locality consists of a series of small sedimentary outcrops lying to the west of Lake Baringo in Kenya's Rift Valley. Preliminary radiometric determinations confirm faunal evidence for a late Pliocene age for the basal part of the succession. In 1966, a fragment of hominid temporal was reported (Martyn & Tobias, 1967) and tentative identification was made attributing the specimen to *Australopithecus*. Although the fossil was located on the surface, there is no reason to doubt the reported provenance and approximate age of between 2 and 3 million years. The fragmentary condition of the specimen makes it difficult to assess its taxonomic affinities but there are several features that suggest differences from *Australopithecus* sensu lato.

Chesowanja, Kenya

Chesowanja represents a very limited outcrop of sediment lying to the east of Lake Baringo. Faunal evidence suggests an age of approximately 1·0 million years (Carney, Hill, Miller & Walker, 1971). A single hominid fossil consisting of the right half of a palate and maxilla with remnants of the cranium has been recovered from this site. The specimen shows dental characters that are typical of *Australopithecus* but the specific affinities of the fossil are not absolutely clear, it appears that the specimen represents a comparatively late example of the East African robust form of this genus (Carney *et al.*, 1971).

Kanapoi, Kenya

A single hominid fossil (Patterson & Howells, 1967) has been reported from this late Pliocene site lying to the west of Lake Rudolf.

The specimen consists of the distal end of a left humerus and it has been referred to cf. *Australopithecus*. The Kanapoi deposits have been estimated to be 4·0 to 4·5 million years old (Patterson, Behrensmeyer & Sill, 1970).

Lothagam Hill, Kenya

Lothagam Hill is an important Pliocene site also lying to the west of Lake Rudolf. A mandibular fragment with M_1 of a small hominid has been recovered from the lowest fossiliferous unit, Lothagam-1 which on faunal evidence appears to be 5·0 million years old (Maglio, 1970). Present tentative identification of the hominid specimen places it within the genus *Australopithecus* and it has been compared to *A. africanus* (Patterson et al., 1970). While the fossil is of considerable interest, in the absence of additional cranial or mandibular material from the Pliocene of Africa, there is every reason to be cautious as to its affinities. A number of authors refer to the Lothagam mandible in their discussions on the early appearance of the genus *Australopithecus* but this overlooks the possibility that the same specimen might represent a Pliocene survival of *Ramapithecus*.

The Omo Valley, Ethiopia

The Omo Valley represents an extensive Plio/Pleistocene formation that has produced invaluable data on this period. The locality has been the scene of intensive investigations since 1967 when an international programme of research was launched. The principal investigators, Professors F. C. Howell and Y. Coppens of the U.S.A. and France respectively, have reported hominid remains from many horizons (Arambourg & Coppens, 1967; Coppens, 1970a, b, 1971; Howell, 1969a, b, 1972a). The earliest deposits that have yielded hominid specimens at the Omo are above a tuff dated by K/Ar at 3·75 million years and the youngest are above a tuff dated at 1·84 million years (Howell, 1972). The major part of the hominid collection from the Omo Valley consists of isolated teeth and tooth fragments although some cranial and postcranial material has also been recovered. Detailed reports on the Omo Valley collection have not yet appeared but preliminary notes indicate the occurrence of two distinct hominid forms throughout the succession. The most common of these appears to represent a robust australopithecine while the second has been compared with the South African gracile australopithecine, *A. africanus* (Coppens, 1970b; Howell, 1972b).

East Rudolf, Kenya

This is a very extensive locality lying to the east of Lake Rudolf in

northern Kenya. The wealth of fossil hominid material known from the area and my own invested interest leads me to expand my account of this locality. In 1968, an initial reconnaissance party noted fossiliferous sedimentary outcrops extending over an area of some 700 miles. Since then detailed investigations have been undertaken as a continuing programme and a total of 87 fossil hominid specimens have been recovered to date (Leakey, 1970, 1971, 1972, in press a, in press b).

The locality can be divided into a series of collecting areas (Maglio, 1972) with the earliest deposits, the Kubi Algi Formation, lying generally to the south and dated at approximately 4·5 to 2·0 million years; the intermediate deposits, the Koobi Fora Formation, lying roughly centrally and dated at approximately 3·0 to 1·5 million years; and the youngest deposits, the Ileret Formation, outcropping to the north and dated at approximately 2·5 to 1·0 million years. There is evidence of a generally continuous succession with only localized disconformities. Since 1968, collecting has been focussed on the deposits that are in the upper half of the sequence and no hominid remains are yet known from levels that could be older than 3·0 million years.

Throughout the sequence, from which hominids have been recovered, there is evidence of the two genera, *Australopithecus* and *Homo*, the former being the most common. A total of 49 specimens of *Australopithecus*, 28 specimens of *Homo*, and 10 specimens of uncertain taxonomic attribution have been collected.

The *Australopithecus* specimens include cranial and postcranial material and appear to represent only one species. Within the species there is a wide range of variation which can in part be explained by sexual dimorphism; the range of dimorphism being greater than that seen in *H. sapiens* but similar to that seen in *Pongo*, *Pan*, and certainly *Papio*. Among the cranial specimens this is shown in the large collection of mandibular specimens and particularly in the large male cranium, KNM-ER 406, and the smaller female demi-cranium, KNM-ER 732 (Leakey, 1970, 1971; Leakey, Mungai & Walker, 1971, 1972). The postcranial collection, which includes many elements of the skeleton, also illustrates this point, but its importance lies in the opportunity it offers to study the locomotor habits of this hominid. It appears that *Australopithecus* was almost certainly bipedal, but there is a growing body of evidence to show that the australopithecine pattern of bipedality differs from that of *H. sapiens*.

The evidence of *Homo* is provided by a series of specimens from Ileret (Leakey, 1972) which appear to show affinities with some of the *Homo habilis* material from Bed II, Olduvai. One beautifully preserved mandible, KNM-ER 992 (Leakey, 1972), demonstrates this point.

During 1971, similar material was recovered from earlier levels of the Koobi Fora Formation (Leakey, in press a, b), so that it appears that this form can be traced back beyond 2·5 million years. Further geological investigations are necessary to establish this point.

In addition to these specimens, a parietal fragment from the upper levels in the East Rudolf sequence (Leakey, in press a, b) strongly suggests the presence of *Homo erectus*. More material is required to firmly establish this.

Artefact occurrences have been reported (Isaac, Leakey & Behrensmeyer, 1971) at East Rudolf associated with the KBS tuff dated by K/Ar at 2·6 million years. A sample of material is also known from below this level although little detailed excavation has yet been undertaken. These early artefacts predate the Olduvai Bed I assemblages quite considerably and show a remarkable sophistication in their manufacture in spite of the very early date. Their occurrence at East Rudolf in the early deposits raises the question of which hominid was making and using stone implements. I feel that the existence of well made stone tools implies the authorship by an example of the genus *Homo*.

Late in 1972 further and unexpected evidence for the early appearance of *Homo* was recovered from the collecting area 131, which is part of the Koobi Fora Formation. The geology of this area has been closely investigated and there is strong evidence to show that the greater proportion of the sedimentary exposures represent deposition prior to that of the KBS tuff which has been dated by K/Ar at 2·6 million years (Fitch & Miller, 1970). A cranium, KNM-ER 1470, showing clear affinities with the genus *Homo*, was recovered from an horizon 37 m below the KBS tuff (Leakey, in press a, b). Although samples of the KBS tuff taken from Area 131 have yet to provide dates, and at present the 2·6 million year date has been obtained from only one locality, Area 105, faunal and palaeomagnetic evidence is strongly in support of a pre 2·6 million year age for this cranium. Further radiometric determinations are being carried out to establish this, and results will be released in due course.

The cranium, KNM-ER 1470, had suffered extensive fragmentation prior to discovery but the many pieces have been reconstructed satisfactorily. The absence of temporal or sagittal crests, the weakly developed supraorbital tori and the endocranial volume of 800 cc plus, combine with other features to make the taxonomic attribution to *Homo* quite clear. Detailed studies are now being made but before these are completed it seems unwise to discuss the specimen in detail and I consequently limit my description to illustrations (Fig. 1).

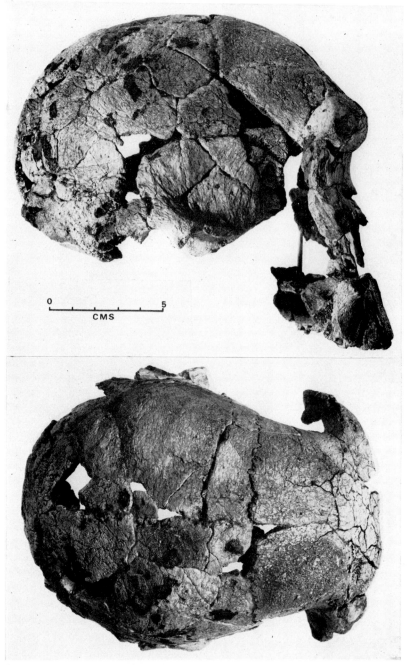

Fig. 1. Cranium KNM–ER 1470. Above, lateral aspect; below, superior aspect.

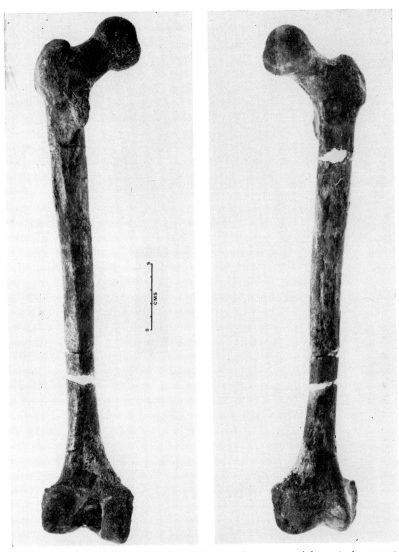

Fig. 2. Left femur KNM–ER 1481. Left, posterior aspect; right, anterior aspect.

Two femora and a fragmented tibia (Leakey, in press a, b) were also recovered from Area 131 during 1972, also from horizons below the KBS tuff. The femora are of particular interest in that they show features not previously seen in other early hominid femora and compare favourably with femora of modern man. The specimens are illustrated in Fig. 2 and a full description will appear after detailed studies have been completed.

The presence of a large-brained bipedally successful hominid at levels earlier than 2·6 million years comes very much as a surprise; it now seems necessary to reassess the presently held models of hominid development.

SUMMARY

Finally I would like to briefly summarize the evidence of the evolution of early man that we now have from East Africa.

Australopithecus sensu lato appears to be represented in deposits ranging from approximately 1·0 million years (Chesowanja, Olduvai) to less than 3·75 million years (Omo Valley); the early specimens have been identified on isolated teeth. There is a claim that the genus is also known at 5·0 million years (Lothagam Hill) but this is far from substantiated at present. *Homo* is less well known and, except at East Rudolf, is absent from sites older than 1·8 million years.

At Olduvai Gorge the contemporary presence of *Homo* and *Australopithecus* is known between 1·8 and about 1·0 million years and at this site the presence of *Homo erectus* in East Africa has also been established. Stone artefacts have been recovered throughout the sequence at Olduvai, the earliest being in deposits slightly older than 1·8 million years.

At East Rudolf the contemporary presence of *Australopithecus* and *Homo* already seen at Olduvai Gorge is confirmed and extends back beyond 2·6 million years. Thus the existence of both these forms is established for a span of time from a little over 2·6 million years to approximately 1·0 million years. At East Rudolf the marked sexual dimorphism of *Australopithecus* is clearly demonstrated and the many specimens of postcranial elements recovered from this locality permit a more certain appraisal of the mode of locomotion of this hominid. The presence of a form of *Homo* apparently similar to *Homo habilis* is extended back to a time prior to Olduvai. The unexpected appearance of a large-brained bipedal *Homo* at levels prior to 2·6 million years illustrates the complex nature of hominid evolution and negates the relatively simple picture previously suggested. Finally the occurrence of relatively sophisticated stone tools within the KBS tuff firmly establishes that 2·6 million years ago man was relatively skilled in this craft.

In conclusion I would like to say that East Africa has yielded a wealth of well documented material and it is to be hoped that future studies of hominid evolution will take cognizance of these collections which provide an important basis for interpretation. The National

Museum of Kenya serves as a repository for the material and welcomes bona fide scientific inquiry.

ACKNOWLEDGEMENTS

I should like to pay tribute to the National Geographic Society, the National Science Foundation, the William Donner Foundation, the National Museum of Kenya and other bodies for providing the financial support for the continuing East Rudolf Research Programme. Various colleagues have provided invaluable help and in particular I would like to thank my wife Dr M. G. Leakey, Dr A. C. Walker, Dr B. A. Wood and Prof. M. H. Day. I also thank the Zoological Society of London for their invitation to participate in the symposium and the Rhodes Trust for meeting my travel expenses.

REFERENCES

Arambourg, C. & Coppens, Y. (1967). Sur la découverte dans le Pleistocène inférieur de la vallée de l'Omo (Ethiopie) d'une mandibule d'australopithécien. *C. r. hebd. Séanc. Acad. Sci. Paris* (D) **265**: 589–590.

Carney, J., Hill, A., Miller, J. A. & Walker, A. (1971). Late australopithecine from Baringo District, Kenya. *Nature, Lond.* **230**: 509–514.

Coppens, Y. (1970a). Localisation dans le temps et dans l'éspace des restes d'Hominidés des formations Pli/Pleistocènes de l'Omo (Ethiopie). *C. r. hebd. Séanc. Acad. Sci.*, Paris **271**: 1968–1971.

Coppens, Y. (1970b). Les restes d'Hominidés des series inférieurs et moyennes des formations Plio-Villafranchiennes de l'Omo en Ethiopie. *C. r. hebd. Séanc. Acad. Sci., Paris* (D) **281**: 2286–2289.

Coppens, Y. (1971). Les restes d'Hominides des series supérieures des formations Plio-Villefranchienne de l'Omo en Ethiopie. *C. r. Acad. hebd. Séanc. Sci., Paris* (D) **272**: 36–39.

Curtis, G. H. (1967). Notes on some Miocene and Pleistocene potassium-argon results. In *Background to evolution in Africa*: 365–369. Bishop, W. W. & Clark, J. D. (eds). Chicago: University Press.

Curtis, G. H. & Hay, R. L. (1972). Further geological studies and potassium-argon dating at Olduvai Gorge and Ngorongoro Crater. In *Calibration of hominoid evolution*. Bishop, W. W. & Miller, J. A. (eds). Edinburgh and Toronto: Scottish Academic Press.

Fitch, F. J. & Miller, J. A. (1970). Radioisotopic age determinations of Lake Rudolf artefact site. *Nature, Lond.* **226**: 226–228.

Hay, R. L. (1963). Stratigraphy of Beds I through IV, Olduvai Gorge, Tanganyika. *Science, N.Y.* **139**: 829–833.

Hay, R. L. (1967). Revised stratigraphy of Olduvai Gorge. In *Background to evolution in Africa*: 221–225. Bishop, W. W. & Clark, J. D. (eds). Chicago: University Press.

Howell, F. C. (1969a). Remains of Hominidae from Pliocene/Pleistocene formations in the lower Omo basin, Ethiopia. *Nature, Lond.* **223**: 1234–1239.

Howell, F. C. (1969b). Hominid teeth from White Sands and Brown Sands localities, lower Omo basin, Ethiopia. *Quarternaria* **11**: 47–64.

Howell, F. C. (1972a). Pliocene/Pleistocene Hominidae in eastern Africa. Absolute and relative ages. In *Calibration of hominoid evolution*. Bishop, W. W. & Miller, J. A. (eds). Edinburgh and Toronto: Scottish Academic Press.

Howell, F. C. (1972b). Recent advances in human evolutionary studies. In *Perspectives on human evolution:* 51–128. Washburn, S. L. & Dolhinow, P. (eds.) New York, Chicago, San Francisco, Atlanta, Dallas, Montreal, Toronto, London, Sydney: Holt Rinehart and Winston Inc.

Isaac, G. Ll., Leakey, R. E. F. & Behrensmeyer, A. K. (1971). Archaelogical traces of early hominid activities, east of Lake Rudolf, Kenya. *Science, Wash.* **173**: 1129–1134.

Leakey, L. S. B. (1959). A new fossil skull from Olduvai. *Nature, Lond.* **184**: 491–493.

Leakey, L. S. B. (1961). New finds at Olduvai Gorge. *Nature, Lond.* **189**: 649–650.

Leakey, L. S. B. & Leakey, M. D. (1964). Recent discoveries of fossil hominids in Tanganyika: at Olduvai and near Lake Natron. *Nature, Lond.* **202**: 5–7.

Leakey, L. S. B., Tobias, P. V. & Napier, J. R. (1964). A new species of the genus *Homo* from Olduvai Gorge. *Nature, Lond.* **202**: 7–9.

Leakey, R. E. F. (1970). Fauna and artefacts from a new Plio-Pleistocene locality near Lake Rudolf in Kenya. *Nature, Lond.* **226**: 223–223.

Leakey, R. E. F. (1971). Further evidence of lower Pleistocene hominids from East Rudolf, North Kenya. *Nature, Lond.* **231**: 241–245.

Leakey, R. E. F. (1972). Further evidence of lower Pleistocene hominids from East Rudolf, North Kenya, 1971. *Nature, Lond.* **237**: 264–269.

Leakey, R. E. F. (in press a). Further evidence of lower Pleistocene hominids from East Rudolf, North Kenya, 1972. *Nature, Lond.*

Leakey, R. E. F. (in press b). Evidence for an advanced Plio/Pleistocene hominid from East Rudolf, Kenya. *Nature, Lond.*

Leakey, R. E. F., Mungai, J. M. & Walker, A. C. (1971). New australopithecines from East Rudolf, Kenya. *Am. J. Phys. Anthrop.* **35**: 175–186.

Leakey, R. E. F., Mungai, J. M. & Walker, A. C. (1972). New australopithecines from East Rudolf, Kenya (II). *Am. J. Phys. Anthrop.* **36**: 235–252.

Maglio, J. V. (1970). Early Elephantidae of Africa and a tentative correlation of African Plio-Pleistocene deposits. *Nature, Lond.* **225**: 328–332.

Maglio, J. V. (1972). Vertebrate faunas and chronology of hominid-bearing sediments east of Lake Rudolf, Kenya. *Nature, Lond.* **239**: 379–385.

Martyn, J. E. & Tobias P. V. (1967). Pleistocene deposits and new fossil localities in Kenya. *Nature, Lond.* **215**: 476–480.

Patterson, B., Behrensmeyer, A. K. & Sill, W. D. (1970). Geology and fauna of a new Pliocene locality in north-western Kenya. *Nature, Lond.* **226**: 918–921.

Patterson, B. & Howells, W. W. (1967). Hominid humeral fragment from early Pleistocene of northwestern Kenya. *Science, N.Y.* **156**: 64–66.

Tobias, P. V. (1965). An early *Australopithecus* and *Homo* from Tanzania. *Anthropologie* **3**: 43–48.

DISCUSSION

CAVE (CHAIRMAN): I think we have all been very privileged to be shown this evidence of Mr Leakey's discoveries, particularly what you or I would call a man and contemporary creatures with crested skulls, which anybody seeing them in the flesh would call apes—hominids if you like. This is quite surprising and affords food for thought.

ZUCKERMAN: Mr Chairman, may I first congratulate Mr Leakey, an amateur and not a specialist, for the very modest and moderate way he has given his presentation. May I also express my personal gratitude, and certainly the gratitude of many others who have worked with him, and his father for the work they have done, not as anatomists, as Mr Leakey pointed out, not as geochemists or anything else, but just as people interested in collecting fossils on which specialists can work. The generous offer that these are available to students is certainly something that we shall take great account of, and I trust use.

Even though, Mr Leakey, you say it is too early to draw conclusions, I do think that you have in fact drawn *the* conclusion. Had it been available before, the last piece of evidence you presented would never have permitted creatures with sagittal crests to be described as *Homo*. Nobody who was able to do an analysis of anatomical correlation, of the kind that Professor Day was talking about, starting with the sagittal crest, could end up with the supposition that one was dealing with a creature whose head was balanced on its vertebral column as in men like us—like the creature you have shown us. They are manifestly different and I think that the view some of us held that there were, more than a million years ago, protohumans or humans as well as these other australopithecine creatures, and—do not let us forget—chimpanzees and gorillas as well—was right; we now really know, and much of this is due to you.

DAY: I should like to add my congratulations on the remarkable achievement that has been made by Richard Leakey in East Africa. I say this because I have been privileged to witness personally what this involves. It is not just a matter of getting into a Land Rover and going and having a look on a hillside. There is a vast pyramid underneath of administrative work, even some political work, fund raising, collecting, organizing and so on. That this is being done for us, as scientists, by people as dedicated to the subject as the Leakey family are, is something for which I think we should be profoundly thankful.

ZUCKERMAN: Mr Chairman, in congratulating Mr Leakey I failed to put to him the question I wanted to ask. Your dating of 2·6 million years— that is the argon–potassium method, is it?

LEAKEY: Yes. The dating at East Rudolf is unfortunately not nearly as advanced as we all would have hoped. The fault lies with no person, but rather with the fact that the samples that we are collecting at East Rudolf are not behaving as they should, and the team at Cambridge, which I visited two days ago, are distinctly upset by the fact that these samples are not behaving in the way that they should behave, and consequently they are not prepared to release numbers to be used as dates until they fully understand what these numbers mean.

The only date that we have is the 2·6 date, which was taken from a tuff (volcanic ash) which is about midway down the total succession. This date is based on the potassium–argon technique, and both Professor Miller and Dr Fitch, who are responsible for the dating, are in the room, and if anyone would like to go specifically into details of that date—if they are worried about it—I know that both of them would be very happy to explain. I am not sure if that is your point or whether you just wanted to know that it was potassium–argon.

ZUCKERMAN: Yes, I take it that it is from the stratigraphic evidence too?

LEAKEY: In addition to the date of that tuff, we have very good geological stratigraphic control now throughout the succession, and we can relate all of the fossils that we are talking about to well-documented stratigraphic sections. There is no doubt at all that all of these important fossils are coming from where we think they are coming from, and these have been checked and counter-checked by different geological teams. We have to be absolutely sure that we are not dealing with deflated horizons or this kind of thing.

A SPEAKER: Could you, on behalf of Mrs Leakey, give us a word or two more about the artefacts? You say that they are very highly developed. Is it possible to say more than that at this time?

LEAKEY: All I would like to say about the artefacts is that when we found them in 1969 and showed them—we only had a sample of 17 in 1969—not only to my mother, but also to other archaeologists who are familiar with the Early Pleistocene developments, they felt a little unhappy with the suggestion that they were earlier than Olduvai. The reason for this is that they showed some characteristics which suggested that they were better made or made by a more skilled technique. This was subsequently helped by a much larger sample from excavation, of over 450 pieces from two sites. It is now possible to say that there is a distinct parallel with the early material at Olduvai, but that there are also differences. These differences might be attributed to the fact that at East Rudolf they were making artefacts with better raw material, but it is equally possible that those making artefacts at East Rudolf were better able to make them, and I think the fact that we have now a specimen of cranial capacity of 800+ at East Rudolf, whereas the Olduvai evidence is still below 650, suggests that this might also be the case. But I do not know, and I do not think that anybody knows. It requires considerably more fossil material to answer your question on that point.

D

NAPIER: May I pick you up on your point about sexual dimorphism? You went into this in detail so far as East Rudolf was concerned, but would you extend this diagnosis to the South African sites? If so, my follow-up question would be: would sexual dimorphism account for the differences between the two types of pelvis which, as Professor Day showed, are extremely distorted, but, distorted or not, are still morphologically different enough to be regarded as of two distinct types?

LEAKEY: What I would say in answer to that is that I do feel strongly that at East Rudolf there is evidence of sexual dimorphism, and the range or degree of this dimorphism can and will be documented. On a superficial look, which is all I am able to give it, it is quite clear that the range is very wide. A number of experts have pointed to the similarity between the Robust australopithecine in East Africa and that in South Africa, and as a complete amateur, and a layman if you will, I would merely say that if such a range of dimorphism exists in East Africa, is it not possible that a similar range would exist in South Africa? I do not know the answer, except to say that I do not think the South African collections have been looked at in the light of the East African collections. I would also say that when they are, there is no reason to suppose that everything that has been said about South Africa need necessarily be wrong.

WEINER: May I ask a simple-minded question, which I am sure can be demolished very quickly by the geologists? Is there any possibility at all that this volcanic ash layer of 2·6 million years really covers younger beds? Is that at all conceivable or is that a silly idea?

LEAKEY: I would answer that by saying that it is not for me to say whether it is a silly idea or not. I think, however, that on other evidence, such as the collections of vertebrate fossils, such as the evidence of paleomagnetic periods of reversal and non-reversal that we now have throughout this succession, there is no reason to think that we are in fact dealing with an elderly tuff covering young beds. I would also say that to me it is geologically inconceivable to have that at East Rudolf without very good evidence for it, and there is no good evidence for this to have been the case. But perhaps I could ask one of my team? We have a very strong Rudolf team in the room, and I think I would ask Dr Vondra just to comment specifically on that point, if that is all right, Mr Chairman.

VONDRA: I would say that it is inconceivable that the tuff is older than the underlying sediments. It is definitely younger than the sediments from which this hominid has been collected. The tuff has been designated the KBS tuff and has been dated at 2·6 million years B.P. It is traceable over a very wide area and the sediments in question occur beneath it in a normal stratigraphic sequence and therefore are older.

WORTHINGTON: Would Mr Leakey say a word or two about the associated fauna? He has mentioned that the vertebrate fauna helps a good deal with the dating and relationships of the human relatives, which implies a substantial amount of evolution of that fauna during the period concerned. It seems to me a little surprising that the *Australopithecus* apparently remain the same, indicating a zero rate of evolution, while the associated fauna, through its rapid evolution, is helping the dating.

LEAKEY: I think that was a very good point, and I am glad that you raised it because it will enable me to come back to *Australopithecus* and I hope help put *Australopithecus* in its right perspective in terms of human development. We do have a very large collection of fossil vertebrate forms from East Africa. We have a particularly large collection from the Omo in south-west Ethiopia, where there are tuffaceous horizons that have been well dated at regular intervals. So we have a very good chronologically documented situation upon which to base our comparisons. One of the most striking features about an examination of this total assemblage is that you do get a number of taxa that appear to have specialized or reached an adaptive plateau prior to the period that we are sampling. When you find them at the earliest levels, either at East Rudolf or the Omo, they can be traced right through this succession with virtually no apparent change in the morphology. This is certainly the case in *Australopithecus*.

There are other forms which appear to be evolving relatively rapidly, and these too can be traced through the succession. I think that if *Australopithecus* was the only species that showed no signs of change, one might indeed be worried. There are a number of pigs, a number of other Primates, a number of carnivores, not to mention the many different bovids that we have, that do different things, and some behave very similarly. It suggests to me that it is not surprising, but is rather indicative of the possibility that *Australopithecus* is a pre-Pleistocene specialization that came into the Pleistocene, well suited to a particular mode of life, and subsequently became extinct.

ZUCKERMAN: In the additional material which you have, have you any other creatures like *Australopithecus*? For example, have you any gorilloid material or chimpanzee material?

LEAKEY: In answer to that I would say that, as the collection stands at present, we do not seem to be finding any evidence of pongid occupation of this area. The evidence all points towards the probability that we are dealing with a savanna situation with some gallery forest or forest influence. The large proportion of the forms that we are obtaining appear, if one uses a modern day analogy, with a grassland situation, and therefore it is not perhaps surprising that we do not have any pongids. We do have a number of other Primates, including the baboon *Theropithecus*, the

true baboon and a number of the colobines. No other higher Primates
at this time are known anywhere from the Eastern African Pleistocene.

JOLLY: You mention in your abstract that this assemblage leads you to
speculate on australopithecine social behaviour. Would you care to enlarge on that?

LEAKEY: Yes, I did say that in my abstract and I would like to enlarge on it.
I was counting on only about 20 people coming to this seminar! I think
what I would say in answer to your point is that there is room for consideration of the possible implications of sexual dimorphism in the
Pleistocene situation. I think if one takes the modern situation and one
looks, as Lord Zuckerman did in his youth, at baboons, one finds that
sexual dimorphism does have social significance, in that a large group
usually has one or two extra large males for defence and social organization.
This I think is generally true in the terrestrial primate world with one or
two exceptions. *Australopithecus*, as represented by the fossils we are
now getting from East Rudolf, shows that the majority of specimens
are not particularly large, but that there are some that are hyper-large,
and that there is sexual dimorphism, with a very distinct difference in
size between what appear to have been the smaller females and the very
large males. I think someone with a lot more bravery than I have might
enter into a speculative study and interpretation of this in terms of
possible social organization. One could have a situation there where
these somewhat dumb Primates were sitting around on savannas, and
because of their lack of behavioural advance, had this very primitive
mechanism of social organization. I do not know, but I merely raise it
as one area of future research on the basis of this much larger sample
that has now become available.

ALLBROOK: As an anatomist from East Africa, I wish to add my warmest
congratulations and thanks to Mr Leakey for showing this material.
    In the light of the cultural finds and the remarkable new hominid
fossils, obviously of considerable cranial capacity, dated at 2·6 million
years, can we now say that Keith's "cerebral rubicon" was crossed, not
comparatively recently, but certainly well back into the Pliocene?

LEAKEY: Well, I should not like to get into this discussion at any great depth,
but I would say that on the material we now have there is very strong
evidence pointing to the probability that these things that we are calling
Australopithecus at East Rudolf and these things that we are not calling
Australopithecus at East Rudolf, developed prior to 2·6 million years.
If you want to call this the Pliocene, I would go along with it, but having
heard the exchange on era names a little earlier, I do not want to be
precise on that. But I do think that there is a very definite body of evidence
now pointing to the fact that the diversification and the adaptation of

these early hominids took place much further back than hitherto has been thought to be the case.

ZUCKERMAN: May I add to your comment. I believe that the question is not so much the Pliocene, but a dateless point. There have always been a few anatomists, I amongst them, who have refused to be bulldozed by a single fossil, or by attempted reconstructions. The evidence has always pointed to one conclusion that no-one could deny, the independent differentiation of a human line well into what used to be called the Pliocene/Pleistocene horizon. What has been revealed to us today is that these new creatures were there first. And as I said before, and let me repeat, had today's discovery been reported in this Society when the *Australopithecus* skull first rested on the speaker's bench in our old meeting room, any amount of time would have been saved. People would not have been turning themselves inside out, as they did with the Piltdown skull, in order to establish anatomical conclusions which were nonsensical. You may not have intended to, but you have demolished all that with your new skull.

LEAKEY: I am quite pleased that I have.

STRINGER: Could you tell us whether there is much evidence of hunting activity from Rudolf?

LEAKEY: The only evidence that we have of hunting, and it is difficult to say whether it is hunting or food gathering, is that at the sites where we have recovered artefacts, both at the 2·6 level and higher up in the succession, there is evidence of food refuse in the form of broken bones associated with artefacts on a floor situation. This might be taken as evidence of hunting. It might be taken as evidence that those tools were being used to do something to the animals that those bones represent. I would not say necessarily that there is evidence of hunting, but there is certainly evidence of tool use on mammalian prey.

CHAIRMAN'S SUMMING-UP

Summing up this session, those of us who have had the good fortune to be here will count it a most memorable occasion, at which novel material of the most far-reaching scientific importance has been presented to us, beautifully illustrated, by Professor Day and Mr Leakey. We are very fortunate to have been the recipients of this information and of the excellent scientific manner in which it has been presented.

Symp. zool. Soc. Lond. (1973) No. 33, 71–165.

SOME LOCOMOTOR FEATURES OF THE PELVIC GIRDLE IN PRIMATES

S. ZUCKERMAN

University of East Anglia, England

E. H. ASHTON

Department of Anatomy, University of Birmingham, Birmingham, England

R. M. FLINN

Department of Anatomy, University of Birmingham, Birmingham, England

C. E. OXNARD

Department of Anatomy, University of Chicago, Chicago, U.S.A.

and

T. F. SPENCE

Department of Anatomy, University of Birmingham, Birmingham, England

SYNOPSIS

The primate pelvis is constructed according to a common mammalian pattern. Certain minor variations occur between subhuman Primates, but the most conspicuous are between sub-human Primates as a group, and man. Certain of these contrasts—e.g. in the proportions of the iliac blade—have been much emphasized in the literature, but their mechanical significance is, in most cases, uncertain.

Interest in the general question of locomotor adaptation in the pelvic girdle was aroused almost 25 years ago by the discovery of an almost complete innominate bone and other parts of the pelvic girdle now assigned to the genus *Australopithecus*, together with more fragmentary remains from related groups. Frequent emphasis upon certain apparently human features of this fossil (e.g. the relative expansion of the iliac blade) has led to a view that *Australopithecus* was habitually bipedal.

In view both of the lack of precision implied in such comparison and also of the ill-defined significance of many features already studied, the present study has been undertaken to compare in sub-human Primates, man and *Australopithecus*, groups of features related to the pattern of forces impressed upon the pelvic girdle and therefore of definable mechanical significance in relation to locomotion.

Four of the characters selected, define the relative siting of the sacro-iliac and hip joints, thus being associated with the transfer of weight from the vertebral column, through the pelvic girdle, to the leg. Five further features relate to contrasts, established in a preliminary quantitative study, between the main blocks of hip muscles. These concern principally the orientation of the lesser gluteal muscles (abductors in man but not in sub-human Primates), and the lever arm of the hamstring muscles in relation to the hip joint (extensors of the hip being the more prominent in sub-human Primates).

Clear-cut contrasts as revealed by both univariate and multivariate study, exist between man and sub-human Primates. Within these latter, differences that are significant, although smaller in scale, exist between certain groups defined on the basis of the pattern of use of the hindlimb.

In the characters relating to transmission of weight, *Australopithecus* is quite like man and contrasts with all sub-human Primates. In features relating to muscular disposition, it tends to contrast with man and approaches the sub-human Primates. In the combination of both groups of characters it is unique.

The evidence provided by these findings in relation to the question of whether or not *Australopithecus* was habitually bipedal, is not unequivocal. If *Australopithecus* had been habitually or occasionally bipedal, its weight would have been carried more efficiently than in any sub-human primate. But any bipedalism must, because of the almost certain lack of any powerful source of abduction of the hip, have been quite different from that typical of *Homo sapiens*. The possibility is not excluded that the overall locomotor pattern of these extinct types might have included components additional to bipedalism, but as yet undefined.

INTRODUCTION

Despite major differences in overall bodily size and in the mode of use of the hindlimb, the mammalian pelvic girdle is constructed according to a common pattern, which, in turn, reflects that typical of lower vertebrate classes. A laterally placed innominate bone articulates ventrally and in the mid-line with the corresponding contralateral unit to form a symphysis, while dorsally, the innominate bones articulate with the alae of the sacrum. A complete bony ring of appreciable stability is thus formed. Each innominate bone comprises three regions: ilium, pubis and ischium. These, again, correspond with similar units in lower vertebrates and join constantly in a triradiate pattern in the acetabulum through which articulation is effected with the head of the femur. The ilium comprises the more dorsal part of the innominate bone, being expanded to a variable extent in a cranial direction by the side of the vertebral column. The ventral elements that are found in placental mammals comprise the ischium and pubis, and these unite to enclose the big obturator foramen. It is the more cranially disposed of these ventral elements (pubes) that contribute most or all to the ventral symphysis.

Features of the pelvic girdle unique to mammals concern, in the main, shape and proportions, Huxley (1879), for instance, noting that a distinctive characteristic of the mammalian pelvis derives from the fact that the part of the ilium, caudal to the sacro-iliac articulation ". . . elongates backwards, carrying with it the pubis and ischium . . ."

The caudal part of the pelvic girdle discharges a protective function for the reproductive organs, while the girdle as a whole gives attachment to several significant muscle blocks. These, although varying in relative proportions and, to some extent, in disposition, again conform to a general pattern throughout mammals. Functionally these muscle blocks fall into two main categories.

The first comprises muscles not directly related to the hip joint and has three principal subdivisions:

(a) the caudal extreme of the erector spinae complex attaching to the dorsal parts of the pelvic girdle,

(b) sheet-like muscles of the anterior abdominal wall attaching in a somewhat variable pattern to the ventral regions of the ilium and pubis, and

(c) a set of muscles responsible for moving the tail and extending to the ischial and pubic rami.

Elements of these are closely related to components that form, in some species, a muscular floor to the pelvic cavity.

The second contains muscles that are concerned with movement of the leg and with the mechanism of the hip joint. The group has five subdivisions:

(a) a set of protractors (flexors), taking origin from the ventral aspect of the iliac blade, sacrum and adjoining vertebrae and inserting into the proximal part of the femur;

(b) a block of retractors (extensors) consisting primarily of the hamstring muscles, taking origin from the ischial tuberosity, spanning, in general, both the hip and knee joints, and inserting into the proximal part of the leg. These have, as synergists, certain units from the gluteal group of muscles which, in turn, take origin from the dorsal aspect of the ilium and the dorsal part of the sacrum to insert generally into the proximal part of the femur;

(c) a group of abductors comprising a variable number of units from the gluteal complex that pass from the dorsal aspect of the ilium to the proximal parts of the femur;

(d) a block of adductors of the thigh. These originate from the lateral aspect of the ischio-pubic ramus and insert linearly into much of the length of the femoral shaft. Although distinct both morphologically and from the viewpoint of homology, the "adductors" are orientated—and notably in cursorial species—so as to work synergistically with the protractors and retractors, i.e. with the muscles responsible for producing the main locomotor movements. Finally,

(e) a group of short muscles takes origin from the regions of the innominate bone surrounding the acetabulum and inserts into the most proximal parts of the femur. These muscles help to maintain the stability of the hip joint.

This basic pattern of bones and muscles is modified by variations in their form and disposition—some being of striking extent. For instance, in certain artiodactyls (e.g. the dromedary—*Camelus*) the dorso-ventral dimension of the ilium and especially its crest are appreciably elongated, the neck of the iliac blade also being extended dorsally. In contrast in, for instance, a large number of quadrupedal Primates (e.g. the African green-bodied monkeys of the genus *Cercopithecus*), the blade of the ilium is elongated parallel to the vertebral column and is more rectangular in form. Again, in such widely divergent forms as the elephant (*Loxodonta*), the ground sloths of the suborder Xenarthra, and man (*Homo*), the blade of the ilium as a whole is much widened to form a conspicuous false pelvis lying cranial to the basin of the true pelvis.

The more caudal region of the pelvic girdle also displays marked contrasts, the symphysis, for instance, being relatively elongated in certain forms (for example in many Old World Primates—e.g. *Papio*— and in the kangaroo—*Macropus*), while in others (e.g. man (*Homo*)), it is relatively short. In certain bats of the suborder Megachiroptera the two pelvic bones may fail to articulate ventrally and there is thus no pubic symphysis.

The ischia are much more elongated caudally in certain cursorial animals (e.g. the horse—*Equus*), much shortened in graviportal types (e.g. the elephant—*Loxodonta*), and may even be fused with caudal vertebrae (e.g. some sloths of the superfamily Bradypodoidea). In Old World Primates of the superfamily Cercopithecoidea the ischia are much expanded, and the heavily cornified skin overlying the ischial tuberosities forms the ischial callosities.

The extremes of morphological variation in the pelvic girdle are seen in aquatic mammals. In the seals of the family Phocidae ". . . the pelvis is small, and of a different form from that of the terrestrial Carnivora. The ilia are exceedingly short, and with much everted upper borders; the pubes and ischia are very long and slender, inclosing a long and narrow obturator foramen, and meeting at a symphysis of very small extent, in which the bones of the two sides are very slightly connected . . ." (Flower, 1885). In other aquatic mammals (whales and dolphins of the order Cetacea together with the dugongs and manatees of the order Sirenia) the girdle has become rudimentary.

It is not easy to assess whether or not certain of these morphological variations—e.g. the proportions of the obturator foramen—are mechanically associated with functional differences in the pelvic girdle such as might derive from contrasts in function of the hindlimb, between, for instance, cursorial forms on the one hand and leaping and clinging types on the other. The significance of other differences can be more readily appraised. For example, the caudal extension of the ischial tuberosities, typical of cursorial forms whose locomotion is characterized by powerful and rapid cranio-caudal movement of the limb, gives the attached hamstring muscles a greater mechanical advantage. They are thus, as retractors of the hip, better able to contribute to the power stroke in locomotion.

Similar variations also exist in the pelvic muscles (e.g. Haughton, 1864, 1867; Howell, 1936, 1938). Most meristic contrasts affect, in the case of muscles, features that are less conspicuous than are those of the associated bones, but some quantitative differences exist between the amounts contributed to the total musculature by each of the principal blocks.

It is in the single form that is habitually and uniquely bipedal—*Homo sapiens*—that conspicuous variations, not only in muscular proportion but also in disposition exist, and in these features man appears to contrast with all other mammalian groups. For instance, m. gluteus maximus is exceptionally strongly developed and is orientated so as to retract (extend) the hip rather than to provide a relatively weak contribution to retraction or to abduction as in many quadrupedal mammals. Again, the lesser gluteal muscles (m. gluteus medius and m. gluteus minimus), in man pass lateral to the hip joint and therefore act as abductors. But in many other mammals, including subhuman Primates, they pass dorsal and caudal to the hip joint, and thus, by medial rotation and possibly retraction of the hindlimb, contribute to the locomotor power stroke.

Correlated with these and other conspicuous features of the musculature of the human hip, are corresponding features of the pelvic girdle, and especially of the innominate bone, that are also strikingly unique. These can be related to aspects of the mechanism of man's upright posture. Some have been analysed in earlier studies (e.g. Weidenreich, 1913; van de Broek, 1914; Waterman, 1929; Reynolds, 1931; Washburn, 1950) and have been summarized and interpreted by Le Gros Clark (1955) as:

"(1) the dorsal extension of the dorsum ilii, which brings the gluteus maximus and gluteus medius muscles into a different alignment with the hip joint; (2) the great dorsal extension of the iliac crest, which provides a more extensive attachment for the muscles used in supporting the trunk in the erect posture, and in particular for the powerful sacro-spinalis muscle; (3) a rotation of the sacral articulation associated with a reorientation of the sacrum as a whole relatively to the vertical axis of the os innominatum in an upright position; (4) a great reduction in the total relative height of the ilium; (5) the formation of an angulated and relatively deep greater sciatic notch associated with an accentuation of the ischial spine; (6) the development of a conspicuous and stoutly built anterior inferior iliac spine associated with the attachment of the iliofemoral ligament and the origin of the large rectus muscle; (7) an abbreviation of the body of the ischium with a corresponding approximation of the tuber ischii to the acetabulum; and as a corollary of the previous items (8) a relative approximation of the sacral articular surface to the acetabulum."

Such contrasts in osteological or in associated myological features are, in some cases, so great and so specifically orientated as to leave little doubt about their mechanical significance. But others, because of the complexities that are associated with mechanical adaptation in bony form, are more susceptible to varying interpretation. This can, however, be made more objective, if, as shown by work of the last decade, focusing principally upon the pectoral girdle (e.g. Ashton, 1972; Ashton, Flinn, Oxnard & Spence, 1971; Oxnard, 1967, 1973), mechanical adaptation is studied, not so much by concentrating upon those features that show the most conspicuous contrasts, but by selecting characters functionally related to the pattern of forces impressed upon the region. In relation to the pectoral girdle, this pattern can be defined relatively simply because, as the girdle is virtually completely suspended by muscles, it is these that are almost uniquely responsible for producing the pattern of impressed forces. But in the pelvic girdle, the picture is complicated because, in addition to the forces impressed upon the region during locomotion by contraction of the associated blocks of muscle, complexes of forces are produced by direct transmission of weight through the articulating bones. Yet others may result from visceral, e.g. reproductive, function. The most significant almost certainly derive from the locomotor functions of the hindlimb (weight bearing and muscular contraction).

Even within the limits of the Order Primates, there is, between different species and especially between genera, a range of locomotor patterns that results in an appreciable diversity of force patterns being impressed upon the pelvic girdle. For instance, in both Prosimii and Anthropoidea, there are many types that are relatively uncomplicated quadrupedal runners, either on the ground (e.g. *Lemur* from the Prosimii; *Erythrocebus* from the Anthropoidea) or in the trees (e.g. *Tupaia* from the Prosimii; *Cercopithecus* from the Anthropoidea). In these animals, the pelvic girdle is subject primarily to forces of compression that are, in general, non-impulsive in character. In contrast, and again in both suborders of Primates, there are genera in which leaping is a significant component of the total locomotor pattern (e.g. *Propithecus* from the Prosimii; *Presbytis* from the Anthropoidea). In such types, the function of the hindlimb is, in propulsion, dominant relative to the forelimb, and the pelvic girdle is subjected to powerful impulsive forces. Again, in other genera, the hindlimb is, for a greater or lesser part of the time, used for suspending the animal (e.g. *Perodicticus* from the Prosimii; *Pongo* from the Anthropoidea). In these types, the pelvic region is subjected extensively to tensile forces. Such forces are also applied to the pelvic region although to a lesser extent and, in a reversed functional pattern, in genera whose principal mode of locomotion is swinging from branch to branch by the arms, the hindlimbs tending to be suspended freely (e.g. *Hylobates*). Again, sometimes a unique pattern of tensile forces is applied to the pelvic region—for instance in such New World monkeys as habitually suspend themselves from a prehensile tail (e.g. *Brachyteles*), tensile forces are transmitted through the caudo-pelvic area rather than through the hip region.

The arrangement of such forces impressed upon the pelvic girdle does not necessarily bear a marked mechanical relationship to, or even statistical correlation with, the pattern of forces imposed upon the forelimb and shoulder such as can be summarized in the "forelimb locomotor classification":—brachiators, semibrachiators (hangers) and quadrupeds (Ashton & Oxnard, 1964a)—there being little correlation in particular species between the patterns of use of the fore- and hind-limbs.

Again, although various intermediate conditions obtain in loco-motor use of the hindlimb, there is, in the case of resultant force patterns impressed in the pelvic region, no strong analogy with conditions that exist in the shoulder where a single and complete spectrum of force patterns ranging from extreme tension to extreme compression can be defined (Ashton & Oxnard, 1964a; Stern & Oxnard, 1973).

Because of man's bipedal posture and gait, the human pelvis is subjected to a pattern of forces very different from that characteristic of quadrupeds. These forces are, as in quadrupedal runners, largely non-impulsive, but the contrasts arise because, in adult man, the leg is almost parallel to the vertebral column rather than at right angles to it as in quadrupeds. This results in differences in the direction and patterns of transmission.

Consideration of morphological variation, functionally associated with such contrasts in the pattern of impressed forces, is central to an interpretation of the features of the pelvic girdle in fossil Primates (Ashton, 1972; Ashton *et al.*, 1971; Oxnard, 1967, 1973). Interest in this problem was aroused some 25 years ago by the discovery in South Africa of an almost complete innominate bone assigned, at that time, to the australopithecine genus *Plesianthropus* (but now included in the genus *Australopithecus*). Certain earlier descriptions of fossils of this group (e.g. Dart, 1925) from the first emphasized characters (e.g. the presumed forward position of the foramen magnum of the skull base compared with the living apes) that were believed to indicate a poise of the head and posture differing from that characteristic of the living apes and approaching that typical of man. More recent studies (e.g. Ashton & Zuckerman, 1951, 1952, 1956) have failed to give support to certain such observations and interpretations, but the innominate bone has been widely regarded from its first and preliminary descriptions by the late Dr. Robert Broom and his colleagues (e.g. Broom, Robinson & Schepers, 1950) as contrasting in essential features with the apes. In apparently resembling man, its features have often been taken as indicating that these fossil creatures had assumed a bipedal posture and gait. A similar assessment was also made of recovered fragments of a pelvis now assigned to *Australopithecus robustus* (originally designated: *Paranthropus crassidens*) (Broom & Robinson, 1952). This general view was, in the years immediately following, reiterated by many scholars concerned with the study of this group, and possibly most forcefully by Le Gros Clark (1955). In that publication, it was maintained that, in combination, a large number of features of the fossil pelvis apparently related "to the mechanical requirements of posture and gait", indicated that "the posture and gait of the Australopithecinae were very different from those of any of the Recent large anthropoid apes" and that "the group had acquired the erect posture and gait distinctive of the family Hominidae". This view was stated to be valid despite a number of noted contrasts (e.g. in the extent of ventral extension of the anterior superior iliac spine) between the australopithecines and modern man. A similar appreciation characterized later works of this author (e.g. Le Gros Clark, 1962) and the same general view has been widely accepted and frequently reiterated in many works accepted as standard (e.g. Coon, 1963). It has also derived, although with some qualification, from certain independent studies of the pelvic girdle of the extinct group. Napier (1964), for instance, directed attention to differences believed to exist between the innominate bone of, on the one hand, the "gracile" australopithecine, formerly styled *Plesianthropus transvaalensis* (but now assigned to *Australopithecus africanus*), and, on the other, the "robust" forms formerly styled *Paranthropus crassidens* (but now assigned to the separate species *robustus* of the genus *Australopithecus*) and advanced the suggestion that, whereas a bipedal gait "closely similar to modern man" was characteristic of the more lightly built types, that obtaining in the "robust" species had not developed to nearly such a stage of perfection.

In addition to the considerable emphasis given to allegedly ᵤ̣ characters in the australopithecine pelvis, the early descriptions referred to a number of features in which the innominate bone of *Australopithecus* departed from the human condition. Broom (Broom *et al.*, 1950) for instance, pointed to such contrasts in the morphology of (a) the anterior superior iliac spine, (b) the acetabular border in relation to the anterior inferior iliac spine, (c) the extent of the auricular surface and (d) the shape and position of the ischial tuberosity. But in general, these have not been the features to which attention has been drawn in assessments of the functional morphology of the australopithecine bone that have led to the thesis that the general configuration of the australopithecine pelvis betokens an upright (bipedal) posture and gait.

Nevertheless, certain workers have expressed some measure of caution. For example, Napier (1964, 1967), while accepting that *Australopithecus* was bipedal, does not hold that the gait—and especially in the robust species—was like that in *Homo sapiens*. Such caution has derived to no small extent from certain features of the iliac blade. These had been detailed in an earlier and little quoted study (Mednick, 1955) in which attention was drawn to facts such as the general orientation of the iliac blade apparently tending, in the Australopithecinae, to be dorsal, as in the sub-human Primates, rather than lateral, as in man. Again a reinforced "pillar" of bone extending cranially from the acetabular border to the iliac crest and lying towards the anterior aspect of the iliac blade, which is believed to be typical of man and apparently related to the transmission of the uniquely human pattern of impressed forces, is, if present, not nearly so strongly developed in the Australopithecinae.

Further, an early biometric study of the australopithecine pelvis (Williams—reported in Zuckerman, 1954) although unavoidably based on limited material, dealing only with certain overall dimensions and, again unavoidably, using only univariate techniques of comparison, showed that supposed human resemblances of the general proportions of the innominate bone were probably less close than had been claimed.

Certain more recent publications have also drawn attention to features in which the australopithecine pelvis differs from that of man and have made further inferences about their probable mechanical significance. For instance, Jenkins (1972) on the basis of a ciné-radiographic study of bipedal walking in the chimpanzee, suggests that differences in the articular surfaces of the hip of *Australopithecus*, man and the chimpanzee may indicate that its type of bipedality was not the same as in modern man, but was intermediate between the human mode of walking and the rolling type of bipedal gait which is all that can be achieved by the chimpanzee. Again, McHenry (1972) on the basis of differences in the ischio-pubic part of the australo-pithecine pelvis, while not disputing that *Australopithecus* was bipedal, adds a further expression of the opinion that its bipedality must have differed from that of man.

From this complex of study (much of a non-quantitative type) and varying interpretation, certain facts appear to emerge: the mor-phology of the innominate bones of robust or gracile australopithecines, although possibly differing in detail, is fundamentally the same; their overall characteristics distinguish them from all living Primates; this anatomical complex of difference appears to include the combination of (1) a wide iliac blade, orientated dorsally, (2) a small sacral surface, (3) an anterior extension of the iliac crest and (4) a small acetabulum.

Although distinctive of the Australopithecinae, the significance of this combination of features, taxonomically or functionally, is not agreed, and this was the first of several considerations that prompted the inception of a further study of the primate pelvis and especially of that of the Australopithecinae.

Another consideration derived from the fact that most of the comparisons between the australopithecine pelvis and that of living man and apes had been based on visual assessment, and uncertainty thus attached to their precision and validity.

Again, uncertainty about the significance of any established con-trasts in pelvic form, stems from the present lack of information about its dependence, possibly through allometric mechanisms operating during phylogenetic and ontogenetic development, upon the overall size of the animal.

Further, even if the purely morphological assessments of similarity and difference prove valid, it is, both in the case of many of the in-dividual characters already stressed in earlier studies and in that of their anatomical compound, difficult to make any critical appreciation of mechanical significance in relation to posture and locomotion.

The present study was undertaken to delineate and prepare a quantitative assessment of the variance of characters of the pelvic girdle, believed to be of direct functional significance in relation to posture and locomotion. These have been derived from a study of the overall pattern of mechanical forces impressed upon the pelvic region and can be subdivided according to two major contributory factors: first, forces due to muscular pull, and second, forces arising directly from the weight bearing function of the pelvis.

The former have been defined as in studies of the pectoral girdle (Ashton, 1972; Ashton & Oxnard, 1964a, 1964b; Ashton, Healy, Oxnard & Spence, 1965; Ashton *et al.*, 1971; Oxnard, 1967, 1973) by preliminary quantitative assessment of contrasts in form and disposition of the associated muscle blocks. The pattern of the second has, at present, and pending the further development of techniques such as stress analysis of the bone (Oxnard, 1967), had to be inferred from a study of general mechanical considerations.

The opportunity has been taken to combine features, thus selected and defined, by the techniques of multivariate analysis, in order to provide a basis for enquiry: first, into the extent to which contrasts between the resultant constellations of functionally significant characters correlate, in extant Primates, with differences in the locomotor use of the hindlimb; secondly, and on this basis, into the extent to which, in mechanically significant features, the australopithecine pelvis resembles or differs from both individual genera of extant Primates and from the groupings into which they can be combined on the basis of function of the hindlimb. From this, an attempt has been made to derive an appraisal of the evidence provided by the pelvic girdle in relation to the general problem of the type of posture and gait that was habitual in the extinct Australopithecinae.

MATERIALS AND METHODS

Materials

Extant genera

Provenance. The study was based upon pelvic girdles of 430 extant Primates representing 13 genera from the Prosimii and 28 from the Anthropoidea (including man). The material was from the Department of Mammals (Osteology) and the Sub-Department of Physical Anthropology of the British Museum (Natural History); the Powell Cotton Museum, Birchington; the Departments of Anatomy and Zoology and the Duckworth Laboratory of Physical Anthropology of the University of Cambridge, and the Department of Anatomy of the University of Birmingham.

The number of specimens of each genus thus available varied between 1 and 112. In the 41 genera, 10 contained fewer than four specimens. Numbers of pelvic girdles studied are listed in Table I.

When a complete pelvic girdle was available, study centred upon the right innominate bone, certain supplementary measurements

E

TABLE I

Genera, locomotor groupings and numbers of specimens used in osteometric study of the primate pelvis and in ancillary study of hip muscle blocks. Serial numbers of genera are as used in text Figs 21, 22, 23, 24, 25 and 28

Sub-order	Genus	Hindlimb locomotor classification	Serial No. of genus	Number of specimens used in study of	
				Pelvic bones	Hip muscle blocks
PROSIMII	*Tupaia*	Quadruped (runner)	1	—	6
	Hapalemur	Quadruped (runner/climber)	2	2	—
	Lemur	Quadruped (runner/climber)	3	20	2
	Lepilemur	Quadruped (runner/climber)	4	5	—
	Cheirogaleus	Quadruped (runner/climber)	5	1	—
	Galago	Quadruped (leaper)	13	5	6
	Euoticus	Quadruped (leaper)	14	9	—
	Propithecus	Quadruped (leaper/clinger)	6	3	1
	Indri	Quadruped (leaper/clinger)	7	1	—
	Loris	Hanger and acrobat	9	4	6
	Nycticebus	Hanger and acrobat	10	2	5
	Arctocebus	Hanger and acrobat	11	4	1
	Perodicticus	Hanger and acrobat	12	8	5
	Daubentonia	Unclassifiable	8	2	—
ANTHROPOIDEA	*Macaca*	Quadruped (regular)	29	20	5
	Cercocebus	Quadruped (regular)	30	9	4
	Papio	Quadruped (regular)	31	12	3
	Mandrillus	Quadruped (regular)	32	1	2
	Cercopithecus	Quadruped (regular)	33	21	6

Erythrocebus	Quadruped (runner)	34	6	6
Aotus	Quadruped (facultative leaper)	15	5	5
Callicebus	Quadruped (facultative leaper)	16	6	4
Callimico	Quadruped (facultative leaper)	26	—	2
Callithrix	Quadruped (facultative leaper)	27	7	6
Leontocebus	Quadruped (facultative leaper)	28	12	5
Saimiri	Quadruped (facultative leaper)	22	12	6
Presbytis	Quadruped (pronounced leaper)	35	16	5
Rhinopithecus	Quadruped (pronounced leaper)	36	4	—
Nasalis	Quadruped (pronounced leaper)	37	1	—
Colobus	Quadruped (pronounced leaper)	38	15	4
Alouatta	Acrobat	20	8	3
Ateles	Acrobat	23	7	5
Brachyteles	Acrobat	24	1	—
Lagothrix	Acrobat	25	6	5
Pongo	Acrobat	41	14	4
Pan	Acrobat	42	20	4
Gorilla	Acrobat	43	20	—
Cacajao	Intermediate	17	6	4
Pithecia	Intermediate	18	5	4
Chiropotes	Intermediate	19	1	—
Cebus	Intermediate	21	7	6
Homo	Biped	44	112	29
Hylobates	Unclassifiable	39	10	5
Symphalangus	Unclassifiable	40	—	3
Australopithecus	Unknown	45	1	—

being made of the sacrum. Measurements of the left innominate bone were substituted when the right hand bone was either incomplete or missing, previous tests having shown there to be no consistent asymmetry.

Measurement was confined to adult specimens in which not only had the triradiate cartilage ossified to give a continuous bony union in the acetabulum between the ischium, pubis and ilium, but in which ossification had also occurred between the ischio-pubic rami. Again in so far as the available samples permitted, attention was also confined to bones in which the secondary centres of ossification overlying (a) the iliac crest and (b) the ischial tuberosity and ramus had also united with the main part of the bone.

Specimens showing any appreciable pathological changes were excluded from the study.

Locomotor grouping. To date, no study of locomotor use of the hindlimb appears to have been published in sufficient detail to make possible a reasonable analysis of variations in the force pattern to which the pelvic region is subject during locomotion. Such analysis is necessary as a basis for functional study of the pelvic girdle. A preliminary assessment was described in an earlier paper (Ashton & Oxnard, 1964a) which was concerned primarily with an analysis of the spectral pattern of tensile and compressive forces on the forelimb. It led to a definition of the force spectrum that relates to the locomotor use of this region in brachiators, semibrachiators (hangers) and quadrupeds. As in the pectoral region, this spectrum ranges, in the forelimb, from habitual extreme tension to extreme compression.

The pattern of forces in the hindlimb does not necessarily correlate with that in the forelimb. For instance, some species on occasion suspend themselves from the arms, which are thus subject to tensile forces. They may also, not infrequently, leap (e.g. *Nasalis*). The pelvic region is then subject to strong impulsive forces of compression. Other genera (e.g. *Lagothrix*), in which suspension is practised to an even greater degree, may seldom if ever leap—the pelvic region thus not being subject to impulsive compressive forces.

It was on the basis of the extent to which the hindlimb is subject or not subject to impulsive forces, that an initial classification of hindlimb use was attempted (Ashton & Oxnard, 1964a). But it was quickly realized that a more comprehensive grouping would be desirable for use in functional analyses of the hindlimb and pelvic girdle. For instance, some genera (e.g. *Ateles*, *Cacajao*), do not use the hindlimb for propulsion during leaping. Instead they may suspend themselves from the lower extremity, the limb and pelvic region thus being subjected to tensile forces. In such cases the forelimb may be subject to tensile forces (e.g. *Ateles*) or to forces of compression (e.g. *Cacajao*).

Again, further complexity is introduced into any grouping procedure because of appreciable variation in locomotor use of the hindlimb that may occur between the different species of the same genus (e.g. some species of *Presbytis* are pronounced leapers, while others are less so).

Any classification based upon the function of either limb can therefore be taken only as a basis for a biomechanical study of anatomical structure in this immediate region and should not be interpreted as being, in any way, an attempt to describe the overall pattern of locomotion typical of the genus.

Subject to such qualification, a grouping based purely upon locomotor use of the hindlimb has been derived and is summarized in Table I.

As with the forelimb classification (Ashton & Oxnard, 1964a), the categories do not have clearly defined boundaries, but an almost continuous spectrum of locomotor method and derived force pattern typical of the forelimb does not emerge nearly so clearly in the hindlimb classification.

Fossil material

Provenance. Descriptions have been published and casts made widely available of only a single fossil pelvis of the Australopithecinae that appears to be sufficiently complete to enable a full comparison of mechanically significant features to be made. This is the specimen of *Australopithecus* (*Plesianthropus*) recovered from the excavation at Sterkfontein (specimen sts 14) and originally described in full by Broom, Robinson & Schepers (1950). A cast of the specimen for the right hand side was, several years ago, made available to one of us (Zuckerman), through the courtesy of Professor J. T. Robinson. The contralateral bone is less complete and although much of the sacrum has been preserved, a full description and casts of this do not appear to be available. It has, however, been featured by Robinson (1964) in diagrams of reconstructions of the complete pelvic girdle of this genus.

Accuracy of cast. Although relatively few dimensions taken directly on the original fossil specimen appear to have been published, such as are available (Broom *et al.*, 1950) have been compared with those taken on this available cast of the fossil. In both major overall dimensions of the bone, and in more detailed measurements of certain specific features, the deviation between corresponding dimensions of the original and its cast was less than 0·5 mm. Subsequent tests using pelves of various extant Primates of similar general size showed that this is much less than one standard deviation from the mean of the measurements used in the present study. Such inaccuracies in measurement are therefore unlikely to affect the results of the comparisons.

Reconstruction. This one available and reasonably complete specimen of the innominate bone lacks an area around the anterior superior iliac spine. But, as noted and depicted in the original accounts (Broom *et al.*, 1950), the corresponding area is present on the contralateral bone, and an accurate reconstruction can therefore be effected.

Again, in the original specimen as recovered from the breccia, the ischio-pubic rami were distorted, the pubic symphysis and adjoining parts of the rami being displaced medially. Further reconstruction was effected by making accurate copies of the available cast. On these the ischio-pubic rami were carefully transected at the points of distortion and then reassembled in a way that gave a smooth and uninterrupted curvature to the arcuate line and to the lateral walls of the true pelvis. As a check upon this procedure, further reconstruction from additional copies showed that slight deviations from this norm in the positioning of the ventral parts of the ischio-pubic rami produced either abrupt changes in curvature or, alternatively, angulation of the arcuate lines such as is quite atypical of any living mammal.

Methods

Selection of morphological features

Overall pattern. Any attempt to relate individual features of the pelvic bone with a particular functional complex is, to some extent, artificial, as all are correlated. Nevertheless, certain features would appear, *a priori*, to be more clearly related to the function of the pelvis in transmitting weight from the trunk to the femora while others seem to have an equally strong mechanical relationship with the pattern of forces produced by contraction of the several associated blocks of muscles.

Characters related to weight bearing. Weight is transmitted through the pelvis to the leg via the sacrum, sacro-iliac articulation and hip joint. In this context, therefore, mechanical relevance attaches to the positions of these two major regions of articulation of the innominate bone, first in relation to each other in both the cranio-caudal and dorso-ventral directions, and secondly in relation to (a) the dorsal

and ventral extremities of the pelvic girdle, and (b) its cranial and caudal boundaries.

Characters related to muscular pull. These characters were defined on the basis of an analysis of contrasts: (a) in orientation of certain muscle groups and (b) in relative proportion of such groups.

Gross contrasts in muscular orientation were defined by inspection of series of dissected specimens, contrasts in proportional composition of the muscle blocks being assessed by a quantitative analysis of their relative percentage weights.

Related bony features were finally defined in consideration of: (a) the size and position of attachments of such muscles and muscle blocks and (b) the length and disposition of the bony lever arms connecting with the associated joints.

Contrasts in muscular orientation relate, principally, to the gluteal muscles, on the one hand, and to the abductors on the other. In sub-human Primates, the lesser gluteal muscles (m. gluteus medius and m. gluteus minimus) (Fig. 1) are disposed dorsally and contribute to the power stroke by retraction and medial rotation; in man (Fig. 2) they pass lateral to the hip joint, and are thus abductors. Again, in sub-human Primates, the cranial part of m. gluteus maximus (Fig. 3) when present (this being equivalent to the anterior portion inserting high on the femur as described by Stern (1971, 1972)) passes lateral to the hip joint, thus becoming an abductor. Only the caudal part and associated slips then act as retractors. But in man, the whole muscle (Fig. 4), although homologous with only the cranial part in the sub-human Primates, assumes a retractor function.

Further, the adductors (with the exception of the hamstring part of m. adductor magnus which is throughout Primates a retractor), whereas in man subserving mainly an adductor function, in sub-human Primates also contribute significantly to both protraction (ventral elements) and retraction (dorsal elements).

Contrasts in proportional composition of the pelvic muscle blocks emerged from a much more extensive enquiry into the morphology (quantitative and qualitative) of the hip muscles. This was based on study of a series of 167 specimens, listed in Table I and representing eight genera from the Prosimii and 25 genera from the Anthropoidea (including man). The human data were supplemented by quantitative information derived from normal subjects included in a work by Theile (1884), an extensive series of statistical tests having confirmed that no incompatible data were being introduced as a result of, for instance, age, or other factors. Muscles together with associated tendons and aponeuroses were dissected out and weighed, the weights were

TABLE II

Relative (percentage) contribution to total hip musculature of individual, functional muscle blocks in man and sub-human Primates

Muscle block	Flexors		Extensors		Adductors		Abductors		Short muscles	
	Mean	S.E. of Mean	Mean	S.E. of Mean	Mean	S.E. of Mean	Mean	S.E. of Mean	Mean	S.E. of Mean
Man (29 subjects)	22·7	0·37	35·4	0·34	22·4	0·37	15·5	0·30	3·9	0·14
Sub-human Primates (averages for 32 genera including 138 specimens)	24·9	0·76	49·5	0·87	17·2	0·69	3·5	0·25	4·9	0·18

compounded into functional muscle blocks; the means and variances of the percentage weights of each functional block were computed for each genus of Primate, and comparisons made by the standard statistical techniques. The results of this preliminary study are summarized in Table II.

Although in the case of each functional block, minor contrasts exist between different prosimians, monkeys and apes, the chief divergence was between man and the sub-human Primates. This contrast applies primarily to the retractors (extensors) and to the abductors.

In sub-human Primates, the retractors contribute some 15% more to the total bulk of the hip musculature than is the case in man, despite the great enlargement of the human m. gluteus maximus. In the sub-human Primates the abductors contribute some 10% less to the total hip muscle bulk than in man. The total relative bulk of protractors scarcely differs while the adductors vary by only some 5%. The short muscles account for less than 5% of the total in both man and subhuman Primates.

The significance of these percentage values is imprecise, as they do not take into account the detailed features of muscle morphology. Nevertheless, they lend some quantative reality to impressions about overall muscular size.

Definition of dimensions

Plane of reference. Ideally the plane of reference for measurements relating to the mechanical functions of the innominate bone should be based upon the entire pelvic girdle, and its position in the body. However, as no cast or detailed description is available for the sacrum, in the case of the one reasonably complete fossil specimen of an innominate bone available for study (*Australopithecus africanus transvaalensis*—sts 14), it was appropriate that a plane should be defined on the innominate bone alone. This was selected as being directed craniocaudally and defined by three osteometric points (Figs 5–11).

1. The most cranial point on the line of junction between the internal surface of the pubis and the medial aspect of the pubic symphysis;

2. the most caudal point on the line of junction between the internal surface of the pubis and the medial aspect of the pubic symphysis; (neither of these points apparently being affected by such imperfections as exist in the symphyseal region of the available fossil);

3. the most dorsal point on the caudal aspect of the auricular surface of the sacro-iliac joint.

Enquiry in the 41 genera of extant Primates used in this study showed that the mean value for the angle subtended by this plane and

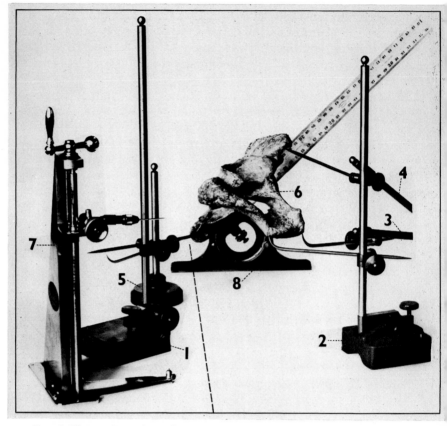

Fig. 5. Human innominate bone orientated for projection of osteometric points. 1, 2, 3. Engineer's surface gauges defining standard osteometric plane. 4, 5. Pointers defining cranio-caudal axis. 6. Innominate bone (human). 7. Projector. 8. Engineer's bevel square.

the mid-sagittal plane was small, averaging only 24° with 90% fiducial limits of 13° and 35°—i.e. the range of variation was also relatively small. The value of this angle, in so far as it can be assessed in the one complete pelvis of *Australopithecus* as measured on the diagram of the reconstruction published by Robinson (1964) was 24°—i.e. almost exactly equal to the average for all the modern Primate genera studied. The corresponding values for man (32°) and for the three great apes (*Pan* 20°, *Gorilla* 19°, *Pongo* 23°) were again all close to the overall mean for extant Primates.

The relative constancy of this angle indicates that dimensions taken on the innominate bone in relation to the defined standard plane are, in all cases, likely to be closely and relatively constantly related

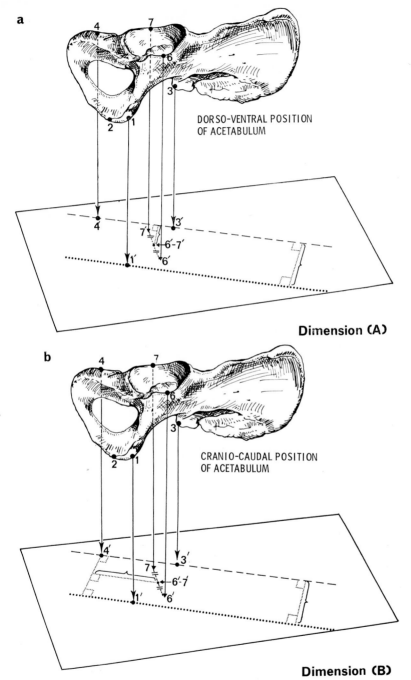

FIG. 6. Geometrical definition of four dimensions describing the relative disposition of the sacro-iliac and hip joints.

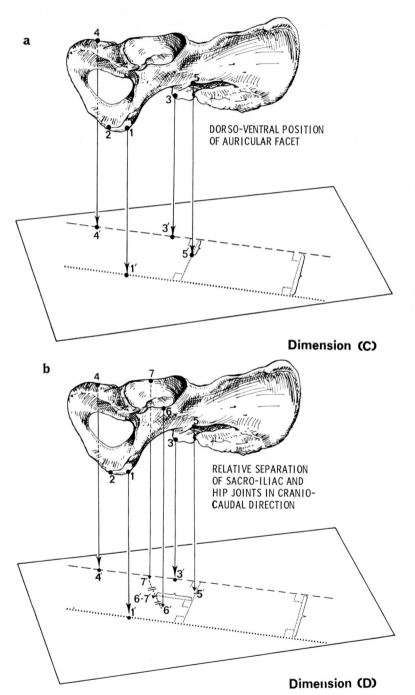

DORSO-VENTRAL POSITION
OF AURICULAR FACET

Dimension (C)

b

RELATIVE SEPARATION
OF SACRO-ILIAC AND
HIP JOINTS IN CRANIO-
CAUDAL DIRECTION

Dimension (D)

Fig. 7. Geometrical definition of dimensions describing the relative disposition of the sacro-iliac and hip joints.

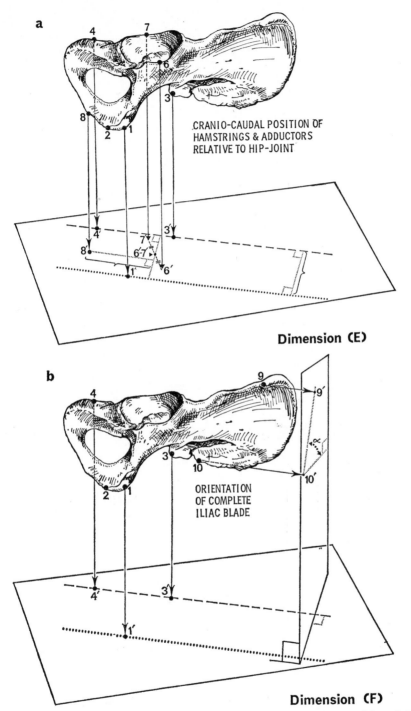

a

CRANIO-CAUDAL POSITION OF
HAMSTRINGS & ADDUCTORS
RELATIVE TO HIP-JOINT

Dimension (E)

b

ORIENTATION
OF COMPLETE
ILIAC BLADE

Dimension (F)

FIG. 8. Geometrical definition of five dimensions relating to the disposition of the hip muscle blocks.

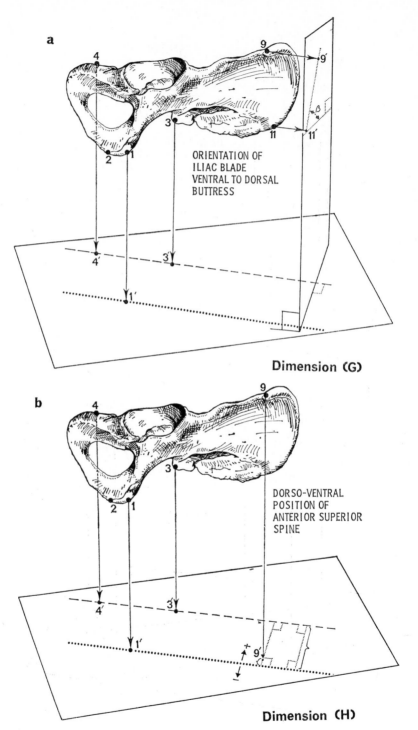

a

ORIENTATION OF
ILIAC BLADE
VENTRAL TO DORSAL
BUTTRESS

Dimension (G)

b

DORSO-VENTRAL
POSITION OF
ANTERIOR SUPERIOR
SPINE

Dimension (H)

Fig. 9. Geometrical definition of dimensions relating to the disposition of the hip muscle blocks.

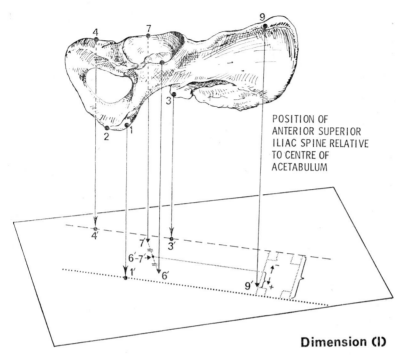

FIG. 10. Geometrical definition of dimensions relating to the disposition of the hip muscle blocks.

to corresponding dimensions that could be taken in relation to, for instance, the mid-sagittal plane of the pelvis.

The cranio-caudal axis of the bone (Fig. 11) was defined by the line joining the most dorsal point on the caudal aspect of the auricular surface of the sacro-iliac joint with the most dorsal point on the cranial (medial) edge of the ischial tuberosity, radiographs of each of the carcasses used in the ancillary analysis of the pelvic muscles (Table I) having shown that this line lay almost parallel to the general direction of the vertebral column.

Osteometric points. (Figs 6–11.) Eleven points on the bone were located and marked lightly in pencil. These were defined as follows, the terms cranial, caudal, dorsal and ventral relating not only to the topography of the entire animal but also to the standard osteometric plane and cranio-caudally directed axis in which the bone was orientated. The points thus selected, all of which were readily apparent on the fossil cast, were:

1. the most cranial point on the line of junction between the internal surface of the pubis and the medial aspect of the pubic symphysis;

F

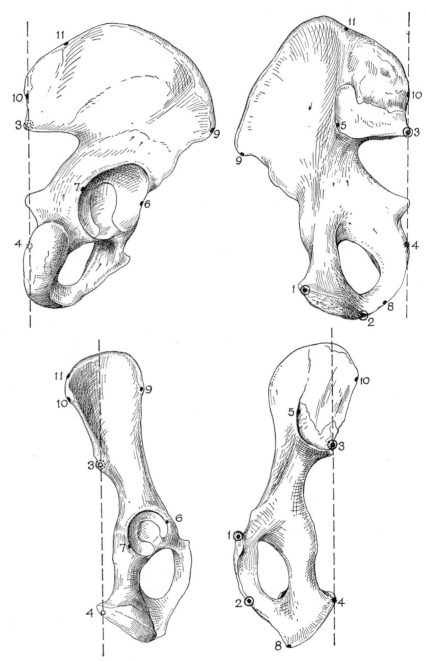

Fig. 11. Osteometric points and the cranio-caudal axis of the innominate bone in man and a quadrupedal monkey (*Papio*).

2. the most caudal point on the line of junction between the internal surface of the pubis and the medial aspect of the pubic symphysis;

3. the most dorsal point on the caudal aspect of the auricular surface of the sacro-iliac joint;

4. the most dorsal point on the cranial (medial) edge of the ischial tuberosity. (This practically coincides with the point at which the edge of the ischial tuberosity meets the ridge extending caudally from the sacro-iliac joint, the ischial spine projecting from its intermediate part);

5. the most ventral point on the surface of the sacro-iliac joint in relation to the standard cranio-caudal axis;

6. the most ventral point on the rim of the acetabulum in relation to the standard cranio-caudal axis. (This practically coincides with the point at which the ventral end of the dorso-ventral diameter of the acetabulum cuts its rim);

7. the most dorsal point on the rim of the acetabulum in relation to the standard cranio-caudal axis. (This practically coincides with the point at which the dorsal end of the dorso-ventral diameter of the acetabulum cuts its rim);

8. the most caudal point on the ischio-pubic ramus in relation to the standard cranio-caudal axis. (This is almost identical with the most caudal point on the edge of the area of muscular insertion on the ischial tuberosity);

9. the most ventro-lateral point on the ridge between the iliac crest and the medial aspect of the iliac blade. (This practically coincides with the anterior superior iliac spine);

10. the most dorsal point on the iliac crest. (This lies close to the posterior superior iliac spine);

11. the point on the ventral surface of the iliac blade at which the iliac crest is cut by the lateral edge of the dorsal buttress of the ilium. (This latter comprises bony ridges lying between the most dorsal extent of the area of origin of m. iliacus and the ventral aspect of the area of ligamentous attachment located cranial to the auricular surface).

Technique of measurement. The bone was clamped with the standard plane in the horizontal position as defined by engineers' surface gauges, whose points after preliminary alignment were applied to the three geometrical points on the bone defining the plane of reference (Fig. 5). Each of the 11 osteometric points was then transferred vertically downwards and recorded by the projector as a pin prick in an underlying sheet of cartridge paper.

Dimensions. The following linear and angular measurements were taken between the defined points to give indications of the functional quantities previously defined as significant in relation to the mechanics of the pelvic girdle. These are illustrated in Figs 6–10.

Dimension (A). The projected distance between the mid-point of the acetabulum as defined in Dimension (B) below and the standard cranio-caudal axis. This measurement describes the dorso-ventral position of the acetabulum. (Fig. 6a).

Dimension (B). The projected distance between the mid-point of the acetabulum (defined as the mid-point of the line joining osteometric point 6 and 7), and the caudal extremity of the bone as defined by osteometric point 4. This dimension is related to the cranio-caudal position of the acetabulum (Fig. 6b).

Dimension (C). The projected distance between points 3 and 5 measured at right angles to the line of the standard cranio-caudal axis. This measure relates to the dorso-ventral position of the auricular facet (Fig. 7a).

Dimension (D). The projected distance between the mid-point of the acetabulum (again defined as in Dimension (B) above) and the line at right angles to the cranio-caudal axis passing through osteometric point 5. This dimension relates to the relative separation, in the cranio-caudal direction, of the sacro-iliac and hip joints (Fig. 7b).

Dimension (E). The projected distance between the mid-point of the acetabulum (as defined in Dimension (B) above) and the position of point 8—this measurement again being taken parallel to the cranio-caudal axis. It describes the position of the caudal extreme of attachment of the hamstring and adductor muscle blocks relative to the hip joint (Fig. 8a).

Dimension (F). The angulation of the complete iliac blade, measured as projected in a plane at right angles, both to the standard plane and to the cranio-caudal axis. It is defined as the angle between the standard plane and the projection on the plane at right angles of a line joining points 9 and 10. This measures the orientation of the entire iliac blade relative to the standard orientation of the bone and hence to the true sagittal plane (Fig. 8b).

Dimension (G). The angulation of the ventral part of the iliac blade; this was defined by an angle, designed and measured in a way similar to Fig. 8a above, but confined to the part of the bone between points 9 and 11. This measures the orientation of the part of the iliac blade ventral to the dorsal buttress—i.e. the part giving origin, on the dorsal surface, to m. gluteus medius and m. gluteus minimus and, on the ventral aspect, to m. iliacus (Fig. 9a).

Dimension (H). The distance between the projected positions of point 9 and point 1, measured, in the standard plane, at right angles to the cranio-caudal axis. This measure was regarded as positive in sign if the projected position of point 9 lay dorsal to the projected position of point 1. It portrays the dorso-ventral position of the anterior superior iliac spine (Fig. 9b).

Dimension (I). The distance, again measured in the standard plane and at right angles to the projection of the cranio-caudal axis between point 9 and the centre of the acetabulum as defined in Fig. 6b above. This measurement relates to the position, in the dorso-ventral direction, of the anterior superior iliac spine relative to the centre of the acetabulum (Fig. 10).

Test of measuring technique

In order to separate out any gross errors in recording dimensions due, for instance, to misreading of scales, bivariate plots were made of pairs of dimensions that appeared to be closely related and which preliminary tests had shown to be quite highly correlated. Most points in such plots fell in reasonably circumscribed bands; measurements lying away from this band were rechecked and invariably proved to be due to mistakes in measurement.

As a further check on the techniques of measurement (including the accuracy of identification of osteometric points), sets of dimensions were redefined and repeated on a series of five similar pelvic girdles from each of the genera *Pan* and *Macaca*, on six from *Euoticus* and on seven from *Callithrix*. These genera were selected as representative of the range of size differences encountered throughout the primate order.

Analysis showed that the contribution to the total variance of each dimension as a result of inconsistency in measurement was quite insignificant ($P < 0.01$) compared with that existing between specimens of the same genus. Such inaccuracy, on average, accounted for only 8% of the total variance and it is therefore unlikely that it could have influenced materially the statistical significance of differences of the magnitude recorded in this study.

Metrical analysis

Compensation for differences in overall size. As variation in size exists not only between individuals within a genus but, also, to a most marked extent, between different primate genera—e.g. between *Leontocebus* and *Gorilla*—it was necessary to compensate by standardizing each dimension against a quantity that could be taken as indicative of the general size of the animal and of its complete pelvic girdle.

The dorso-ventral dimension of the pelvis as defined by the shortest projected distance, measured in the standard plane, between osteometric point 1 (p. 99; Figs 6–10) (this being approximately equivalent to the most ventral point on the bone) and the standard cranio-caudal axis (Fig. 11), is a major dimension of the innominate bone that reflects its overall size and which also appears unrelated to any of the bio-mechanical features examined in this study. Also, analysis showed that this dimension correlates closely ($r = 0\cdot9$) with the crown-rump dimensions, obtained in the case of man from Bean (1922), in that of most primate genera from Elliot (1913) and from Napier & Napier (1967), in that of *Pongo* from Schultz (1941) and those of *Pan* and *Gorilla* from field records made available to us by the Curator of the Powell Cotton Museum (Birchington).

The projected dorso-ventral dimension of the pelvis thus appears to reflect fairly not only the general size of the pelvic girdle but also that of the animal as a whole, although, as pointed out by Biegert & Maurer (1972), the overall size of limbs is not necessarily constantly related to that of the trunk.

Two techniques were used to compensate for the difference in size. First, each linear (although not angular) measurement was expressed as an index, the projected dorso-ventral dimension of the innominate bone (reflecting overall size) being used as the denominator. Second, for each dimension (linear and angular) a regression was calculated, the projected dorso-ventral dimension of the pelvis being the in-dependent variate. After logarithmic transformation of the data (necessary to eliminate curvature in the regression lines indicative of an allometric relationship between each functional dimension of the pelvis and overall size as reflected in its dorso-ventral diameter) it was found that, for each variate, the regression lines for each genus had, within the limits of sampling error, similar slopes (i.e. similar allometric constants applied). Again, within the limits of sampling error, such lines were superimposed. It was, therefore, possible for each dimension, to compute an overall regression for all specimens of extant Primates used in the study. The slope of this overall regression varied from dimension to dimension indicating a different allometric relationship between each and the overall size.

For each dimension in each specimen in each genus, a correction was then applied in accordance with the slope of the relevant overall regression line in order to assess the probable value that would have obtained in that specimen had the overall size of its pelvic girdle (as depicted by the projected dorso-ventral dimension) been the same as the average for all the 430 specimens of extant Primates used in this study.

In the subsequent analyses, parallel computations were carried out using: (a) indices and (b) regression adjusted data. In all parts of both univariate and multivariate studies, the results obtained not only paralleled each other, but were so constantly similar that identical conclusions emerged from the results of either method of analysis.

This similarity was further emphasized when, at a later stage of computation, an overall correlation matrix within groups was derived. This showed that in each of the nine variates, the correlation between corresponding regression adjusted values and logged indices, was always greater than 0·9.

In the present study, therefore, it has been possible to present only the results obtained from analyses using the technique of regression adjustment—this being selected as generally giving a somewhat better degree of discrimination than was obtained from comparison between indices. The method is also logically more rigorous in cases, such as the present dimensions, where there are significant deviations from an isometric relationship with overall size.

Sexual differences. It was feasible to make an initial separation into male and female sub-groups only in the case of man and the great apes—for all of which adequate series of specimens were available. The sexes of the chimpanzee and gorilla skeletons used in this study were known from field records in the Powell Cotton Museum. The orang skeletons were sexed on the basis of anatomical examination of the customary cranial and dental features which tend to be more prominent in the male. In other genera of sub-human Primates in which sexual dimorphism is prominent (e.g. *Papio*) there were insufficient specimens available to make a separation into sex sub-groups feasible.

Human pelves were often not associated with skulls or parts of the postcranial skeleton. They were sexed by an examination of features recognized in standard anatomical texts as showing appreciable sexual dimorphism and including the contour of the brim of the true pelvis (more circular in the female), the size of the subpubic angle (larger in the female), the orientation of the ischial tuberosities and ischial spines (more everted in the female), the width of the sacrum (greater in the female), and the characters of the sciatic notches (shallower in the female).

Although the locomotor dimensions studied in the present analysis are not specifically related to any of the features normally listed as showing marked sexual dimorphism in man, in both man and each of three genera of great apes, the available samples were adequate to establish certain of their apparent sexual differences as significant statistically. But all contrasts were small in scale compared with those existing between different primate genera and locomotor groups. There-fore, and additionally because of the generally small numbers of specimens (most of them unsexed) available in other genera, males and females were, in each genus, pooled for all parts of the main data analysis. *Differences between species or races.* Lack of material in genera of sub-human Primates in which clearly defined species are recognized prevented any enquiry into contrasts that may exist in the dimensions examined, between the several constituent species and subspecies. In man, however, where a total of 112 specimens was measured, it was feasible to divide not only into sex subgroups but also into major geographic divisions ("races") of *Homo sapiens* (Caucasoid, Negroid, Mongoloid, Capoid and Australoid). Again, within certain of these, further subdivision was practicable; for instance the available sample of Caucasoids contained the following definable subgroups: mixed "modern", Ancient Egyptian, Mediaeval and Iron Age. Again, even though these ultimate subdivisions contained relatively limited numbers of specimens, statistical significance was shown to attach to certain of the differences that exist: (a) between sexes within racial subgroups, (b) between subgroups within major geographic subdivisions and (c) between major geographic subdivisions. However, as with sexual differences in both the great apes and man, these were always small in scale compared with the intergeneric differences and, therefore, in subsequent analyses, all racial as well as sex subgroups of *Homo sapiens* were pooled.

Basic statistical analysis and transformation of data. For each adjusted dimension in each genus (transformed back for this phase of analysis to linear measure), basic statistical data comprising the mean, standard deviation and standard error of the mean were computed. In each of the 31 genera for which four or more specimens were available plots were made of the standard deviation against the mean, and regression lines were fitted. In all cases, the regression coefficient was positive, the standard deviation increasing with the mean, and the data being in most instances, sufficient to show that the positive slope of the line was significantly greater than zero. In order to equalize the variances between genera, basic data were recomputed, using the values that had previously been transformed logarithmically prior to regression adjustment to compensate for differences in overall bodily size. Further plots and derived regression coefficients indicated that, in all cases, the previous increase in variance in the genera, in phase with increases in the mean, had been eliminated—i.e. the variances had been homogenized.

Computations were, therefore, throughout, based upon the logarithmically-transformed data that had previously been regression-adjusted to compensate for variation in overall size.

Precision of basic data. In the 31 genera for each of which four or more specimens were available, counts were made of the number of instances among the nine variates (logarithmically transformed and regression adjusted), in which the standard error of the mean was less than 5% of the mean, together with the total number in which it was less than 1%. Of the total of 279 sets of basic data (nine dimensions in each of 31 genera), the standard error of the mean was less than 5% of the mean in 261 instances (94% of the total) and was less than 1% of the mean in 126 cases (45% of the total).

Univariate comparison. Two sets of analyses were carried out in which each osteometric dimension was treated as a separate entity and in which account was thus not taken of statistical correlation with other dimensions in the series of nine.

First, for each variate, comparisons were made by means of an analysis of variance, between the 42 genera (including *Australopithecus*) used in the study. As in all instances, the differences proved to be highly significant statistically ($P < 0.001$), the significance of differences between selected pairs of genera was assessed by means of t tests, the standard error of each mean being derived from a common standard deviation computed from the total variance within genera. As the variance within the 42 genera had been homogenized by the logarithmic transformation it could thus be taken as applying equally to those genera for which only minimal numbers of specimens were available.

Secondly, contrasts between locomotor groups were defined by computing for each of the nine regression-adjusted dimensions, an overall mean and its standard error for each locomotor group derived from the series of means of its constituent genera. Even though relatively few genera were available in many locomotor groups, the precision of these supplementary basic data, as judged by the size of the standard error in relation to its corresponding mean, was sufficiently good to make comparison meaningful.

As an analysis of variance for each dimension showed that in both the Anthropoidea and Prosimii the overall pattern of contrast between all the locomotor groups was statistically significant ($P < 0.001$), comparisons between means for selected pairs of genera were, again, accordingly carried out by means of t tests.

Multivariate comparison.

(a) Basic procedures. In order to determine the overall pattern of similarity and difference between the different species of living and extinct Primates as portrayed by groups of dimensions examined in this study, further analysis was carried out using techniques which make proper allowance for the statistical correlations that exist between the various pelvic dimensions.

The basic multivariate statistic computed was the squared generalized distance between groups (Mahalanobis, 1936), calculated using the computational technique described by Gower (1966). This, although making full allowance for statistical correlation between dimensions, does not take account of their geometrical (i.e. anatomical) inter-relationship. But, in the context of the present problem, this factor does not assume a significance comparable with that which obtains in a study designed to define contrasts in overall shape and size (Ashton, 1972).

Two further analyses were undertaken in order to provide a readily assimilable method of presenting a majority of information contained in the matrix of generalized distances in multidimensional space. First, the technique of cluster analysis was applied to the 903 squared generalized distances between the 42 genera in order to group these hierarchically in accordance with degrees of similarity. Of the large numbers of types of cluster analysis now available, a single linkage analysis was selected. This, although not eliciting all possible information from the data, is of established statistical stability (Gower & Ross, 1969).

Secondly, canonical variates (Hotelling, 1936) were derived from the matrix of squared generalized distances again using the computational technique described by Gower (1966). Bivariate plots of the positions of each of the 42 genera were made, relative to those canonical axes which, as judged first from the latent roots of the associated matrix and second from an inspection of the coordinates of each genus in that axis, effected a marked degree of discrimination between the groups.

Although, by definition, the information contained in the axes as serially numbered, declines progressively—the majority of the information thus being contained in plots of the first few axes—some measure of discrimination was found to be effected by even the higher axes. Some information is thus always excluded if bivariate plots of only the early canonical axes are made.

For graphical presentation of the data, plots were made of the position of each genus in relation to pairs of axes, normally axes 1 or 2 and, in those cases where axis 3 effected marked discrimination, axes 1 and 3 in addition. Approximately 90% fiducial limits, described by circles of radius 2·14 standard deviation units, were then drawn in around each mean. Where several genera from a single locomotor group are juxtaposed, and their 90% fiducial limits therefore overlap, the boundaries of the overall territories only have been presented for clarity.

In each canonical axis, an assessment was made of the contribution effected by each osteometric dimension to the overall pattern of discrimination. This was effected by means of examination, first of the crude values of the corresponding loading factors as listed in the computer print-out, and secondly by a parallel examination of the products of each loading factor and the spread of generic means of each dimension as summarized by the standard deviation between these means.

(b) Computation. Computations were carried out on the KDF9 Computer in the Computer Centre of the University of Birmingham, using a programme that provided, for each analysis, basic statistical data for each variate of each genus, an overall correlation matrix within groups, a matrix of squared generalized distances, the sums of squares accounted for by each canonical variate (latent roots), the coordinates of group means in each canonical axis, together with the associated weighting (loading) factors and constant terms appropriate to centring upon an overall mean of zero.

The data, prior to processing by this programme, had been punched in a standard form on 80 column IBM cards, the accuracy of transfer having been checked both by electro-mechanical verification and by visual comparison of the original data with a list printed from the punched cards.

The programme, in addition to printing out the table of squared generalized distances, also provided for this part of the output to be punched, again in a standard format, on to a further pack of 80 column IBM cards. These, after checking by electro-mechanical interpretation and subsequent visual comparison with the printed matrix of squared generalized distances, were then subjected to cluster analysis using a programme prepared from the instructions provided by Gower & Ross (1969).

(c) Effect of inclusion of small groups. The initial multivariate analysis (squared generalized distances and canonical coordinates) was carried out including the entire group of 42 genera. A further analysis was then carried out including only the 31 genera for each of which four or more specimens were available, theoretical considerations indicating that the overall pattern of discrimination might be influenced if appreciable numbers of very small samples were included in the analysis (Gower, 1968). The final 11 genera (including *Australopithecus*) for each of which fewer than four specimens were available, were then positioned both in each canonical axis and then in the matrix of squared generalized distances by appropriate computations using the weighting factors provided in the original print-out. These calculations, although simple in concept, were, again, sufficiently extensive to warrant the use of a further computer programme.

Comparison showed that the overall results obtained by these two approaches were so similar as to warrant inclusion in each member of the group of canonical analyses finally carried out, of all 42 genera, irrespective of the number of specimens available in each.

(d) Final multivariate compounding of dimensions. Two sets, each containing three multivariate analyses, were finally carried out using different combinations of anatomical dimensions.

The first set of analyses was designed to elicit contrasts between genera and was thus based upon statistical data derived from each of the 42 individual genera. The second set was designed to elicit contrasts between locomotor groups and was based, as in the corresponding univariate analyses, on statistical data derived from contrasts in the mean values of the constituent genera of each locomotor group.

In each set, the first analysis included all nine dimensions (logged to homogenize intragroup variance and regression adjusted to compensate for differences in overall bodily size). It was undertaken to give an indication of the overall pattern of contrast in the functionally significant osteological features directly related both to joint position and to muscular disposition.

The second analysis of each set included the four dimensions related to the positions of the sacro-iliac and hip joints. Its results thus bear upon the mechanism associated with the transmission of weight from the vertebral column to the limbs. It compounded Dimensions A–D detailed on page 102 and in Figs 6 & 7.

The third analysis in each of the two sets was designed to establish the overall pattern of contrast in those osteological features related to the proportions and disposition of the principal blocks of muscle acting upon the hip joint. It thus reflected such features as the mechanical leverage of the muscle blocks responsible for propulsion or for stabilization of the pelvic girdle. It compounded Dimensions E–I detailed on pages 102–103 and in Figs 8–10.

A final canonical analysis was carried out in which data only for man and the great apes were included, but for which, coordinates of *Australopithecus*, in each canonical axis, were subsequently computed and tabulated, as also were its squared generalized distances from each of the four extant groups. This method of inserting the fossil specimen rather than including it in the initial overall analysis, was used in this one instance because, although adequate samples of bones were available for man and for each of the great apes, tests proved that, in accordance with the theoretical consideration detailed by Gower (1968) and probably because of the small number of groups involved in the analysis, the overall pattern of discrimination was distorted by the inclusion of one group for which only a single representative specimen was available.

This canonical analysis, including only man and the great apes, was also based upon a study of indices—the dorso-ventral dimension of the pelvis reflecting overall bodily size being used, in each case, as the denominator, and the data again being transformed logarithmically in order to homogenize intragroup variances. This was necessary because it was not possible to use the regression calculated for the 41 extant genera to correct for differences in overall size as only four of these genera (and actually four with the biggest bodily size) were involved.

RESULTS

Univariate study

Individual genera

General pattern of contrast. In most instances, the principal contrast in individual variables lay between man on the one hand and sub-human Primates on the other. In some cases, certain genera of the latter differed appreciably from the main group which, in any case, varied significantly within itself. In a minority of instances, certain variations among the main group of sub-human Primates were greater in scale than the contrast between man and the average of this latter group. *Australopithecus* fell in a variable position, resembling man in certain instances; contrasting with man and resembling different sub-human Primates in others.

Dimensions relating to joint position.

Dimension (A). The siting of the acetabulum (Fig. 12) in the dorso-ventral direction as portrayed by Dimension (A) (depicted in Fig. 6a and described on page 102) is more ventral in man than in any sub-human Primate. Appreciable variation exists among the sub-human primates, Old World monkeys and apes generally showing a somewhat more ventral position of the acetabulum (i.e. approaching man more closely) than obtains in New World monkeys or prosimians. Among the latter, the hanging Lorisinae show a position of the acetabulum more dorsal than in any other primate genus.

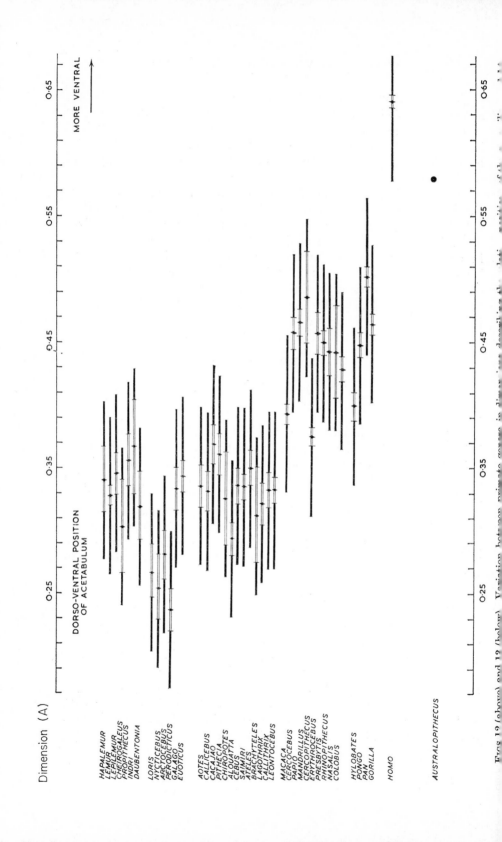

Figs. 18 (above) and 19 (below). Variation between primate genera in dimension ...

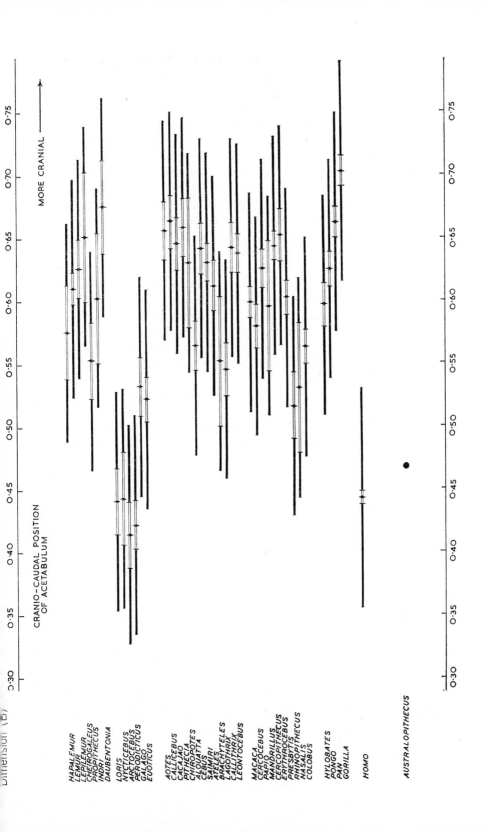

CRANIO-CAUDAL POSITION
OF ACETABULUM

MORE CRANIAL ⟶

Dimension (b)

0·30   0·35   0·40   0·45   0·50   0·55   0·60   0·65   0·70   0·75

HAPALEMUR
LEMUR
LEPILEMUR
CHEIROGALEUS
PROPITHECUS
INDRI
DAUBENTONIA

LORIS
NYCTICEBUS
ARCTOCEBUS
PERODICTICUS
GALAGO
EUOTICUS

AOTES
CALLICEBUS
CACAJAO
PITHECIA
CHIROPOTES
ALOUATTA
CEBUS
SAIMIRI
ATELES
BRACHYTELES
LAGOTHRIX
CALLITHRIX
LEONTOCEBUS

MACACA
CERCOCEBUS
PAPIO
MANDRILLUS
CERCOPITHECUS
ERYTHROCEBUS
PRESBYTIS
RHINOPITHECUS
NASALIS
COLOBUS

HYLOBATES
PONGO
PAN
GORILLA

HOMO

AUSTRALOPITHECUS

0·30   0·35   0·40   0·45   0·50   0·55   0·60   0·65   0·70   0·75

In this feature, *Australopithecus* lies at the extreme of the human range, being approximately intermediate between the mean for man and the most extreme of the Old World monkeys and apes (e.g. *Pan*). The fossil is, however, closer to the mean for man than to that of any genus of the New World monkeys or prosimians.

Dimension (B). The situation of the acetabulum (Fig. 13) in the cranio-caudal direction as portrayed by Dimension (B) (depicted in Fig. 6b and described on page 102) is more caudal in man than in the majority of sub-human Primates. In contrast to the previous dimension, the hanging Lorisinae (*Loris, Nycticebus, Arctocebus* and *Perodicticus*), in this respect, are clearly similar, on average, to man and contrast with the remaining sub-human Primates. The hopping Galaginae and leaping Colobinae are approximately intermediate between, on the one hand, man and the Lorisinae, and on the other, the remaining Primates.

Australopithecus falls (a) relatively close to the mean for man and the Lorisinae, (b) within the 90% fiducial limits for the intermediate hopping and leaping forms and (c) just outside the 90% fiducial limits of all other Primates. It is well removed from the mean position of any ape.

Dimension (C). The positioning of the auricular facet (Fig. 14) in the dorso-ventral direction as portrayed by Dimension (C) (depicted in Fig. 7a and described on page 102) is more ventral in man than in any other Primate. There is relatively little variation in this feature among genera of the sub-human Primates, except that the great apes tend to be somewhat closer to man than do most other sub-human Primate genera. *Brachyteles* appears to have the auricular facet placed exceptionally dorsally, but here, the result might have been influenced by sampling errors as only a single specimen of this rare genus was available.

In this feature, *Australopithecus* is most similar to the great apes and especially to *Pongo* and *Gorilla*, lying in a position approximately mid-way between the remaining sub-human Primates on the one hand and man on the other.

Dimension (D). The relative separation of the sacro-iliac and hip joints (Fig. 15) in the cranio-caudal direction as portrayed by Dimension (D) (depicted in Fig. 7b and described on page 102), is smaller in man than in the majority of sub-human Primates (with the exception of *Callithrix* and *Leontocebus*). However, the main contrast here is between the apes plus the prehensile and semi-prehensile tailed Cebidae, on the one hand, in which the separation of the joints is greater, and man, together with the remaining sub-human Primates, on the other, where the separation is less.

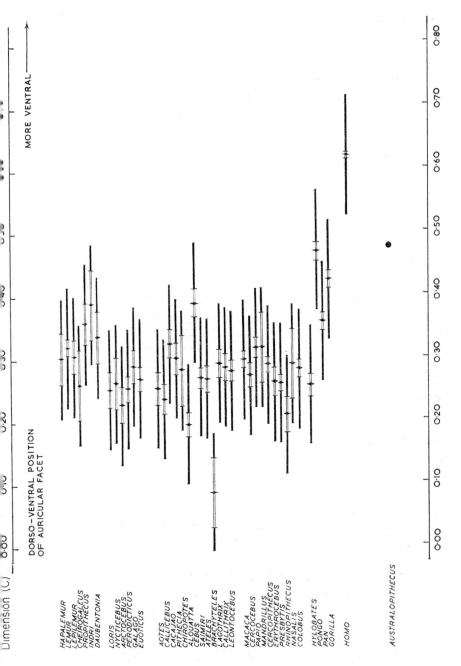

Fig. 14. Variation between primate genera in dimensions describing the relative positions of the sacro-iliac and hip joints—means and 90% fiducial limits. The band embracing two standard errors on each side of the mean indicates the variance of sample means.

Dimension (D)

FIG. 15. Variation between primate genera in dimensions describing the relative positions of the sacro-iliac and hip joints—means and 90% fiducial limits. The band embracing two standard errors on each side of the mean indicates the variance of sample means.

Australopithecus in this respect is similar to man, and thus, although lying towards the extreme of their variation, also resembles several other genera of sub-human Primates. But it contrasts with the great apes and prehensile and semi-prehensile tailed New World monkeys.

Summary (joint position). In so far as can be judged from a combination of univariate comparisons, account not being taken of statistical correlation between dimensions, it would seem that the principal consistent finding concerning the relative positions of the sacro-iliac and hip joints is that the specimen of *Australopithecus*, although in

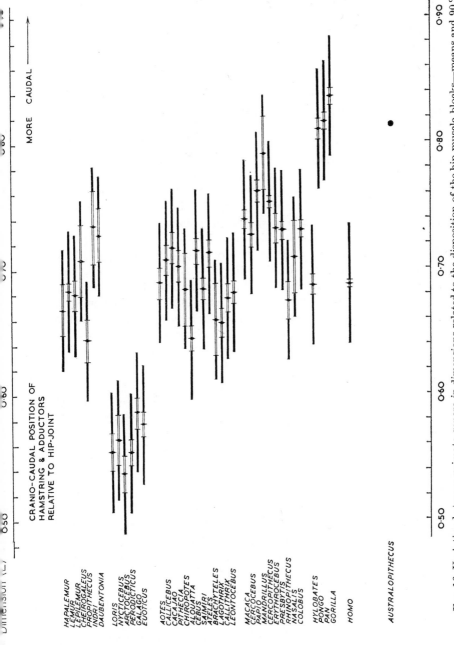

Fig. 16. Variation between primate genera in dimensions related to the disposition of the hip muscle blocks—means and 90% fiducial limits. The band embracing two standard errors on each side of the mean indicates the variance of sample means.

certain dimensions agreeing with various individual sub-human primate genera, tends, overall, to resemble man more closely than the sub-human Primates. In three dimensions out of four it shows a consistent contrast with the great apes.

Dimensions relating to muscular leverage and disposition.

Dimension (E). The position of attachment of the hamstring and adductor musculature (Fig. 16) in the cranio-caudal direction as portrayed by Dimension (E) (depicted in Fig. 8a and described on page 102) does not show an overall contrast between man and the sub-human primates. But the Lorisiformes are distributed as a distinct cluster, having a more cranial attachment of these muscles than in any other primate genus. At the other extreme are the great apes. Man lies, in respect to this character, in an intermediate position along with the remaining sub-human primate genera.

Australopithecus is, in this quantity, similar to the great apes, contrasting with man and practically all other sub-human Primates.

Dimension (F). In relation to the transverse plane, the entire iliac blade is orientated (Fig. 17), as described by Dimension (F) (depicted in Fig. 8b and described on page 102), in most Primate genera with the exception of certain Prosimii together with Callithricidae, more dorsally than in man. This dorsal orientation is most marked in the Pongidae, the Colobinae and the prehensile and semi-prehensile tailed Ceboidea all of which are widely divergent in this respect from man.

The orientation of the entire iliac blade in *Australopithecus* is more dorsal than in man. Although the plane of the iliac blade is, in this extinct genus, not directed so dorsally as in the apes, its orientation resembles that of a large number of sub-human primate genera.

Dimension (G). In relation to the transverse plane, the orientation of the iliac blade ventral to the dorsal buttress (Fig. 18), as portrayed by dimension (G) (depicted in Fig. 9a and described on page 102) shows a pattern of contrast somewhat similar to that revealed by the previous dimension, with the exception that in the Cercopithecinae the orientation of this part of the iliac blade is appreciably more dorsal than is that of the entire blade. The principal contrast thus becomes between, on the one hand, Old World monkeys and apes, and, on the other, the remaining Primates including man.

The position of *Australopithecus* is intermediate between, on the one hand, man, and, on the other, the Old World Primates together with the prehensile and semi-prehensile tailed members of the Ceboidea.

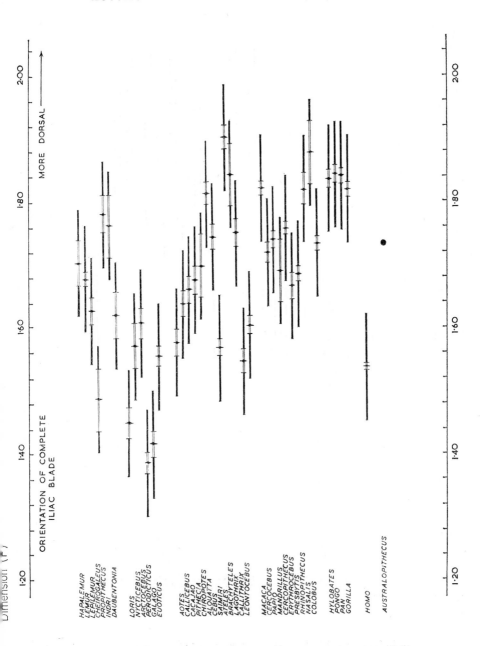

Fig. 17. Variation between primate genera in dimensions related to the disposition of the hip muscle blocks—means and 90% fiducial limits. The band embracing two standard errors on each side of the mean indicates the variance of sample means.

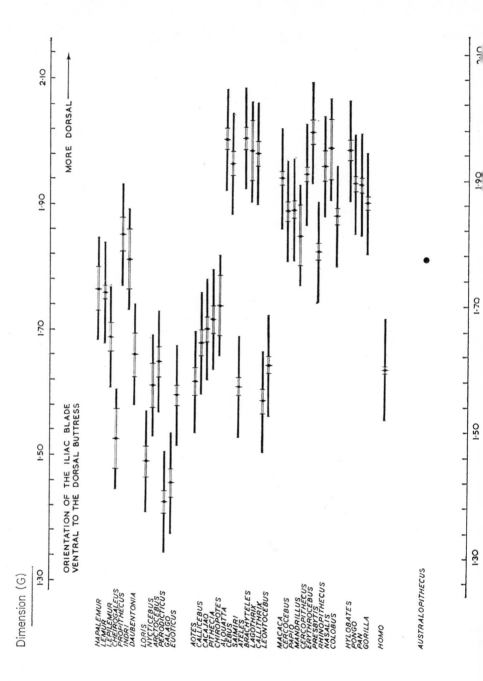

Figs. 18 (above) and 19 (opposite). Variation between primate genera in dimensions related to the disposition of the hip muscle blocks—means and 90% fiducial limits. The band embracing two standard errors on each side of the mean indicates the variance of sample means.

Dimension (H)

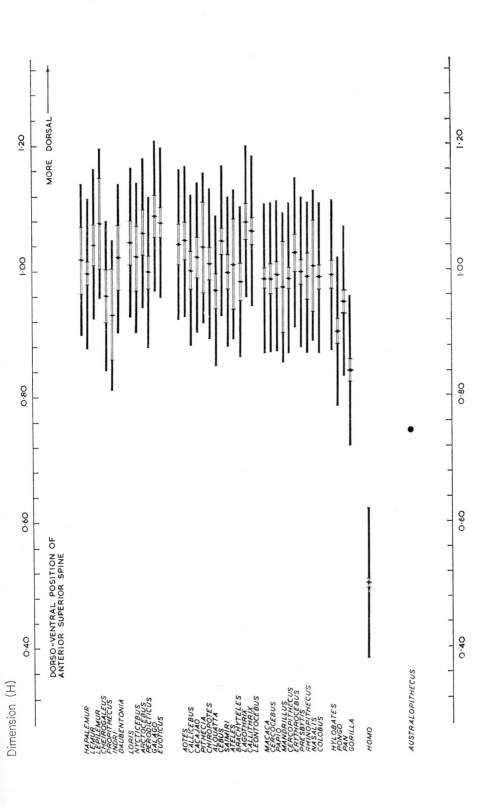

MORE DORSAL ⟶

DORSO-VENTRAL POSITION OF
ANTERIOR SUPERIOR SPINE

0·40   0·60   0·80   1·00   1·20

HAPALEMUR
LEMUR
LEPILEMUR
CHEIROGALEUS
PROPITHECUS
INDRI
DAUBENTONIA

LORIS
NYCTICEBUS
ARCTOCEBUS
PERODICTICUS
GALAGO
EUOTICUS

AOTES
CALLICEBUS
CACAJAO
PITHECIA
CHIROPOTES
ALOUATTA
CEBUS
SAIMIRI
ATELES
BRACHYTELES
LAGOTHRIX
CALLITHRIX
LEONTOCEBUS

MACACA
CERCOCEBUS
PAPIO
MANDRILLUS
CERCOPITHECUS
ERYTHROCEBUS
PRESBYTIS
RHINOPITHECUS
NASALIS
COLOBUS

HYLOBATES
PONGO
PAN
GORILLA

HOMO

AUSTRALOPITHECUS

0·40   0·60   0·80   1·00   1·20

Dimension (H). The position of the anterior superior iliac spine, relative to the innominate bone as a whole (Fig. 19), described by Dimension (H) (depicted in Fig. 9b and described on page 102) is more ventral in man than in any sub-human primate. Between the members of this latter group, there is appreciably less variation than emerged in the analysis of Dimensions (A)–(G). Of the various sub-human Primates, *Gorilla* lies closest to man.

In this feature *Australopithecus* is intermediate between man and the sub-human Primates lying outside the upper 90% fiducial limit of the former. It is somewhat closer to the mean of *Gorilla* and falls just within its lower 90% fiducial limit.

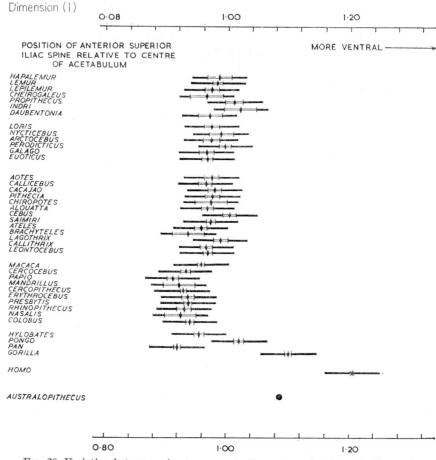

FIG. 20. Variation between primate genera in dimensions related to the disposition of the hip muscle blocks—means and 90% fiducial limits. The band embracing two standard errors on each side of the mean indicates the variance of sample means.

Dimension (I). The position of the anterior superior iliac spine in relation to the acetabulum (Fig. 20), reflected in Dimension (I) (depicted in Fig. 10 and described on page 103), is more ventral in man than in any other sub-human primate. As in Dimension (H), variation among the sub-human Primates is relatively less than in most other dimensions used in this study. The mean for Gorilla and to a lesser extent that for *Pongo* lie appreciably closer to the human mean than do those of any other genus of sub-human primate.

In this feature, *Australopithecus* lies close to *Gorilla*. Its anterior superior iliac spine lies much more dorsally than obtains even in individuals at the dorsal extreme of the human range.

Summary (muscular attachment). Again, in so far as can be judged within the methodological limitations of a univariate study, the principal consistent finding relating to the position of attachment of the blocks of muscles acting upon the hip joint is that, in each feature, *Australopithecus* differs from man and consistently resembles many groups of sub-human Primates including, always, one or more of the great apes.

Hind-limb locomotor groups

Dimensions relating to joint position (Table III). For each dimension, the variance within each locomotor group is high, reflecting partly the relatively small numbers of genera within each and possibly also the semi-arbitrary nature of their definition. Nevertheless, certain contrasts appear to emerge.

In the Anthropoidea, in three of the four dimensions associated with joint position: Dimension (A) (dorso-ventral position of acetabulum), Dimension (B) (cranio-caudal position of acetabulum) and Dimension (C) (dorso-ventral position of auricular facet), apart from the big contrasts between man and sub-human Primates already described, the only locomotor grouping which appears to distinguish itself even marginally in the relatively high variance introduced by the grouping process is that comprising leapers (facultative or pronounced). The remaining Dimension (D) (cranio-caudal separation of sacro-iliac and hip joints) differs in that, in the Anthropoidea, it effects an additional separation of hindlimb acrobats.

In the Prosimii, in relation to these dimensions, it is (a) the hind-limb acrobats and hangers and (b) the leapers and clingers, that show some contrast with other groups. This is brought about by Dimensions (A), (B) & (C), Dimension (D) not producing an effective contrast.

This pattern of metrical contrasts means that in the hindlimb leapers of the Anthropoidea and in the (1) hindlimb acrobats and hangers

Variation between means of hindlimb locomotor groups in each of nine logged and
are as in

Dimension	(A)		(B)		(C)		(D)	
	Mean	S.E. of Mean	Mean	S.E. of Mean	Mean	S.E. of Mean	Mean	S.E. of Mean
PROSIMII								
Quadruped (runner/climber)	0·33	0·011	0·61	0·017	0·30	0·014	0·69	0·024
Quadruped (leaper/clinger)	0·38	0·005	0·59	0·015	0·39	0·005	0·71	0·005
Quadruped (leaper)	0·34	0·005	0·53	0·005	0·28	0·010	0·65	0·015
Hanger and acrobat	0·26	0·011	0·43	0·008	0·25	0·009	0·67	0·009
Unclassifiable	0·32	—	0·67	—	0·34	—	0·71	—
ANTHROPOIDEA								
Quadruped (regular)	0·44	0·222	0·60	0·009	0·30	0·008	0·71	0·009
Quadruped (runner)	0·46	—	0·65	—	0·26	—	0·68	—
Quadruped (intermediate)	0·35	0·008	0·64	0·006	0·33	0·024	0·70	0·008
Quadruped (facultative leaper)	0·33	0·000	0·64	0·007	0·27	0·010	0·61	0·009
Quadruped (pronounced leaper)	0·45	0·005	0·55	0·018	0·26	0·019	0·70	0·010
Acrobat	0·39	0·033	0·61	0·022	0·30	0·050	0·80	0·021
Unclassifiable	0·40	—	0·60	—	0·26	—	0·82	—
Biped	0·64	—	0·44	—	0·63	—	0·61	—
Unknown (*Australopithecus*)	0·58	—	0·47	—	0·48	—	0·57	—

Dimension (A): Position of acetabulum in the dorso-ventral direction.
Dimension (B): Position of acetabulum in the cranio-caudal direction.
Dimension (C): Position of auricular facet in the dorso-ventral direction.
Dimension (D): Relative separation of the sacro-iliac and hip joints in the cranio-caudal direction.

III

regression-adjusted dimensions of Primate pelvis. Hindlimb locomotor groupings

Table I

(E) Mean	S.E. of Mean	(F) Mean	S.E. of Mean	(G) Mean	S.E. of Mean	(H) Mean	S.E. of Mean	(I) Mean	S.E. of Mean
0·69	0·009	1·48	0·156	1·64	0·057	1·04	0·017	0·98	0·007
0·70	0·030	1·77	0·010	1·80	0·025	0·96	0·025	1·01	0·010
0·58	0·000	1·48	0·080	1·48	0·070	1·09	0·005	0·96	0·000
0·54	0·018	1·50	0·053	1·50	0·054	1·04	0·012	0·98	0·008
0·73	—	1·62	—	1·62	—	1·02	—	0·97	—
0·74	0·013	1·75	0·024	1·86	0·018	0·99	0·002	0·94	0·012
0·73	—	1·66	—	1·98	—	1·03	—	0·93	—
0·70	0·005	1·69	0·018	1·74	0·061	1·01	0·017	0·98	0·007
0·69	0·006	1·59	0·015	1·59	0·016	1·06	0·006	0·96	0·003
0·71	0·012	1·78	0·043	1·88	0·038	1·00	0·004	0·93	0·005
0·73	0·033	1·83	0·017	1·92	0·016	0·96	0·025	0·98	0·023
0·69	—	1·84	—	1·95	—	1·00	—	0·95	—
0·69	—	1·53	—	1·60	—	0·51	—	1·21	—
0·82	—	1·73	—	1·77	—	0·75	—	1·09	—

Dimension (E): Position of attachment of hamstring and adductor muscles in the cranio-caudal direction.

Dimension (F): Orientation of the entire iliac blade relative to the transverse plane.

Dimension (G): Orientation of the iliac blade ventral to the dorsal buttress and relative to the transverse plane.

Dimension (H): Position of anterior superior iliac spine relative to entire innominate bone and measured in the dorso-ventral direction.

Dimension (I): Position of anterior superior iliac spine relative to the acetabulum and measured in the dorso-ventral direction.

and (2) leapers and clingers of the Prosimii, the sacro-iliac and hip joints are positioned relatively more dorsally and caudally within the pelvis. Their joint surfaces are, in these groups, also relatively nearer to each other.

Dimensions relating to muscular leverage and disposition (Table III). Again, and probably for reasons similar to those connected with the dimensions reflecting joint position, the pattern of contrast between locomotor groups is not clear-cut, although again, some apparently noteworthy differences emerge.

In the Anthropoidea, although no big contrasts emerged between the locomotor categories in the case of Dimension (E) (cranio-caudal position of attachment of hamstrings and adductors), in each of Dimensions (F) (orientation of entire iliac blade), (G) (orientation of iliac blade ventral to buttress), (H) (dorso-ventral position of anterior superior iliac spine in relation to entire bone) and (I) (dorso-ventral position of anterior superior iliac spine relative to acetabulum), the leapers (pronounced and facultative) attained, even with the relatively high variance introduced by the process of pooling into locomotor categories, a measure of distinction. The hindlimb acrobats were also separated by Dimensions (F) (orientation of entire iliac blade) and (G) (orientation of iliac blade ventral to buttress). These were, of course, additional to the contrasts between man and sub-human Primates already established.

In the Prosimii, leapers and clingers were, to some extent, distinct in each case, as also were the acrobats and hangers in the case of Dimension (E) (cranio-caudal position of attachment of hamstrings and adductors) and Dimension (G) (orientation of iliac blade ventral to buttress).

The morphological interpretation of these metrical contrasts is that in the leapers of the Anthropoidea, although the iliac blade is twisted more ventrally than in other locomotor groups, it is placed more dorsally, resulting in the anterior superior iliac spine also assuming a more dorsal position. In the Prosimii, the converse appears to obtain. In acrobats of the Anthropoidea, the iliac blade is orientated more dorsally, while the reverse appears to be the case in the acrobats and hangers of the Prosimii. In addition, in these latter, the attachment of the hamstring muscles lies more cranially.

Multivariate study

Individual genera

Canonical analysis of nine regression-adjusted dimensions.

(a) Axis 1 (Figs 21 & 22). The overall separation produced by this axis is about 17 standard deviation units (S.D.U.). The Lorisiformes

(4·5 S.D.U. in extent) lie at the extreme positive end of this axis. The remainder of the Prosimii less the Indridae, together with the more primitive New World monkeys (i.e. all those forms except the Atelinae, Alouattinae and *Cebus* that have a prehensile or semi-prehensile tail), lie in the interval between +1·0 and +5·2 S.D.U. *Cheirogaleus* is somewhat of an outlier of this group at +6 S.D.U. The two available genera of Indridae lie between 0·0 and −1·0 S.D.U. All the Old World monkeys and apes, together with the prehensile and semi-prehensile tailed New World monkeys, lie between −2·0 and −7·0 S.D.U.

Australopithecus lies towards the extreme negative end of the spectrum of Old World forms at −5·3 S.D.U., while *Homo* is placed on the opposite pole of that group at −1·1 S.D.U.

Three dimensions contribute principally to the overall pattern of discrimination in the direction of the first canonical axis: the dorso-ventral position of the acetabulum (Dimension (A)), the cranio-caudal position of attachment of the hamstring and adductor muscles relative to the hip joint (Dimension (E)), and the orientation of the iliac blade ventral to the dorsal buttress (Dimension (G)).

Axis 2 (Fig. 21). This axis separates man by 6·5 S.D.U., and *Australo-pithecus* by 5 S.D.U., from the non-human Anthropoidea, both forms lying outside the 90% limits of that group. A unique separation between three groups comprising: (a) the sub-human Primates, (b) man and (c) *Australopithecus* is thus provided by axes 1 and 2 in combination. Some separation is also produced by axis 2 between the great apes, that lie between 0·0 and −3·0 S.D.U., and the remaining Old World Primates lying between 0·0 and +3·0 S.D.U. This axis also aids in the separation of the Indridae from the Lemuriformes.

In axis 2, four variates contribute substantially to the overall discrimination. These are: orientation of the iliac blade ventral to the dorsal buttress (Dimension (G)); orientation of the entire iliac blade (Dimension (F)); position of the anterior superior iliac spine relative to the centre of the acetabulum (Dimension (I)), and the dorso-ventral position of the acetabulum (Dimension (A)). Two further dimensions make a contribution that is noticeable, although smaller in scale. These are the relative separation of the sacro-iliac and hip joints in the cranio-caudal direction (Dimension (D)), together with the cranio-caudal position of origin of the hamstring and adductor muscles relative to the hip joint (Dimension (E)).

Axis 3 (Fig. 22). This axis produces separation between Old World monkeys, apes and New World monkeys (prehensile tailed and semi-prehensile tailed), the Old World monkeys being somewhat more

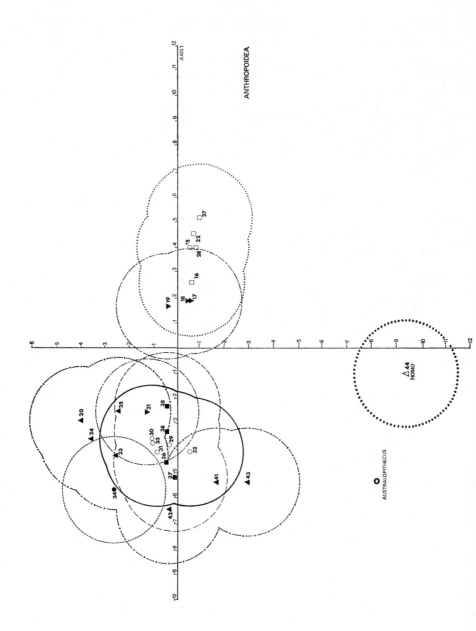

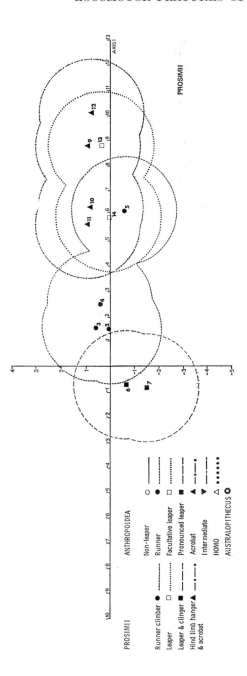

FIG. 21. Bivariate plot (canonical axes 1 & 2) of relative positions of primate genera as derived from canonical analysis of nine logged and regression adjusted dimensions. Ninety per cent boundaries of different locomotor groups are indicated. Serial numbers of genera are listed in Table I. Canonical co-ordinates for genera whose hindlimb locomotor pattern could not be included in the classificatory scheme are: *Daubentonia*: 3·03, 0·16; *Hylobates*: —4·46, 2·72.

G

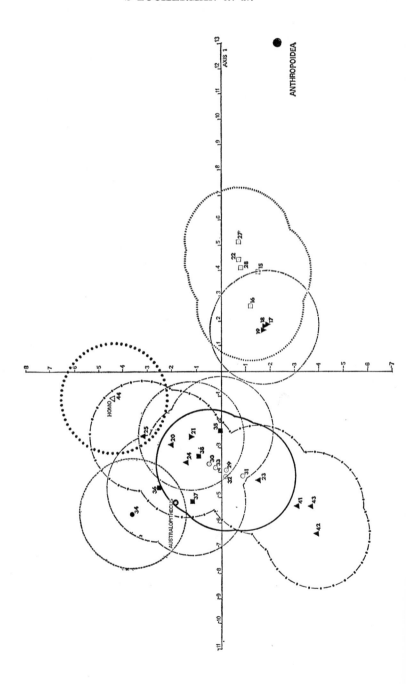

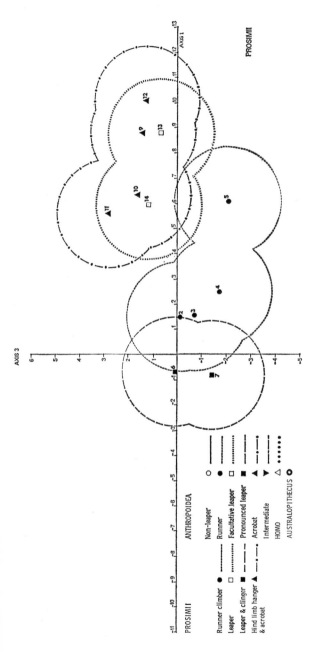

FIG. 22. Bivariate plot (canonical axes 1 & 3) of relative positions of primate genera as derived from canonical analysis of nine logged and regression adjusted dimensions. Ninety per cent boundaries of different locomotor groups are indicated. Serial numbers of genera are as listed in Table I. Canonical co-ordinates for genera whose hindlimb locomotor pattern could not be included in the classificatory scheme are: *Daubentonia*: 3·03, −3·04; *Hylobates*: −4·46, 0·04.

positive and the great apes somewhat more negative than is the bulk of New World forms. The axis further accentuates some of the distinction already brought out by axes 1 & 2—e.g. between the Lemuriformes and the Lorisiformes.

Australopithecus and man are separated by 2·5 S.D.U. from other genera, lying at +4·4 and +1·9 S.D.U. respectively from the origin in the direction of this axis.

In axis 3, the principal discrimination is effected by two dimensions: orientation of the iliac blade ventral to the dorsal buttress (Dimension (G)), and orientation of the complete iliac blade (Dimension (F)). Two further dimensions also contribute to an appreciable extent: origin of the hamstring and adductor muscles relative to the hip joint (Dimension (E)), and the relative separation of the sacro-iliac and hip joints—again measured in the same direction (Dimension (D)).

Axis 4 (not figured). This axis produced a partial separation between the Galaginae (positive) and the Lorisinae (negative) more distinct than that produced by any of the first three canonical axes. In addition, axis 4 accentuates separations already achieved between: (a) the prehensile tailed plus semi-prehensile tailed New World monkeys, (b) the great apes, and (c) the majority of Old World monkeys.

Australopithecus and man are not separated from Old World Primates by this axis, although they are separated from each other by 3·0 S.D.U.

In axis 4, the principal discrimination is effected by dimensions depicting the dorso-ventral position of the acetabulum (Dimension (A)), the orientation of the complete iliac blade (Dimension (F)) and the position of the anterior superior iliac spine relative to the centre of the acetabulum (Dimension (I)).

Generalized distance and cluster analysis. The minimum spanning tree derived from and summarizing the matrix of squared generalized distances associated with all nine canonical axes, confirms all the separations noted from bivariate canonical plots and, in addition, suggests that the information contained in the later axes is not only consonant with that relating to axes 1–4 but, in fact, separates such groupings in such a way as to make them even more obvious.

Australopithecus is separated from man by 8·3 units of generalized distance (U.G.D.), this being of the same general order as its distance from such extant sub-human Primates as approximate most closely. The nearest members of the latter to *Australopithecus* are *Mandrillus* and *Nasalis* that lie 8·8 and 8·9 U.G.D. away respectively. The nearest ape is *Pongo* at 9·2 units. This result confirms that only marginal alterations to the overall picture are effected by the information contained in the fourth and higher canonical axes.

In so far as can be judged from the sums, throughout the nine canonical axes, of products of loading factors and standard deviations between generic means, all dimensions contribute, although to a varying extent, to the overall pattern of discrimination. The dimension making the biggest contribution to the overall separation between genera as summarized by the generalized distances is that defining the orientation of the iliac blade ventral to the dorsal buttress (Dimension (G)). Five other dimensions also contribute: orientation of the entire iliac blade (Dimension (F)); position of origin of the hamstring and adductor muscles relative to the hip joint (Dimension (E)); dorso-ventral position of the acetabulum (Dimension (A)); position of the anterior superior iliac spine relative to the centre of the acetabulum (Dimension (I)) and cranio-caudal position of the acetabulum (Dimension (B)). The three remaining dimensions: dorso-ventral position of the auricular facet (Dimension (C)), separation of the sacroiliac and hip joints in the cranio-caudal direction (Dimension (D)) and dorso-ventral position of the anterior superior iliac spine (Dimension (H)) make an appreciable although smaller contribution.

Canonical analysis of four pelvic dimensions relating to joint position.

Axis 1 (Fig. 23). This axis produces a distinction from other Primates, of *Australopithecus* and man, in that both lie at the negative end of the spectrum between −4·0 and −5·3 s.d.u. from the origin. The range for other Primates extends from −4·3 s.d.u. to +4·9 s.d.u. However, smaller separations are evident between (a) all Old World monkeys and apes (lying between −0·4 and −4·3 s.d.u.); (b) all New World forms, together with the Prosimii (lying between 0 and +4·9 s.d.u.). This axis also produces a specific separation of the Lorisinae (lying between +3·7 and +4·9 s.d.u.) from all other Primates.

In axis 1, a single dimension, relating to the dorso-ventral position of the acetabulum (Dimension (A)), makes a marked contribution to the discrimination. It is supported, although to a much smaller extent, by the measure depicting the cranio-caudal position of the acetabulum (Dimension (B)).

Axis 2 (Fig. 23). The major separation produced by this axis is of, on the one hand, man and *Australopithecus* lying at +6·4 and +5·9 s.d.u. respectively, from, on the other, all remaining Primates whose range in this axis extends from −3·4 to +2·4 s.d.u.

Two dimensions are principally responsible for discrimination effected in the direction of canonical axis 2. They depict the relative

S. ZUCKERMAN *et al.*

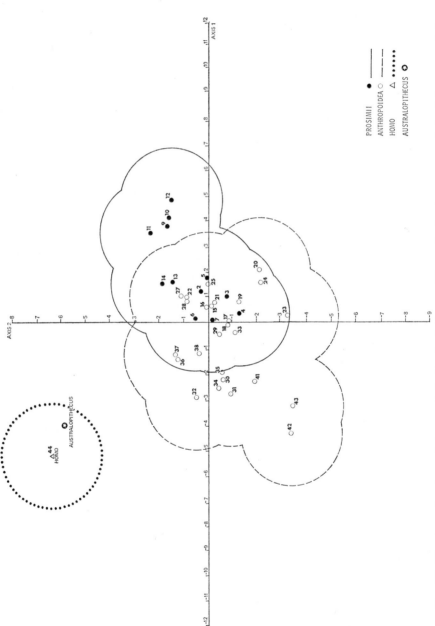

FIG. 23. Bivariate plot (canonical axes 1 & 2) of relative positions of primate genera as derived from canonical analysis of four logged and regression adjusted dimensions describing the relative disposition of the sacro-iliac and hip joints. Ninety per cent boundaries of the Anthropoidea and Prosimii are indicated. Serial numbers of genera are as listed

separation of the sacro-iliac and hip joints in the cranio-caudal direction (Dimension (D)), together with the cranio-caudal position of the acetabulum (Dimension (B)).

Axes 1 & 2 (Fig. 23). The combination of canonical axes 1 & 2 gives the minor additional information that, within the major group of Primates, there seems to be a spectrum extending from the apes through the Old World monkeys to the New World monkeys, thence to the Lemuriformes and, finally the Lorisiformes.

Generalized distance and cluster analysis. The minimum spanning tree derived from cluster analysis of the matrix of squared generalized distances associated with the four dimensions relating to joint position, confirms the general pattern of separation portrayed by canonical axes 1 & 2. It also confirms the similar separations of both *Australopithecus* and man from the next nearest primate genera: *Mandrillus* and *Nasalis*.

Australopithecus is separated from man by only 2·6 U.G.D., this being much less than that from any sub-human primate. Of this group, the closest to *Australopithecus* is again *Nasalis* lying at 5·5 U.G.D. and *Mandrillus* at 5·6 U.G.D. The nearest great ape is *Pongo* lying at 8·5 U.G.D. This result again confirms that only marginal alterations to the overall picture are effected by information contained in higher axes.

Although all dimensions appear to make an appreciable contribution to the overall pattern of discrimination, the contribution from one, depicting the dorso-ventral position of the acetabulum (Dimension (A)), is prominent.

Canonical analysis of pelvic dimensions—five dimensions relating to muscular disposition.

Axis 1 (Figs 24 & 25). This axis, spanning some 15 S.D.U., separates most clearly the Lorisiformes (+ 5·3 to + 9·1 S.D.U.) from the majority of the Lemuroidea (less Indridae) with which are grouped the non-specialized genera of New World monkeys (lying between + 0·8 and + 5·5 S.D.U.). The Old World monkeys and apes, together with the acrobatic prehensile and semi-prehensile tailed New World monkeys, lie between − 1·4 and − 5·2 S.D.U. The Indridae occupy an intermediate position centring around − 1·0 S.D.U.

Australopithecus, at − 3·4 S.D.U., lies towards the centre of the Old World forms, while *Homo* at + 0·6 S.D.U., is intermixed with the non-specialized New World monkeys, some 4 S.D.U. away from *Australopithecus*.

It is the orientation of the iliac blade ventral to the dorsal buttress (Dimension (G)) that contributes principally to the overall discrimination

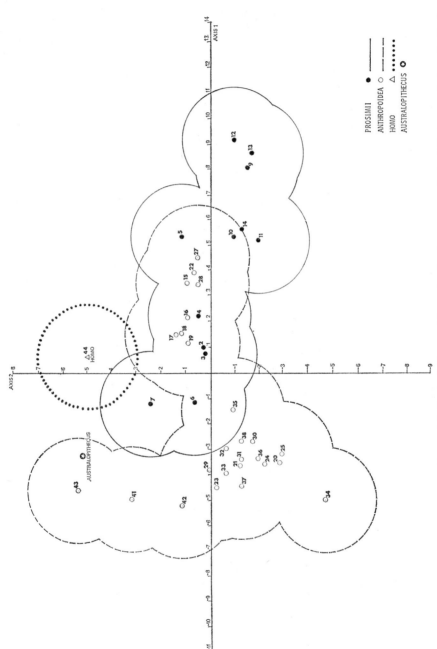

FIG. 24. Bivariate plot (canonical axes 1 & 2) of relative positions of primate genera as derived from canonical analysis of five logged and regression adjusted dimensions related to the disposition of hip muscle blocks. Ninety per cent boundaries of Anthropoidea and Homo are indicated. Serial numbers of genera are as listed in Table I. Canonical co-ordinates for genera Anthropoidea and Prosimii are indicated.

in the direction of canonical axis 1, although contributions are also made by the dimensions depicting the cranio-caudal position of attachment of the hamstring and adductor muscles relative to the hip joint (Dimension (E)), together with the dorso-ventral position of the anterior superior iliac spine (Dimension (H)).

Axis 2 (Fig. 24). This axis repeats the separation between, on the one hand, the Lorisiformes and, on the other, the Lemuriformes and the majority of the non-specialized New World monkeys. The third major grouping, comprising Old World Primates together with prehensile tailed and semi-prehensile tailed New World monkeys, lies in an intermediate position overlapping the two former groups. Within this third group, there are marginal separations between the great apes and the general complex of other Old World forms.

It is contrasts in orientation of the iliac blade ventral to the dorsal buttress (Dimension (G)) that again contribute maximally to the discrimination parallel to axis 2. Appreciable supporting contributions are also made by dimensions depicting the orientation of the complete iliac blade (Dimension (F)), the position of origin of the hamstring and adductor muscles relative to the hip joint and measured in the cranio-caudal direction (Dimension (E)), and the position of the anterior superior iliac spine relative to the centre of the acetabulum (Dimension (I)).

Axis 3 (Fig. 25). This axis further distinguishes the Indridae by enhancing their distinctions from the remaining Lemuriformes and the unspecialized New World monkeys. But the major separation that it effects is of man (− 6·0 units) from all other forms, the nearest of which is *Lagothrix* (lying at − 2·8 units). *Australopithecus* remains distinct from man, at − 0·7 units, being well within the general range of all sub-human Primates.

In axis 3, the principal discrimination is effected by variation in the orientation of the iliac blade ventral to the dorsal buttress (Dimension (G)), by contrasts in the position of origin of the hamstring and adductor muscles relative to the hip joint (Dimension (E)) and in the position of the anterior superior iliac spine relative to the acetabulum (Dimension (I)).

Axes 1, 2 & 3 (Figs 24 & 25). Axes 1 & 2 in combination make a good discrimination between man and other Primates. *Australopithecus*, however, is not specially differentiated from the sub-human Primates by any axis, lying, in fact, close to the centre of the group of means for the three genera of great apes.

Discrete or semi-discrete groups separated by the first three canonical axes in combination are: (a) *Homo*, (b) Lorisiformes, (c)

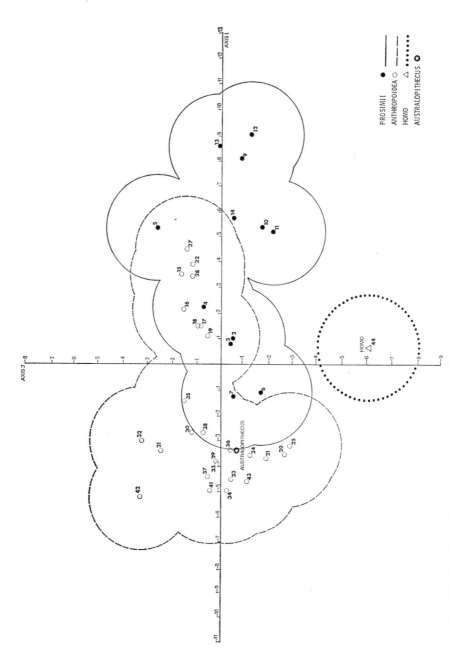

FIG. 25. Bivariate plot (canonical axes 1 & 3) of relative positions of primate genera as derived from canonical analysis of five logged and regression adjusted dimensions related to the disposition of hip muscle blocks. Ninety per cent boundaries of Anthropoidea and Prosimii are indicated. Serial numbers of genera are as listed in Table I. Canonical co-

Lemuriformes (less the Indridae) together with unspecia.
World monkeys, (d) Indridae and (e) Old World forms plu
hensile tailed and semi-prehensile tailed New World monkey
this last group, *Australopithecus*, together with the extant gɪ‸‸ apes,
can be partially distinguished.

Generalized distance and cluster analysis. The single linkage cluster
analysis of squared generalized distances between genera, when
summarized as a dendrogram and minimum spanning tree, re-
emphasizes the basic pattern of separation shown by plots of canonical
axes. It shows especially that *Australopithecus* lies close to the great
apes, of which the closest is *Gorilla* lying only 2·4 U.G.D. away. This is
much less than its separation from man (7·0 U.G.D.).

All measurements appear to make a noteworthy contribution to
the total pattern of discrimination as summarized by the matrix of
squared generalized distances, although that made by the orientation
of the iliac blade ventral to the dorsal buttress (Dimension (G)) is
pronounced. The orientation of the complete iliac blade (Dimension
(F)), the point of attachment of the hamstring and adductor muscles
relative to the centre of the acetabulum (Dimension (E)), and the
position of the anterior superior iliac spine relative to the centre of
the acetabulum (Dimension (I)) are somewhat less so.

*Canonical analysis of nine pelvic dimensions: man and apes only:
logged indices.*

Axis 1 (Fig. 26, Table IV). Axis 1 effects principally a separation
between man and the great apes, the former lying some 8·4 S.D.U. in the
negative direction, the latter between 1·5 and 4·2 S.D.U. in the positive
direction. This axis also distinguishes *Australopithecus* from all groups,
the fossil lying approximately 2·8 S.D.U. from the origin in the negative
direction.

The maximum contribution to the overall discrimination in this
direction of canonical space is made by dimensions depicting the point
of origin of the hamstring and adductor muscles relative to the hip
joint (Dimension (E)), the position of the anterior superior iliac spine
relative to the centre of the acetabulum (Dimension (I)), the orientation of the iliac blade ventral to the dorsal buttress (Dimension (G)),
and the separation of the sacro-iliac and hip joints in the cranio-
caudal direction (Dimension (D)).

Axis 2 (Fig. 26, Table IV). This axis effects a separation between
the three great apes, the mean for *Gorilla* lying approximately 2·8
S.D.U. in the positive direction, the mean for *Pongo* lying 1·0 S.D.U.
from the origin also in the positive direction, and that for *Pan* being

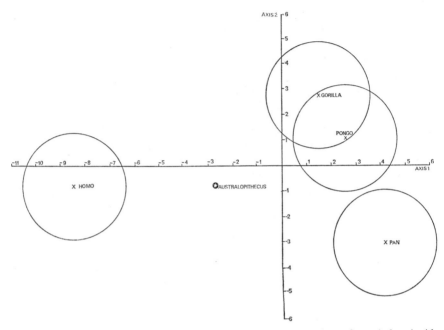

Fɪɢ. 26. Bivariate plot (canonical axes 1 & 2) of relative positions of certain hominoid genera as derived from canonical analysis of nine logged indices—the great apes and man, only, being included in the analysis. Approximate 90% fiducial limits for each extant genus are indicated.

positioned some 3·1 s.ᴅ.ᴜ. in the negative direction. Both man and *Australopithecus* lie on the negative side of the origin at −0·7 and −0·8 s.ᴅ.ᴜ. respectively—i.e. in the intermediate part of the range for the great apes.

Discrimination in this direction of canonical space is produced principally by two dimensions portraying

1. the position of the anterior superior iliac spine relative to the centre of the acetabulum (Dimension (I)) and,

2. the position of the acetabulum in the dorso-ventral direction (Dimension (A)).

Generalized distance analysis. The matrix of squared generalized distances emphasizes first, the relatively close position of the great apes to each other, the maximum separation (*Pan–Gorilla*) being only 6·5 ᴜ.ɢ.ᴅ., and second the marked contrast that each shows with man from whom the distance varies between 10·6 and 12·8 ᴜ.ɢ.ᴅ. *Australopithecus* lies 5·6 ᴜ.ɢ.ᴅ. from man, its separation from the great apes varying between 5·8 ᴜ.ɢ.ᴅ. (*Gorilla*) and 7·4 ᴜ.ɢ.ᴅ. (*Pan*). Thus,

TABLE IV

Canonical co-ordinates (axes 1, 2 and 3) of great apes, man and Australopithecus derived from computations including only the great apes and man and compounding (a) nine locomotor dimensions of primate pelvis; (b) four dimensions describing the relative disposition of the sacro-iliac and hip joints; (c) five dimensions describing the relative disposition of the hip muscle blocks. Analyses were based on logged indices

Genus	Analysis of nine locomotor dimensions of primate pelvis			Analysis of four dimensions describing the relative disposition of the sacro-iliac and hip joints			Analysis of five dimensions describing the relative disposition of the hip muscle blocks		
	Axis 1	Axis 2	Axis 3	Axis 1	Axis 2	Axis 3	Axis 1	Axis 2	Axis 3
Pongo	2·7	1·0	1·7	1·9	1·8	−0·5	1·9	0·9	0·1
Pan	4·2	−3·1	−0·5	2·2	−2·1	−0·3	3·7	−2·1	0·0
Gorilla	1·5	2·8	−1·2	2·3	0·4	0·7	0·6	1·9	−0·1
Homo	−8·4	−0·7	0·1	−6·4	−0·1	0·0	−6·2	−0·8	0·0
Australopithecus	−2·8	−0·8	−0·1	−5·4	−1·3	0·0	−0·3	2·5	−2·0

although the fossil appears from a plot of the first two canonical axes (these containing most of the meaningful information) to be intermediate between man and the apes, this analysis of overall contrasts as portrayed by generalized distance shows that the fossil is, in actual fact, uniquely placed, being widely different from both man and the living great apes.

Canonical analysis—four pelvic dimensions relating to joint position: man and apes only: logged indices.

Axis 1 (Table IV). Axis 1 separates man from the great apes, the mean for the former lying 6·4 S.D.U. from the origin and in the negative direction, while those for the latter lie in the positive direction: 1·9 to 2·3 S.D.U. from the mean. *Australopithecus* lies 5·4 S.D.U. from the origin in the negative direction—i.e. close to man but deviating slightly in the general direction of the great apes.

The maximum contribution to discrimination in the direction of canonical axis 1 is made by two dimensions, these depicting the relative separation of the sacro-iliac and hip joints in the cranio-caudal direction (Dimension (D)) and the dorso-ventral position of the acetabulum (Dimension (A)). Further contribution is also made by the dimension defining cranio-caudal position of the acetabulum (Dimension (B)).

Axis 2 (Table IV). Axis 2 effects only a small measure of separation between the great apes, *Pongo* and *Pan* at the extremes being 3·9 S.D.U. apart.

This small measure of discrimination appears to be effected by two dimensions depicting the dorso-ventral position of the acetabulum (Dimension (A)) and the dorso-ventral position of the auricular facet (Dimension (C)).

Generalized distance analysis. The matrix of squared generalized distances again emphasizes that the three great apes lie close to each other, their overall separation being only 3·9 U.G.D. But their overall contrast with man varies between 8·6 and 8·9 U.G.D. The matrix of generalized distances also shows the close similarity of *Australopithecus* to man in respect to the four variables relating to joint position, only 1·6 U.G.D. separating them, while the most similar ape (*Pan*) is separated from *Australopithecus* by 7·7 U.G.D.

Canonical analysis of five pelvic dimensions relating to muscular disposition: man and apes only: logged indices.

Axis 1 (Table IV). Axis 1 separates man from the great apes, the mean for the former lying 6·2 S.D.U. from the origin and in the negative direction, those for the former varying in position from 0·6 to 3·7 S.D.U. in the positive direction from the origin.

Australopithecus lies, in this axis, close to the apes, but deviates marginally in the human direction, lying only 0·3 S.D.U. from the origin, although in the negative direction.

The maximum contribution to discrimination in the direction of axis 1 is effected by two dimensions depicting first the cranio-caudal position of origin of the hamstring and adductor muscles relative to the hip joint (Dimension (E)) and second, the position of the anterior superior iliac spine relative to the centre of the acetabulum (Dimension (I)). An appreciable contribution is also made by the dimension that depicts the orientation of the ventral part of the iliac blade (Dimension (G)).

Axis 2 (Table IV). This axis effects an appreciable separation between the great apes whose means lie between 2·1 units in the negative direction (*Pan*) and 1·9 units positively (*Gorilla*). *Pongo* falls in an intermediate position at +0·9 units. Axis 2 effects a further separation between *Australopithecus* and man, the former lying 2·5 S.D.U. in the positive direction—a value that makes it, in fact, more positive than the mean value for any ape—while man lies 0·8 S.D.U. negatively.

In axis 2, it is the variation in the position of the anterior superior iliac spine relative to the acetabulum (Dimension (I)) that is mainly responsible for effecting discrimination, although the orientation of the complete iliac blade (Dimension (F)) together with the cranio-caudal origin of the hamstring and adductor muscles relative to the hip joint (Dimension (E)), play an appreciable ancillary role.

Axis 3 (Table IV). Axis 3, although not differentiating between any of the extant Hominoidea (both man and each ape lying less than 0·1 units from the origin), separates *Australopithecus* which lies 2·0 S.D.U. from the origin in a negative direction.

In axis 3, the principal dimension contributing to discrimination is the orientation of the iliac blade ventral to the dorsal buttress (Dimension (G)), although supplementary contributions are made by the orientation of the entire iliac blade (Dimension (F)), the position of the anterior superior iliac spine relative to the centre of the acetabulum (Dimension (I)) and the position of origin of the hamstring and adductor muscles relative to the hip joint (Dimension (E)).

Generalized distance analysis. The matrix of squared generalized distances again emphasizes the close relationship in respect to the combination of these five dimensions of the great apes, the maximum separation (*Pan–Gorilla*) being only 5·0 U.G.D. It also emphasizes the much bigger separation between the apes and man, the smallest distance (*Homo–Gorilla*) being 7·4 U.G.D., and the biggest (*Homo–Pan*) being 10·0 U.G.D. The matrix also shows that *Australopithecus* is, in this respect, and as inferred from the study of individual canonical axes, and especially axis 1, more like the apes than man. The separation of *Australopithecus* from *Homo* is, in fact, 7·1 U.G.D. The shortest distance from the apes is 2·2 U.G.D. (*Australopithecus–Gorilla*) while the biggest is 6·1 U.G.D. (*Australopithecus–Pan*). The corresponding distance from *Pongo* is 3·4 U.G.D.

Hindlimb locomotor groups

Analysis of means of 42 genera. (Figs 21 & 22.) In descriptions of the principal contrasts emerging from bivariate canonical plots and from generalized distance matrices, the strong separation of man from the sub-human Primates has frequently emerged. Within the latter, some of the notable contrasts relate principally to taxonomic groups (e.g. the Ponginae are often distinct). Although, in some instances, such distinctions automatically have locomotor significance in relation to the hindlimb (e.g. Galaginae), in other instances, groups have separated out that, while not homogenous taxonomically, sometimes display certain similarities in the patterns of locomotor use of the hindlimb and pelvic region (e.g. the combination of the Lemuridae and the non-prehensile tailed New World monkeys—all of which are quadrupeds).

If attention is focused purely upon locomotor groupings— irrespective of the scale of contrast that emerges in the analysis— certain differences appear, some being illustrated in Figs 21 & 22. In these figures the contrasts between the 42 genera in canonical axes 1 & 2 and 1 & 3 as produced by a compound of all nine locomotor dimensions examined in this study are plotted. The overall boundaries of the locomotor groups formed by the coalescence of the circles depicting the 90% fiducial limits of the individual constituent genera are superimposed.

The main contrasts within both Anthropoidea and Prosimii are between the several groups in which varying amounts of leaping occur. The pronounced leapers lie, in both suborders, towards the negative direction in canonical axis 1, whilst the facultative leapers are at the corresponding positive extreme. Intermediate types from the Anthropoidea occupy an intermediate position close to the origin.

Genera from the Anthropoidea, grouped as runners, contrast with the correspondingly named group from the Prosimii, these, in the former, lying on the negative side of the origin and, in the latter, on the positive side. Genera from the unique prosimian group in which the hindlimb is sometimes associated with leaping and sometimes with clinging, fall in a position analogous to that occupied by those members of the Anthropoidea, whose hindlimb has an "acrobatic" function.

The grouping of genera depicted in the dendrogram, derived from the clustering of squared generalized distances (Fig. 27), correlates to an appreciable extent with locomotor grouping. But there are a number of exceptions, certain groupings appearing, to some extent, to cut across locomotor categories, and to correlate with systematic taxa. For example, *Euoticus* (a leaper) links with *Arctocebus* and *Nycticebus* (hindlimb acrobats and hangers), whilst an analogous situation obtains between *Galago* (a leaper) on the one hand, together with *Loris* and *Perodicticus* (acrobats and hangers) on the other. All six genera, however, combine at the next level to form a single cluster which thus includes all the available representatives of the taxon Lorisiformes.

Analysis of means of locomotor groups. (Table V.) In the analysis directed primarily towards eliciting contrasts between locomotor groups rather than between their constituent genera, the contrast between man and sub-human Primates is accentuated as are those between *Australopithecus* and man, on the one hand, and between *Australopithecus* and the sub-human Primates on the other.

Contrasts between locomotor groups of the sub-human Anthropoidea and of the Prosimii are correspondingly reduced, but are again increased in scale (although with some reorientation between axes) when the analysis is repeated excluding man and *Australopithecus*.

Plots of the position of each of the locomotor groups relative to canonical axes 1 & 2 confirm, and to some extent clarify, the general pattern of contrast between locomotor groups that emerged from analysis of differences between 42 individual genera.

In both Anthropoidea and Prosimii there are parallel contrasts between leapers and non-leapers; again hindlimb "acrobats" from the two groups occupy contrasting positions in the spectrum although the position of the prosimian group styled "leapers and clingers" occupies a position analogous to that assumed by the "acrobats" from the Anthropoidea.

In axis 1, the dorso-ventral position of the anterior superior iliac spine (Dimension (H)) makes the dominant contribution, whereas in both axis 1 and axis 2, it appears to be variation in the cranio-caudal position of the acetabulum (Dimension (B)) that is also significant.

146   S. ZUCKERMAN *et al.*

TABLE V

Variation between hindlimb locomotor groups in canonical axes 1, 2 and 3 as derived from combination of nine logged and regression adjusted pelvic dimensions. Variance within each locomotor group was derived from contrasts between means of constituent genera. Hindlimb locomotor groupings are as in Table I.

	Axis 1	Axis 2	Axis 3
PROSIMII			
Quadruped (runner/climber)	3·7	1·3	1·2
Quadruped (leaper/clinger)	1·5	0·5	−0·8
Quadruped (leaper)	3·0	5·8	1·3
Hanger & acrobat	−0·4	8·1	−1·1
Unclassifiable	4·9	0·8	0·6
ANTHROPOIDEA			
Quadruped (regular)	1·7	−2·9	0·0
Quadruped (runner)	3·9	−6·0	1·3
Quadruped (intermediate)	4·0	−0·4	0·1
Quadruped (facultative leaper)	3·9	0·5	3·5
Quadruped (pronounced leaper)	0·6	−2·3	0·0
Acrobat	3·2	−1·8	−3·3
Unclassifiable	4·7	−1·8	−4·4
Biped	−22·2	0·2	−2·2
Unknown (*Australopithecus*)	−12·5	−2·0	3·8

DISCUSSION

The present analysis appears to be the first in which pelvic dimensions have been purposefully selected as being directly related to the principal biomechanical and especially locomotor functions of the pelvic girdle. Although, in this respect, unique, the results remain complementary to those derived from biometric studies undertaken by earlier workers and based upon standard anthropometric dimensions, designed to give a picture, however incomplete, of overall form and proportions. Such work has been reviewed in detail by Martin (1959). It does not, of course, indicate how such differences have arisen nor what they signify. Some could be due to differences in allometric growth pattern, Barham (1971), Coleman (1971) and Biegert & Maurer (1972) having shown that allometric factors operate during development and growth of the pelvic girdle. This concept is extended by the present finding that, in relation to the dimensions analysed, allometric constants are, for each dimension, consistent between different genera. But they vary in size from dimension to dimension, such contrasts thus resulting in changes in proportion in accordance with variation in overall size. But even if allometric factors were responsible for part of such osteometric contrasts as have emerged in the present study, it is apparent that their adult end product has selective significance in that the pattern of contrast within both the Anthropoidea and Prosimii can be correlated functionally with variation in locomotor function of the hindlimb and pelvic region. It thus also correlates with the pattern of forces to which the pelvic region is subject during locomotion.

In so far as *Australopithecus* has, in the overall complex of features studied, emerged as occupying a unique position, quite distinct not only from man but also from sub-human Primates (and not necessarily intermediate between them), an appraisal of the probable biomechanical significance of its similarities and differences (a) in individual features, (b) in their functional groupings and (c) in the entire complex, is a basis for assessment of its probable pattern of locomotion. Further relevant information derives from an assessment of the biomechanical significance of differences between various locomotor categories of extant sub-human Primates. These genera, despite an established ability in many cases to adopt, on occasion, a type of bipedal stance and gait, normally and regularly exhibit a mode of locomotion in which weight is transmitted through all four limbs, which thus usually assume a supporting (quadrupedal) function.

The basic form of a quadruped (to which other locomotor types—including human bipedalism—can be readily related) is, as frequently emphasized in classical and standard descriptions (e.g. d'Arcy Thompson 1917; Gray, 1944, 1953), essentially a flexible rod (the vertebral column) beneath which soft tissues are suspended, and which is supported by the limbs and limb girdles at points near its cranial and caudal ends. The centre of gravity of the whole lies ventrally and varies in its cranio-caudal disposition. The biomechanical system thus formed is therefore one of a resultant force directed ventrally through the centre of gravity and distributed as parallel derived forces at the shoulder and hip joints. The transmission of forces between the vertebral column and the shoulder joint is effected mainly by a muscular mechanism; in the case of the hip, it is by bony connection. Within this system, there obtains, throughout the sub-human Primates, a pattern in which the sacro-iliac and hip joints are relatively widely spaced in the cranio-caudal direction, and in which they, and especially the sacro-iliac joint, tend to lie towards the dorsal aspect of the pelvic girdle. This arrangement might be regarded as mechanically generalized because it is, from the point of view of weight transmission, not only appropriate to quadrupedalism, but also not inconsistent with (i.e. not mechanically ineffective in) a situation in which the animal may suspend itself from the hindlimb, from the forelimb or from both, thus setting up patterns of tensile forces in the hindlimb and pelvic region.

Some characteristics of a generalized quadruped may, in fact, be accentuated in species that not infrequently suspend themselves from the hindlimb. For instance, in the Lorisinae, most of whose members hang from the hindlimb more frequently than do most other primates, the acetabulum is more dorsal than in any other primate group. Again, among Anthropoidea, genera such as *Ateles* and *Pongo*, both of which again frequently use the hindlimb for suspension of body weight, display a more marked separation in the cranio-caudal direction of the sacro-iliac and hip joints than obtains in any other group.

The "quadrupedal" characteristics of the pelvic girdle—even when certain of these extreme variants are superimposed—is not wholly inconsistent with some type of bipedal gait, as is evidenced by the fact that large numbers of primate genera have now been recorded (as summarized by, for example, Ashton & Oxnard, 1964a and by Napier & Napier, 1967), as adopting, on occasion, a bipedal stance or gait. But such anatomical configuration, mechanically efficient for the transfer of weight ventrally to all four limbs when these, as in the quadrupedal pose, are at right angles to the line of the trunk, is not necessarily the best for habitual caudal transmission of weight to the hindlimb when, as in the bipedal position, it lies approximately parallel to the line of the trunk.

Because of man's habitual upright posture and "striding" bipedal gait, the compressive, caudally-directed forces due to weight bearing contrast with those typical of quadrupeds. Transmitted by the vertebral column, these forces pass to the sacrum, thence via the sacro-iliac articulation to the innominate bone, and thence via the hip joint to the femur and leg. This, when extended, is a solid column of bone stiffened by the "locking" of the knee joint due to medial rotation of the femur relative to the tibia during the final stages of extension, the eventual transfer of weight being through the resilience of the ankle and the arches of the foot. The mechanical efficiency of the part of this system lying in the pelvic region is, in man, enhanced as a result of the relatively close approximation, in the cranio-caudal directions, of the sacro-iliac and hip joints, and by their relatively central (i.e. more ventral) siting in relation to the bone as a whole.

Such contrast in these features, between man and all sub-human Primates, while, to some extent, definable visually, has emerged clearly in the present biometrical study, both in comparisons of individual dimensions relating to the disposition of the sacro-iliac and hip joints, and in those relating to the combined group of dimensions. In this complex of characters, *Australopithecus* has emerged from both univariate and multivariate study as being very similar to man and quite dissimilar from all sub-human Primates. It thus appears to have been characterized by an arrangement of the pelvic articulations which, as and when a bipedal posture might have been assumed would, as in man, have enabled body weight to be transmitted, more effectively than can occur in such circumstances in extant sub-human Primates.

Such a finding does not necessarily mean that *Australopithecus* was, in fact, habitually bipedal or that, as and when it might have assumed a bipedal gait, this was necessarily like that typical of man. But it also appears true that if the habitual posture and gait of *Australopithecus* did not normally involve a type of bipedalism—i.e. if the creature was habitually some form of quadruped—the genus would contrast with all living sub-human primate quadrupeds in the close approximation and ventral disposition of the sacro-iliac and hip joints.

It can be argued that, on the one hand, there are no obvious biomechanical reasons why approximation and ventral displacement of the sacro-iliac and hip joints should be inconsistent with quadrupedalism; on the other, it could be postulated that some concealed reason might exist because man's gait, and especially after early infancy, is quite inefficient if he attempts to assume a quadrupedal position. But such inefficiency could equally result from factors other than the relative disposition of the sacro-iliac and hip joints. It might, for instance, be connected with the anatomical characteristics of other regions of the body (e.g. the relatively short forelimb), with factors such as neurological coordination, or, more immediately, with the proportions and disposition of the principal blocks of muscle acting upon the hip joint.

Man does, in fact, in this last respect, differ markedly from all living sub-human Primates, as emerged from preliminary study (undertaken primarily as a means for determining functionally significant bony dimensions) of the relative proportions of the several muscle blocks acting upon the hip joint. For instance, the extensors and medial rotators of the human hip joint (i.e. the muscles contributing mainly to the power stroke in locomotion) are relatively less powerful than in quadrupeds. Their effectiveness is also diminished by reduction of the lever arm between their area of origin from the ischio-pubic region of the innominate bone and the axis of the hip joint. In contrast, the human lesser gluteal muscles, attached to the more lateral part of the dorsal aspect of the iliac blade, are orientated so as to be abductors of the leg. Their functioning, in this respect, is enhanced by the lateral orientation of the human iliac blade as distinct from its dorsal positioning in quadrupeds. Abduction of the hip is thus more powerful in man than in sub-human Primates where the lesser gluteal muscles are medial rotators and possibly extensors of the hip and thus contribute to the propulsive stroke.

This latter feature would seem to be of crucial significance in the production of a specifically human type of bipedal gait. This is characterized by its "striding" pattern (Napier, 1964, 1967), the limbs moving uniformly in a dorso-ventral (postero-anterior) direction. A prerequisite of this type of movement is that the transverse axis joining the acetabular fossae remains relatively horizontal during each phase of the walking cycle, tilting of the pelvis towards the unsupported side during the "swing" phase thus, of necessity, being to a large extent, eliminated. This is brought about, in man, by contraction of the abductors of the contralateral ("stance") side working eccentrically (i.e. with their attachment to the greater trochanter of the femur becoming the origin and that to the dorsal surface of the iliac blade the insertion).

No mechanism of comparable power for stabilizing the unsupported side of the pelvis exists in any other extant primate, for in sub-human Primates, the only power of abduction derives from the small part of m. gluteus maximus that passes lateral to the hip joint, and perhaps the most lateral aspects of the lesser gluteal muscles that derive, as in the great apes and especially *Gorilla*, from the extreme lateral (ventral) part of the expanded iliac blade. As a result, the unsupported side of the pelvis tends, as and when these creatures attempt to walk bipedally, to tilt downwards. But this is compensated by an inclination of the trunk towards the stance side, whilst the forward movement of the leg passing through the swing phase is accentuated by a forward rotation of the trunk and pelvic girdle around the "fixed" limb. A lateral movement of the limb is thus produced in the swing phase, and a rolling type of gait results. This mechanism, as manifest in the chimpanzee, has been studied critically by Jenkins (1972).

In so far as the unique features of the human lesser gluteal muscles, significant in the mechanism of man's specific type of gait, are functionally related to and enhanced biomechanically by the configuration of associated osteological features—these latter can, when viewed in appropriate combination, be taken, in a fossil type, as reflecting the function of the attached muscles. Most significant are the lateral orientation of the iliac blade together with its ventral extension. The result, obtained in this study, that *Australopithecus* is, in both individual features relating to muscular disposition as also in their multivariate compound, quite dissimilar from man and much more nearly like many sub-human Primates (and especially the great apes), suggests that any pattern of bipedal walking that it might have assumed, must, as realized by some earlier workers (e.g. Mednick, 1955; Napier, 1964, 1967) and noted rather more strongly by Jenkins (1972), almost certainly have differed from that typical of *Homo sapiens*.

Particularly significant in this context are the features of the iliac blade relating to the orientation of m. gluteus medius and m. gluteus minimus. The dorsal orientation of the iliac blade in *Australopithecus* as compared with man, together with a much more dorsal positioning of the anterior superior iliac spine, indicate that m. gluteus medius and m. gluteus minimus were not orientated so as to fall sufficiently far lateral to the axis of the hip joint to be such efficient abductors as in man. In fact, it seems unlikely that they could have provided strong abductive power such as is necessary to stabilize the pelvis in a human type of bipedal walking. Confirmation of these anatomical findings, derived solely from studies of the innominate bone (although supported by the finding that the orientation of the innominate bone relative to the sagittal plane as measured by half the angle subtended by the sacrum at the pubic symphysis, is virtually constant throughout the Primates) is provided by Robinson (1964) in a review of certain aspects of the anatomy of the Australopithecinae. Robinson's paper illustrates a reconstruction of the entire pelvic basin of the specimen of *Australopithecus* upon which the present study is based. In this model, the almost complete sacrum (for which no detailed description, even now, appears to be available) is in place. In addition, dorsal rather than lateral orientation of the iliac blade is most obvious. Further evidence supporting the findings of the present study of the innominate bone is that, in *Australopithecus*, the half angle subtended by the sacrum at the pubic symphysis, as depicted by Robinson, is identical with the mean for the 41 extant primate genera measured in this study. Similar reconstructions carried out and described by McHenry (1972) for certain more recently discovered specimens of the Australopithecinae, but for which descriptions and casts are not yet widely available, are in agreement with these observations.

It is, of course, possible that in *Australopithecus*, some other muscular mechanism, as yet undefined, had developed to provide a unique and powerful source of abduction of the hip and thus provided a stabilizing mechanism during bipedal locomotion. A suggestion to this effect has, in fact, been advanced by Lovejoy & Heiple (1970) who point to an apparent elongation of the neck of the femur that could give a greater mechanical advantage to such muscles as contributed to abduction. But in the absence of full statistical substantiation of this point and because, in any case, any such mechanism would *a priori* seem incapable of enhancing the power of abduction to a level even approaching that which obtains in *Homo sapiens*, these observations would, if anything, seem to reinforce the view that any bipedal locomotion practised by *Australopithecus* must have been quite unlike that characteristic of *Homo sapiens*.

These results of univariate and multivariate study of each of two groups of biomechanical features (the first related to the position of the sacro-iliac and hip joints; the second to the disposition of the several functional blocks of pelvic muscles), together with their separate functional assessment, enable some interpretation to be made of findings that emerged from combination, by multivariate techniques, of all nine locomotor osteological features examined in the present study. These findings were similar in the analyses including a comprehensive range of genera from both the Anthropoidea and Prosimii, and in those in which only man and the great apes were considered. The most prominent was that *Australopithecus* occupied a unique position differing markedly from both man and the sub-human Primates rather than being intermediate between them. From a purely morphological and biometric viewpoint such a position appears, of course, readily explicable on the grounds that in one group of characters (those relating to joint position) the fossil resembles man, while in the group related to the disposition of pelvic muscle blocks, it is like certain sub-human Primates and differs from man. Its offset (i.e. unique) position in canonical space (as distinct from one directly intermediate between man and the sub-human Primates such as might be incorrectly inferred from examination of early canonical variates only) also correlates with the correspondingly unique combination of morphological features. From this, a similarly unique biomechanical system might be inferred. The four features relating to the relative position of the two joint surfaces are, in providing the fossil with a relatively efficient mechanism for the transmission of weight in a bipedal position, positively conducive to a bipedal posture and gait, although they are also apparently not mechanically inconsistent with quadrupedalism. The five features relating to muscle disposition show that a bipedal gait, if either habitual or occasional, must, in *Australopithecus*, have differed from that obtaining in *Homo sapiens*. Further, there would seem to be no obvious reason why such muscular disposition need necessarily have been inconsistent with, for instance, an occasional or habitual quadrupedal gait.

It is, of course, conceivable that the habitual posture and gait of *Australopithecus* might have been unique by displaying a combination of quadrupedalism and bipedalism, the latter, although differing from the type of bipedalism characteristic of *Homo sapiens*, being more frequent and in certain respects more efficient, than in any living sub-human primate.

A further possibility, while not excluded by the present findings, might be rendered less likely because of the expressly human disposition of the sacro-iliac and hip joints. This is, that the pattern of use of the hindlimb might have approximated to that found in certain species of living Primates in which, although basically quadrupedal, the hindlimb may also assume an "acrobatic" function, operating both with greater mobility and in a much more extensive quadrant of space—perhaps as in climbing—than does the hind-limb of a true quadruped. This interpretation is suggested by the pattern of contrast revealed in this study between the different locomotor categories of living sub-human Primates. This pattern appears to be best summarized by the dendrogram (Fig. 27) derived from the minimum spanning tree associated with the matrix of squared generalized distances. Although not coinciding fully with the locomotor categories based on the function of the hindlimb and used in this study, and thus paralleling the qualifications that attach to attempts to delineate locomotor categories on the basis of hindlimb function, it does confirm that the major group lying nearest to *Homo* comprises the Old World Primates, including the apes, together with those New World Primates in which a prehensile or semi-prehensile tail is habitually used in locomotion. In some members of this group—and principally in the great apes—the hindlimb has additionally an acrobatic function, i.e. has a greater three-dimensional mobility. But the group also includes the Indridae, and they are pronounced leapers. A quadruped—*Mandrillus*—also lies close, although this result might have been influenced by the fact that only a single specimen was available for this genus.

Despite the fact that, as with man, *Australopithecus* is also very different from any of even these groups of sub-human Primates—although in a different way as indicated by its falling in a different plane of canonical space—they are the only ones that lie near to the fossil. It could thus be permissible to infer that the Australopithecinae might have had, as, for instance, in both the living great apes and in man, a hindlimb with a relatively greater mobility in three dimensional space than obtains in quadrupeds. In pattern and extent, this need not, of course, have been necessarily identical with that typical of any of these

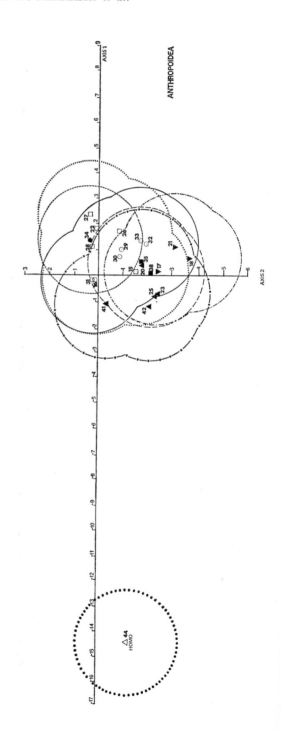

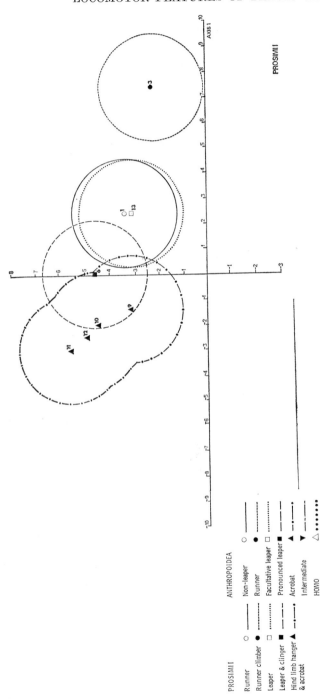

Fig. 28. Bivariate plot (canonical axes 1 & 2) of relative positions of primate genera as derived from canonical analysis of percentage contribution to the total hip musculature of the flexors (protractors), extensors (retractors), adductors, abductors and short muscles. Ninety per cent boundaries for each locomotor group are indicated. Serial numbers of genera are as in Table I. Canonical co-ordinates for genera whose hindlimb locomotor pattern could not be included in the classificatory scheme are: *Hylobates*: −1·60, −0·34; *Symphalangus*: −0·42, −1·35.

present-day groups as suggested by the fact that, within the group of extant Primates with an "acrobatic" hindlimb, the distance between the extreme members (*Gorilla* and *Alouatta*) is virtually the same as that between *Australopithecus* and *Gorilla*. Such inference would remain consonant with a view that the locomotor use of the hindlimb might have been composite, involving possible quadrupedalism, bipedalism and maybe other types of activity, such as an "acrobatic" function.

A relevant ancillary finding that derives from canonical analysis of the nine regression adjusted variables is that the general group of extant Primates closest to *Australopithecus* (Old World monkeys, apes and prehensile tailed New World monkeys, together with the Indridae), are animals which, although not habitual bipeds, frequently (e.g. during rest, feeding, etc.) adopt an orthograde or semi-orthograde posture. This is in contrast to the more distal group in the canonical space (non-prehensile tailed New World monkeys and prosimians other than the Indridae) which, although on occasion bipedal, (e.g. in the trees—*Pithecia*; on the ground—the larger species of *Galago*), normally assume (e.g. when feeding or at rest) a pronograde position of the trunk.

Further weight attaches to such tentative generalizations—and especially to those about the mobility of the hip of *Australopithecus*—because the relationships among genera revealed by multivariate analysis of the blocks of associated hip muscles (Fig. 28) correlate with patterns of locomotor contrast, which although limited in scale, are broadly parallel to those revealed by the analysis of related bony dimensions. Although, possibly because of the smaller scale of contrast, certain resemblances and differences that emerged from the corresponding analysis of bony dimensions are concealed, others that previously emerged as conspicuous, remain pronounced. Man's relatively closer resemblance to the great apes than to other Primates, is, for instance, again apparent.

Thus, despite the seemingly human overall proportions of the iliac blade (e.g. Broom *et al.*, 1950; Le Gros Clark, 1955, 1962), and despite the finding that in the relative positions of the sacro-iliac and hip joints, *Australopithecus* is similar to man, there are other functionally significant features of its pelvic girdle that contrast with man and agree with sub-human Primates. This means that the locomotor proclivities of *Australopithecus* still cannot be deduced with certainty. Its locomotor pattern might, in fact, have been unique, but all that can be said with certainty is that morphologically the pelvic girdle of these extinct creatures was, in its complex of functionally significant features, markedly different from all living Primates—including man.

Such doubt is reinforced by studies of other skeletal regions where a combination of similarities to and differences from both man and the sub-human Primates has also emerged. Thus, while the occipital condyles are placed much further dorsally than in man, and in this respect are quite like the sub-human Primates (Ashton & Zuckerman, 1951, 1952, 1956), the angulation of the occipital condyles relative to the cranial base is similar to that in man and contrasts with the apes (Moore, Adams & Lavelle, 1973).

Again, a clavicle attributed to the group and described by Leakey, Tobias & Napier (1964), although human in its curvature, displays a longitudinal torsion quite different from that in man where the arm mainly operates in the lower quadrant and resembles that characteristic of apes where the movements of the arm are in the upper quadrant (Oxnard, 1968a). Further, a fragmentary scapula first described by Broom *et al.* (1950) and attributed to the genus *Plesianthropus* (now *Australopithecus*), has emerged from univariate analysis of its features as being, and especially in, for instance, the set of the glenoid cavity together with the shape and orientation of the acromion and scapular spine, intermediate between man and the orang (Oxnard, 1968b). Multivariate study, in so far as it can be applied in an analysis of such relatively small fragments, has, as in the present analysis, in fact shown that the fossils occupy a unique position, contrasting with both man and the apes, rather than being intermediate between them (Oxnard, 1969).

A further example is provided by the curvature of the phalanges— this being a feature in which morphologically and biometrically these fossils resemble the African great apes rather than man. But an experimental stress analysis has shown that the architecture of these bones is such as to indicate an efficiency for a type of function apparently quite different from that habitual in the African great apes (Oxnard, 1973).

Examples of this type are multiplying—the most recent analysis of the talus (Oxnard, 1972) reinterpreting earlier multivariate data (Day, 1967; Day & Wood, 1968, 1969) which had appeared to indicate that, in the features of this region, the australopithecine fossils were intermediate between man and the living African apes, has shown that they really fall in a position quite distinct from both.

The mechanical significance of such exceptional combinations of some human features with others that are more like the conditions that obtain in sub-human Primates is, in practically all these skeletal regions, difficult to interpret, and the present study of the innominate bone and its associated muscle blocks has re-emphasized the uncertainty that attaches to such assessments. As ambiguity exists in practically all the examples that are available, and as a common interpretation apparently cannot be derived, even greater uncertainty than appears from the present analysis alone, attaches to inferences about overall locomotor patterns in the Australopithecinae.

SUMMARY

A study has been made of pelvic girdles from a total of 430 extant Primates representing 13 genera from the Prosimii and 28 from the Anthropoidea (including man).

Comparison has been made with reconstructions of an almost complete fossil innominate bone assigned to the genus *Australopithecus* (formerly *Plesianthropus*) and claimed, in many publications, as providing strong evidence in support of the view that the Australopithecinae were bipedal.

Anatomical features were selected so as to relate to the pattern of forces habitually impressed upon the pelvic region, thus permitting biomechanical interpretation of any contrasts established between genera.

Four features were defined as associated with the mechanism for transmitting weight from the vertebral column, through the sacrum and innominate bone to the thigh. They described the positions of the sacro-iliac and hip joints relative both to the bone as a whole and to each other.

Five further features were defined as relating to proportion and disposition of the several blocks of hip muscles—differences in these between man and sub-human Primates having been established in a preliminary quantitative study. These characters concerned: first, the abductors of the hip—these muscles being more prominent in man, and secondly, the extensors of the hip—these being proportionately greater in sub-human Primates.

Preliminary tests showed that inconsistency in measurement contributed only an insignificant amount to the total variance. Again, so far as could be judged from the Hominoidea, sexual and racial differences were small in scale compared with contrasts between genera.

Further preliminary sorting of the data indicated the existence of allometric relationships between each pelvic dimension and the dorso-ventral dimension of the bone—this measure being, throughout the Primates, highly correlated with crown-rump length and therefore reflecting overall size. Differences due to allometry and to contrasts in overall size were eliminated by logarithmic transformation and by regression adjustment—this technique demonstrating a slightly greater contrast between primate genera than was obtained from the less rigorous method based upon calculation of indices.

Univariate comparison showed that, subject to certain variation, there is, in the case of characters related both to weight-bearing and to muscular pull, an appreciable contrast between man and many groups of sub-human Primates.

In features relating to weight-bearing, *Australopithecus* is, in general, like man and unlike most sub-human Primates. The sacro-iliac and hip joints lie, in the fossil, relatively close to each other and are more ventrally disposed than in sub-human Primates, thus providing a mechanism, more efficient than in monkeys and apes, for the transmission of weight in the upright position.

In features relating to muscular pull, *Australopithecus* tends to contrast with man and agrees with many sub-human Primates, and especially the great apes. The iliac blade of the fossil is orientated more dorsally than in man (thus making it unlikely that the lesser gluteal muscles were abductors), while the lever arm of the extensors of the hip is, as in sub-human Primates, relatively long.

When the four dimensions relating to weight bearing are compounded by multivariate techniques (generalized distance and canonical coordinates), *Australopithecus* again emerges as being quite similar to man and dissimilar to sub-human Primates.

When the five features relating to muscular pull are similarly grouped, *Australopithecus* falls with certain sub-human Primates and differs from man.

When all nine features are similarly compounded, *Australopithecus* emerges as unique, differing from both man and sub-human Primates. Of the latter, the group approximating most closely to *Australopithecus* comprises genera in which the hindlimb sometimes supports, sometimes suspends the animal and generally operates in many planes of space.

Australopithecus again emerges as unique when multivariate analysis is confined to the Hominoidea (including *Australopithecus*).

It is impossible to advance an unequivocal interpretation of these findings.

If *Australopithecus* were habitually, or occasionally, bipedal, its weight would have been carried more efficiently than occurs when any extant sub-human primate attempts to walk bipedally.

Any such bipedalism must have differed from that typical of *Homo sapiens* because of the lack of powerful abduction of the hip joint resulting in ineffective lateral support of the pelvic girdle.

The possibility is not excluded that the overall locomotor pattern of *Australopithecus* included, in addition to bipedalism, components as yet undefined.

Difficulty in assessing the probable overall locomotor pattern of the Australopithecinae is increased by the existence of unique combinations of morphological characters in other relevant skeletal regions.

ACKNOWLEDGEMENTS

The work described in this paper was carried out in the Department of Anatomy, University of Birmingham, between 1964 and 1972. The collection and initial processing of the osteometric data together with the preliminary myological study were supported by grant HD00450 from the United States Public Health Service. The continued collaboration of Oxnard was made possible by N.I.H. grant HD02852, by NSF Grant GS 30508 and by contributions from the Louis Block Bequest, the Dr W. C. and C. A. Abbot Memorial Funds of the University of Chicago, and from the Wenner–Gren Foundation for Anthropological Research. We are obliged to the several Museums and University Departments noted on page 81 for permitting access to their collections of Primate skeletal material and to Professor J. T. Robinson for providing a cast of the innominate bone of *Australopithecus*. Text figures 1–20 were originally prepared by the late Mr W. J. Pardoe. They were among the last artistic and graphical works of one who, during the years, had contributed much to anatomical illustration. The remainder of the text figures are the work of Miss Margaret Howdle.

REFERENCES

Ashton, E. H. (1972). *Know thine anatomy.* Inaugural lecture published by University of Birmingham: 1–48.
Ashton, E. H., Flinn, R. M., Oxnard, C. E. & Spence, T. F. (1971). The functional and classificatory significance of combined metrical features of the primate shoulder girdle. *J. Zool., Lond.* **163**: 319–350.
Ashton, E. H., Healy, M. J. R., Oxnard, C. E. & Spence, T. F. (1965). The combination of locomotor features of the primate shoulder by canonical analysis. *J. Zool., Lond.* **147**: 406–429.
Ashton, E. H. & Oxnard, C. E. (1964a). Locomotor patterns in Primates. *Proc. zool. Soc. Lond.* **142**: 1–128.
Ashton, E. H. & Oxnard, C. E. (1964b). Functional adaptations in the primate shoulder girdle. *Proc. zool. Soc. Lond.* **142**: 49–66.
Ashton, E. H. & Zuckerman, S. (1951). Some cranial indices of *Plesianthropus* and other Primates. *Am. J. phys. Anthrop.* (n.s.) **9**: 283–296.

Ashton, E. H. & Zuckerman, S. (1952). Age changes in the position of the occipital condyles in the chimpanzee and gorilla. *Am. J. phys. Anthrop.* (n.s.) **10**: 277–288.

Ashton, E. H. & Zuckerman, S. (1956). Age changes in the position of the foramen magnum in hominoids. *Proc. zool. Soc. Lond.* **126**: 315–325.

Barham, W. W. (1971). A longitudinal study of the growth of the chimpanzee bony pelvis. *Proc. 3rd int. Congr. Primat., Zurich 1970.* **1**: 169–175.

Bean, R. B. (1922). The sitting height. *Am. J. phys. Anthrop.* **5**: 349–390.

Biegert, J. & Maurer, R. (1972). Rumpfskelettlänge, Allometrien und Körper-proportionen bei catarrhinen Primaten. *Folia primat.* **17**: 142–156.

Broek, A. J. P. van de (1914). Studien zur Morphologie des Primatenbeckens. *Morph. Jb.* **49**: 1–118.

Broom, R. & Robinson, J. T. (1952). Swartkrans ape-man *Paranthropus crassidens*. *Transv. Mus. Mem.* **6**: 1–123.

Broom, R., Robinson, J. T. & Schepers, G. W. H. (1950). Sterkfontein ape-man *Plesianthropus*. *Transv. Mus. Mem.* **4**: 1–117.

Coleman, H. (1971). Comparison of the pelvic growth patterns of chimpanzee and man. *Proc. 3rd int. Congr. Primat., Zurich 1970.* **1**: 176–182.

Coon, C. S. (1963). *The origin of races*. London: Cape.

Dart, R. A. (1925). *Australopithecus africanus:* The man-ape of South Africa. *Nature, Lond.* **115**: 195–199.

Day, M. H. (1967). Olduvai hominid 10: a multivariate analysis. *Nature, Lond.* **215**: 323–324.

Day, M. H. & Wood, B. A. (1968). Functional affinities of the Olduvai hominid 8 talus. *Man.* **3**: 440–455.

Day, M. H. & Wood, B. A. (1969). Hominoid tali from East Africa. *Nature, Lond.* **222**: 591–592.

Elliot, D. G. (1913). A review of the Primates. *Monogr. Am. Mus. nat. Hist.* **1**: (3 vols): 1–317, 1–382, 1–262.

Flower, W. H. (1885). *An introduction to the osteology of the Mammalia*. London: Macmillan.

Gower, J. C. (1966). A Q-technique for the calculation of canonical variates. *Biometrika* **53**: 588–590.

Gower, J. C. (1968). Adding a point to vector diagrams in multivariate analysis. *Biometrika* **55**: 582–585.

Gower, J. C. & Ross, G. J. S. (1969). Minimum spanning trees and single linkage cluster analysis. *Appl. Statist.* **18**: 54–64.

Gray, J. (1944). Studies in the mechanics of the tetrapod skeleton. *J. exp. Biol.* **20**: 88–116.

Gray, J. (1953). *How animals move* (Royal Institution Christmas Lectures 1951). Cambridge: University Press.

Haughton, S. (1864). Notes on animal mechanics: 2 On the muscles of some of the smaller monkeys of the genera *Cercopithecus* and *Macacus*. *Proc. R. Ir. Acad.* **8**: 467–471.

Haughton, S. (1867). Notes on animal mechanics: 7 On the muscular anatomy of the *Macacus nemestrinus*. *Proc. R. Ir. Acad.* **9**: 277–287.

Hotelling, H. (1936). Relations between two sets of variates. *Biometrika* **28**: 321–377.

Howell, A. B. (1936). The phylogenetic arrangement of the muscular system. *Anat. Rec.* **66**: 295–316.

H

Howell, A. B. (1938). Morphogenesis of the architecture of hip and thigh. *J. Morph.* **62**: 177–218.

Huxley, T. H. (1879). On the characters of the pelvis in the Mammalia, and the conclusions respecting the origin of mammals which may be based on them. *Proc. R. Soc.* **28**: 395–405.

Jenkins, F. A. (1972). Chimpanzee bipedalism: cinéradiographic analysis and implications for the evolution of gait. *Science, N.Y.* **178**: 877–879.

Leakey, L. S. B., Tobias, P. V. & Napier, J. R. (1964). A new species of the genus *Homo* from Olduvai Gorge. *Nature, Lond.* **202**: 7–9.

Le Gros Clark, W. E. (1955). The os innominatum of the recent Ponginae with special reference to that of the Australopithecinae. *Am. J. phys. Anthrop.* (n.s.) **13**: 19–27.

Le Gros Clark, W. E. (1962). *The antecedents of man* (2nd edition). Edinburgh: University Press.

Lovejoy, C. O. & Heiple, K. G. (1970). A reconstruction of the femur of *Australopithecus africanus. Am. J. phys. Anthrop.* (n.s.) **32**: 33–40.

Mahalanobis, P. C. (1936). On the generalized distance in statistics. *Proc. natn. Inst. Sci. India* **2**: 49–55.

Martin, R. (1959). *Lehrbuch der Anthropologie.* 3rd ed. Stuttgart: Fischer.

McHenry, H. (1972). *Study of the post cranium of* Australopithecus. Ph.D. thesis: University of Harvard.

Mednick, L. W. (1955). The evolution of the human ilium. *Am. J. phys. Anthrop.* (n.s.) **13**: 203–216.

Moore, W. J., Adams, L. M. & Lavelle, C. L. B. (1973). Head posture in the Hominoidea. *J. zool. Lond.* **169**: 409–416.

Napier, J. R. (1964). The evolution of bipedal walking in the hominids. *Archs Biol., Liège* **75**: 673–708.

Napier, J. R. (1967). The antiquity of human walking. *Scient. Am.* **216**: 56–66.

Napier, J. R. & Napier, P. H. (1967). *A handbook of living Primates.* London & New York: Academic Press.

Oxnard, C. E. (1967). The functional morphology of the primate shoulder as revealed by comparative anatomical, osteometric and discriminant function techniques. *Am. J. phys. Anthrop.* (n.s.) **26**: 219–240.

Oxnard, C. E. (1968a). A note on the Olduvai clavicular fragment. *Am. J. phys. Anthrop.* (n.s.) **29**: 429–432.

Oxnard, C. E. (1968b). A note on the fragmentary Sterkfontein scapula. *Am. J. phys. Anthrop.* (n.s.) **28**: 213–218.

Oxnard, C. E. (1969). Evolution of the human shoulder: some possible pathways. *Am. J. phys. Anthrop.* (n.s.) **30**: 319–332.

Oxnard, C. E. (1972). Some African fossil foot bones: a note on the interpolation of fossils into a matrix of extant species. *Am. J. phys. Anthrop.* (n.s.) **37**: 3–12.

Oxnard, C. E. (1973). *Form and pattern in human evolution: some mathematical, physical and engineering approaches.* Chicago: University of Chicago Press.

Reynolds, E. (1931). The evolution of the human pelvis in relation to the mechanics of the erect posture. *Pap. Peabody Mus.* **11**: 255–334.

Robinson, J. T. (1964). Adaptive radiation in the australopithecines and the origin of man. In *African ecology and human evolution:* 385–416. Howell, F. Clark & Bourlière, F. (eds) London: Methuen.

Schultz, A. H. (1941). Growth and development of the orang-utan. *Contr. Embryol.* **29**: 57–110.

Stern, J. T. Jr (1971). Functional myology of the hip and thigh of cebid monkeys and its implications for the evolution of erect posture. *Biblthca Primat.* **14**: 1–318.

Stern, J. T. Jr (1972). Anatomical and functional specializations of the human gluteus maximus. *Am. J. phys. Anthrop.* (n.s.) **36**: 315–339.

Stern, J. T. Jr & Oxnard, C. E. (1973). *Primate locomotion: some links with evolution and morphology.* Basel: Karger.

Theile, F. W. (1884). Gewichtsbestimmungen zur Entwickelung des Muskelsystems und des Skelettes beim Menschen. *Nova Acta Acad. Caesar. Leop. Carol.* **46**: 133–471.

Thompson, W. d'Arcy. (1917). *On growth and form.* Cambridge: University Press (new edition 1942).

Washburn, S. L. (1950). The analysis of primate evolution with particular reference to the origin of man. *Cold Spring Harb. Symp. quant. Biol.* **15**: 67–78.

Waterman, H. C. (1929). Studies on the evolution of the pelvis of man and other Primates. *Bull. Am. Mus. nat. Hist.* **58**: 585–642.

Weidenreich, F. (1913). Über das Hüftbein und das Becken der Primaten und ihre Umformung durch den aufrechten Gang. *Anat. Anz.* **44**: 497–513.

Zuckerman, S. (1954). Correlation of change in the evolution of higher Primates. In *Evolution as a process:* 300–352. Huxley, J. S., Hardy, A. C. & Ford, E. B. (eds) London: Allen & Unwin.

DISCUSSION

BARNICOT: You have pointed out that certain features of the australopithecine pelvis may indicate that they were not bipedal in quite the same way as modern man. May one ask whether this type of bipedalism was one from which the modern form of bipedalism could have evolved?

ASHTON: Well, this question is intriguing, but its answer can only be speculative. Personally I prefer to stress facts, and in this I agree with Professor Day and Mr Leakey. The type of problem to which I think one can profitably direct one's attention is: was the type of bipedalism like that in man? If not, what type of bipedalism might it have been? These are the questions that we have been asking, and here, I think I would go a little further than Professor Day and agree with some of Professor Napier's earlier publications, in which it was suggested that the bipedalism of the australopithecines was probably of a rolling type rather than a striding type—I trust I have not misinterpreted him.

NAPIER: I think the point I would make is that, as Professor Day showed this morning, there are two types of pelvis there from South African sites. We do not need to get taxonomically heated over this issue and call one *Paranthropus* and the other *Australopithecus*, although I am prepared to come out in clear and say that that is how I would see them myself. There are two types of pelvis, whether or not they are sexually dimorphic, which I think is most unlikely; both of them, as Professor Day has pointed out, are rather distorted. But the one that you have described this afternoon is the least distorted and the more complete of the two. I am rather surprised that you have not mentioned the Kromdraai pelvis, which is so very distinctive from the Sterkfontein one. My view about the gait of *Australopithecus* has been for some years that it was not exactly like modern man's. It was a form of bipedalism, although striding, so far as I could see, was absent. My mention of waddling, shuffling, rolling was really with reference to the other pelvis, which may belong to *Australopithecus robustus*, or *Paranthropus*, whatever you like to call it. So the waddling and shuffling really refer to the Kromdraai pelvis and not the Sterkfontein specimen.

ASHTON: Thank you; obviously one would clearly like, as I mentioned— I think twice in the course of presenting the paper, Mr Chairman—to have the opportunity of applying biomechanical analysis of this type to other fossil pelves. The one to which Professor Napier has just alluded is another fascinating specimen, but one which, as I see it, is not sufficiently complete to be amenable to this type of biometric study, and it would be rather dangerous to attempt to reconstruct sufficiently to make possible an application of the methods that I have outlined. Possibly it would be appropriate at this stage if I were to re-emphasize that the onus of this study has not been so much to study the probable posture and gait of the Australopithecinae, as to develop a biomechanical method for enquiring into the functional significance of the features of any fossil pelvis that is found. Professor Oxnard has shown, in the case of the scapula, that this concept can be applied outside the Primates, certainly through the mammals and probably beyond.

YOUNG (CHAIRMAN): If I might interrupt there, I think many of us were especially pleased to see the emphasis you placed on the particular functions of the pelvis of the different groups. We do not only want to know different muscle groups, but what they are likely to be doing.

DAY: That was one of the points I was about to make, Mr Chairman, but you have made it for me, I do not want to go any further into the question of the Sts 14 pelvis, but I think this pelvis can be reconstructed properly, and until this is done, I think we can again generate a great deal of controversy that is unproductive and unnecessary. Three reconstructions have been produced today, any one of which might be right, but all three of which may well be wrong, and therefore why should we argue?

So far as the results that you have shown us are concerned, and a very large number of results were presented in a very short time so it was difficult to take in the significance of each, but as I understand it, the weight-bearing relationships gave a relationship towards man, and muscular ones towards the pongids, shall we say. However, at one point a combination of the two in a very dramatic slide left us with the Australopithecinae and *Homo* still reasonably close together.

ASHTON: But only in the first two canonical axes.

DAY: You then amplified that and said in fact that at a later stage you concluded with an intermediate position. To me this is extremely interesting, because in my much smaller and much more limited canonical analysis, of a restricted area of the femoral neck, I obtained the same sort of answer. I have an intermediacy in this situation. I am not suggesting that they are parallel, but I do take your point that if this is the case, then these things are different. So we go round in a circle; we did know before we started that they were different. We have to be very careful as to what sort of progress we are making if we are merely confirming by elaborate metrical and statistical means things that are fairly obvious anatomically.

ZUCKERMAN: I think Professor Day is quite incorrect in saying that "we all knew". I embarked on these studies, and encouraged my colleagues into them, precisely because people were saying that this particular innominate bone was more a human than an ape innominate bone. That is why we launched upon this enquiry—to try to see what specific characteristics we could find, and to see if they were human or simian.

We learnt this morning that there are other post-cranial bones of a different type from those belonging to the new skull. Frankly I am less interested in the australopithecine remains than are the anatomists who regarded them as the remains of some ancestor of Man. I would much sooner wait until those new bones, about which Mr Leakey told us this morning, are available for the kind of detailed study we have been carrying out. But I know of course that the modern statistical analysis which Professor Ashton has been describing is very arduous. I myself do not mind now where *Australopithecus* falls, on which circle, on which line. The matter is irrelevant.

ASHTON: Surely: our main object was, as I mentioned a moment ago, to define a method for enquiring into the biomechanical significance of the features of these bones.

YOUNG (CHAIRMAN): Thank you very much, Professor Ashton. I should also thank Lord Zuckerman and the other members of this team. They really have shown us how you can begin a biomechanical study—an example for the future.

COMPARATIVE ANATOMY AND EVOLUTION OF THE FOREBRAIN

Chairman: J. Z. Young

CHAIRMAN'S INTRODUCTION

I should like to say one or two words about Sir Grafton Elliot Smith. I never knew him myself, but as his successor but one, I have naturally been very interested in his career, and many have recollections of him at University College.

Lord Zuckerman said this morning that he was not sure whether the Rockefeller Foundation had given the Anatomy Department—they did in fact. In 1919 the Rockefeller Foundation came over to England with money to spend, so one understood, and Sir Grafton Elliot Smith was clever enough to realize that this was a good opportunity. They gave a very large sum of money to University College and University College Hospital to build the Medical School, to endow various chairs and to build the Anatomy Department.

Because of his interest in Egyptology, the architect, who I understand was a Professor Simpson, took an Egyptian model for his building. I wonder how many of you walking down Gower Street have ever noticed this. There is a lot of decoration, all with Egyptian motifs and looking at one entrance it might almost be the doorway of a tomb. Inside there are pillars on the landing and the staircase. I expect many of you have had your hands on that stair rail and never noticed that it had Egyptian motifs too.

We pass on now to another aspect of Sir Grafton Elliot Smith's work, hinted at by Professor Day when he spoke of eyeball study; Elliot Smith's work was perhaps primarily in the brain and the study of cerebral structure, and some of his most striking work was done actually with the naked eye. (I expect Dr Powell will be speaking about this in his paper.) But a great deal of it was done in the ventral parts of the brain too, and he really provided us with a scheme for studying the comparative anatomy and hence the proper morphology of the human brain many years ago. It is on that theme that we planned this session,

with Dr Webster to talk about the striatum and thalamus of birds and
reptiles, Professor Irving Diamond to talk about the same parts of the
brain in more simple Primates and Dr Powell on the cerebral cortical
end of it. Professor Diamond is from Duke University Department of
Psychology. This is interesting because although one does not at once
associate Elliot Smith's name with psychology, I have heard that he
was known to say that the thing he most enjoyed when he went to
Cambridge was meeting Dr Rivers, who was a psychologist. Lord
Zuckerman confirmed that this morning. So Elliot Smith did have this
interest in psychology, and I am sure he would have been fascinated by
what Professor Diamond is going to tell us.

Sir Grafton Elliot Smith I am sure would have been delighted with
this session as with the other parts of the Symposium. He had really
sought—you could feel it in reading his papers—for a scheme for
understanding these really very difficult problems of homology in the
brain. I think he would be delighted at the progress that has been
made.

Symp. zool. Soc. Lond. (1973) No. 33, 169–203.

THALAMUS AND BASAL GANGLIA
IN REPTILES AND BIRDS

K. E. WEBSTER

Department of Anatomy, University College London, Gower Street, London,
England

SYNOPSIS

The vertebrate forebrain has long been assumed to conform to a fundamental pattern. All its subdivisions are, with the possible exception of the neocortex, presumed to be identifiable in all brains, although developed to differing degrees. With the exception of the olfactory system, all inputs to the telencephalon are by way of the dorsal thalamus. In mammals, the nuclear configurations of the dorsal thalamus include two important sub-groups, each in turn related to a major sub-division of the telencephalon: the intralaminar system on the one hand—projecting onto the corpus striatum—and the principal nuclei on the other—related to the neocortex. In all other vertebrates the failure to identify a significant homologue of the neocortex has led to the general conclusion that the larger part of the dorsal thalamus of birds and reptiles is comprised of "intralaminar nuclei". Earlier studies of the projection of the non-mammalian thalamus onto the cerebral hemisphere indicated little more than the existence of the projection: the results could be interpreted to mean that either the projection was distributed diffusely throughout the whole striatal complex or restricted to a small part of it. More recently it has become apparent that both these views are erroneous and evidence is accumulating of a highly organized system of projections, comparable in its complexity with the projections of the mammalian principal nuclei. It is possible that the most significant difference in the forebrain organization of birds and reptiles compared with mammals lies in the mechanics of expansion of the hemispheres: in the former to form a solid mass protruding into the lateral ventricle, and in the latter to produce a folded sheet.

INTRODUCTION

In the Arris and Gale lectures of 1909 Elliot Smith (1910) reviewed the then available evidence and speculations concerning the evolution of the mammalian neocortex. He concluded that the neocortex is a uniquely mammalian feature, notwithstanding the presence in the dorsal part of the forebrain hemispheres of all amniotes of a cortical structure of some kind (see Fig. 1). This view has since become a touchstone for comparative neurologists, and so has its corollary—that in reptiles and birds the telencephalon shows a marked development of the basal regions that form, as a result, a striking bulge into the lateral ventricle. This promontory, probably first described by John Hunter, was for some time known as "Hunter's eminence". Unfortunately this harmless eponym was later displaced by the term "striatum" to which were appended numerous prefixes to deal with the variety of cyto-architectural subdivisions of which the eminence is composed. The use of the term "striatum" was unfortunate because it carries with it a series of preconceived ideas relating to the mammalian striatum. That Elliot Smith was aware of this is revealed perhaps, by a long neglected remark in the second of the Arris and Gale lectures:

> "Birds again present such a highly specialized and diversely modified cerebrum that they have quite left the path leading to the formation of a neopallium and evolved a mechanism peculiar to their own class, which does much the same kind of work as the neopallium". (Elliot Smith, 1910: 151).

By the word "mechanism", Elliot Smith could have alluded to nothing but the so-called corpus striatum of birds. And the sentence opens in such a way as to suggest that he may have thought that something similar might well apply to reptiles—the group that he discusses

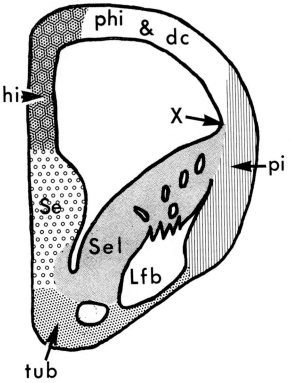

Fig. 1. Transverse section of a hypothetical forebrain from which the form and composition of all amniote telencephalons can be derived. Site X is that from which it is possible to imagine neuronal proliferation to occur either to expand the cortex (dc) or to create the dorsal ventricular ridge (ds in Fig. 2). See text. Key to abbreviations p. 200.

briefly in the paragraph immediately preceding this extract. That Elliot Smith may have equivocated about applying this concept to reptiles is understandable—the living representatives of this Class all possess a dorsal or "general" cortex, and the "striatal" complex is not nearly so elaborated as that of birds. As will be noted later, Elliot Smith's (1919) more complete account of the reptilian forebrain and analysis of the reptilian/mammalian condition he forced into a conventional inter-pretation. Briefly, this interpretation involves a comparison with the mammalian condition as follows:

The enlargements constituting the basal telencephalon of mammals include the corpus striatum and amygdaloid complex. The former may be thought of as "somatic" whilst the latter is part of the limbic system and is closely related to the olfactory pathways. The corpus striatum may be divided into two parts: a medial segment containing large neurons (the globus pallidus or "palaeostriatum") and an outer small-cell region (the striatum or "neostriatum"). The amygdala, in this convention, is called the "archistriatum". In reptiles and birds it therefore becomes imperative to discover three fundamental subdivisions in the striatal elevation, which may, indeed, be done by grouping the smaller subdivisions. In order to demonstrate this it is necessary first to consider the normal cytoarchitecture of the relevant regions of the reptilian and avian forebrain.

THE CYTOARCHITECTURE OF THE STRIATAL ELEVATION

Reptiles

The morphology of the reptilian cerebral hemispheres has been extensively reviewed by Ariëns Kappers, Huber & Crosby (1936) and, more recently by Goldby & Gamble (1957). Before describing the striatal eminence it is important to note that all the structures shown in Fig. 1 are present in the reptilian telencephalon. Special attention is directed towards the dorsal cortex, which comprises little more than a single lamina of pyramidal cells between the para-hippocampal cortex medially and the piriform cortex laterally (Fig. 2, dc). Its exact status has been disputed for many years and few authors have been inclined to compare it with mammalian neocortex. This latter has been des-cribed as a "primordial" element in the rostral part of the dorsal cortex of reptiles (see pp. 173, 183, 193).

The striatal mass in these forms may be divided into a rostral and caudal part. The boundary may be either vague or clearcut, and in some species is marked by an obliquely transverse sulcus—the sulcus

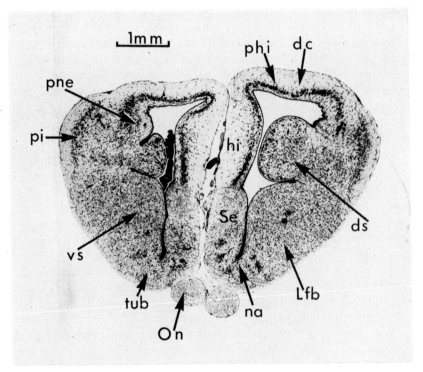

Fig. 2. Transverse section of the telencephalon of the tortoise (*Testudo graeca*). The left of the photograph is more rostral than the right. (Cresyl violet stain). Key to abbreviations p. 200.

strioarchistriatica of Ariëns Kappers (1921). The caudal region (the archistriatum) shows much interspecies variation in its complexity—it shows little elaboration in, for instance, Crocodylia, but in other forms it contains a well-circumscribed nucleus (the nucleus sphaericus) which receives a large input from the accessory olfactory bulb (Heimer, 1969).

The rostral part of the striatal eminence comprises about two thirds of the whole. It may be divided into a dorsal and a ventral part, which in some species (e.g. Chelonia) are separated by a distinct sulcus as shown in Fig. 2. The ventral division shows relatively little variation from species to species. It contains mainly medium size or small neurons, mixed with larger perikarya that are scattered among the radiating fibres of the lateral forebrain bundle. The ventral striatum abuts on the septum by virtue of an extension known as the nucleus accumbens (Fig. 2, na). The ventral division is the "palaeostriatum" of Ariëns

Kappers (1921), and is sometimes divided into a medial "olfactory part" and a lateral "somatic part". As Goldby & Gamble (1957) point out, there exists little real evidence for or against this idea.

The dorsal subdivision is sometimes referred to as the dorsal ventricular ridge, or hypopallial ridge (Johnston, 1923), in which case, it must be pointed out, the ridge is considered to comprise the whole rostro-caudal length of the striatal eminence, i.e. to include the "archistriatum" (see Fig. 5, arch). The rostral part of the dorsal ventricular ridge exhibits considerable interspecies variation in its cytoarchitecture, but always contains more than a single nucleus, as may be seen in Fig. 2, which is from a specimen of *Testudo graeca*, one of the less elaborate forms. Riss, Halpern & Scalia (1969) have compared the cytoarchitecture of several turtles with that of a caiman and concluded that all the subdivisions found in the latter are in fact present in turtles. In some species, and particularly in *Sphenodon* (Elliot Smith, 1919; Cairney, 1926; Durward, 1930), this region of the striatum contains a well-defined cell lamina that is continuous with either the piriform cortex in its caudal part or with the dorsal cortex rostrally. Rostrally the continuity forms a distinct ridge that has become known as the "primordium neopallii" (Crosby, 1917) (see Fig. 2, pne). Elliot Smith (1919) accepted this term, partly because he considered that, unlike the remaining part of the dorsal cortex, this region receives ascending fibres from the lateral forebrain bundle (the fibre system ramifying throughout the dorsal striatum). Elliot Smith stressed the importance of this continuity, and considered the dorsal striatum to be a "pallial derivative". He therefore referred to this apparent infolding as the hypopallium. It is of some interest that in spite of this he went on to compare the greater part of the hypopallial ridge with the mammalian striatum and the ventral (palaeo-) striatum with the globus pallidus. Johnston (1915, 1916b, 1923) went so far as to use these terms in describing the chelonian striatal complex. These comparisons (including that of the caudal striatum or "archistriatum" with the amygdala) therefore neatly dispose of the subdivisions of the striatal elevation of reptiles in terms of the structures found in the mammalian telencephalon. But can this be done for the complex striatal structure found in birds?

Birds

The dorsal corticoid area of birds is in general poorly developed and may represent little more than large but rudimentary para-hippocampal and periamygdaloid fields, except rostro-dorsally where the cortex is inseparable from the superficial Wulst (see p. 175). The accessory hyperstriatum and possibly all the Wulst may be considered cortical derivatives (see Craigie, 1940 a, b; Pearson, 1972).

In birds the enormous striatal eminence is split into a series of subdivisions proceeding from a rostro-dorsal position to the caudo-ventral part of the hemisphere (Fig. 3). The subdivisions have been

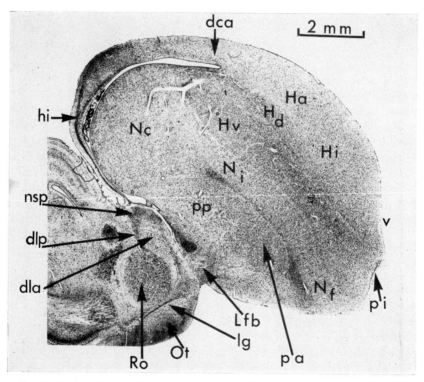

Fig. 3. Parasagittal section of the forebrain of the pigeon (*Columba livia*). The letter "v" on the extreme right of the photograph lies in the shallow depression of the vallecula. (Cresyl violet stain). Key to abbreviations p. 200.

reviewed by Ariëns Kappers *et al.* (1936) and by Pearson (1972). The nomenclature used below is that of Huber & Crosby (1929). Although there are interspecies variations in the topography and relative development of the various nuclei, these subdivisions have been recognized in all species of birds so far examined (see, for example, Rose (1914); Stingelin (1958)). The most complete set of illustrations is provided by Karten & Hodos (1967). For convenience and in anticipation of results to be discussed later, the subdivisions will be grouped into two major categories—those belonging to the external striatum, and those comprising the internal striatal complex.

The external striatum

(a) The Wulst comprises a dorsal region of the hemisphere from which it is separated laterally by a shallow groove—the vallecula—on the surface of the brain. It contains three subdivisions—the outer hyperstriatum accessorium (Ha) and the inner hyperstriatum dorsale (Hd), with the intercalated nucleus (Hi) between them (Fig. 3). The nucleus intercalatus is particularly variable, as was noted by Rose (1914). It is, for example, very well developed in owls (Nauta & Karten, 1970). According to Stingelin (1958) it is absent from the Wulst of *Columba*. This is an error: the intercalated nucleus is present but relatively poorly developed in this group (Karten & Hodos, 1967; Hunt & Webster, 1972, and unpubl. obs.). As noted previously, the dorsal corticoid area becomes continuous with the hyperstriatum and is difficult to distinguish from it (Fig. 3, dca): Edinger, Wallenberg & Holmes (1903) referred to the accessory hyperstriatum as "frontal cortex".

(b) The hyperstriatum ventrale (Fig. 3, Hv) which is sometimes further subdivided into ventral and dorsal parts.

(c) The neostriatum (Nc, Ni, Nf), which has a large caudal extension, the neostriatum caudale (Nc), that is easily recognized in Fig. 3. The neostriatum caudale contains a subdivision which Rose (1914) referred to as "field L".

(d) The ectostriatum, (Fig. 4B, E), forming an isolated nucleus deep in the striatal complex.

(e) The archistriatum is usually considered to be an olfactory or "visceral" region, but the work of Zeier & Karten (1971) indicates that its rostral and dorsal portions are somatic sensori-motor regions.

The internal striatum

(a) The palaeostriatum augmentatum. This is a region of relatively small neurons arranged in the form of an inverted, shallow cup around the palaeostriatum primitivum (see below) and continuous rostrally with the paraolfactory nucleus. It can be seen in Fig. 3 (pa).

(b) The palaeostriatum primitivum (Fig. 3, pp) contains relatively few neurons, but these are large. It abuts ventromedially on the entopeduncular nucleus (Fig. 4B, ent).

Comparison of this complex formation with the simple condition found in mammals, and even that seen in reptiles, is at first sight difficult. The basic triad was, in spite of this, found by many authors (see,

for example, Ariëns Kappers, 1921). In these schemes, it became customary to compare the archistriatum with the amygdaloid complex, the
palaeostriatum primitivum with the globus pallidus, and all the remainder with the striatum (caudate nucleus and putamen). Edinger *et al.*

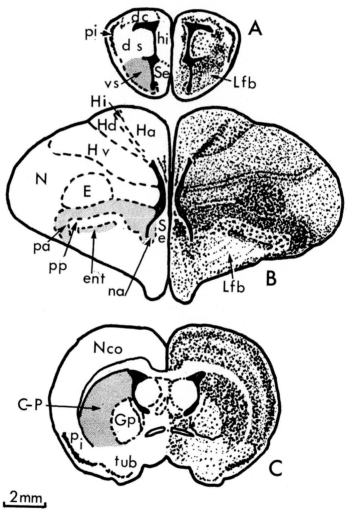

Fig. 4. Semi-diagrammatic drawings of transverse sections through the telencephalon of: A, tortoise; B, pigeon; C, rat. The stippled areas on the left of the figures
represent the distribution of acetyl cholinesterase. The distribution of dopamine is
identical except that no monoamines are found in the entopeduncular nucleus (B, ent).
Key to abbreviations p. 200.

(1903), however, placed the hyperstriatum accessorium in the category of cortex. Support for the comparison between the palaeostriatum primitivum and globus pallidus has recently been produced by Fox, Hillman, Siegesmund & Sether (1966), who point to the striking ultrastructural resemblances between these nuclei.

Considered solely in terms of cytoarchitecture, however, the avian striatum remains enigmatic, except that the palaeostriatum augmentatum et primitivum presents an appearance that recalls the lentiform nucleus (i.e. the putamen plus globus pallidus) of some mammals. This apparently trite observation becomes more interesting and significant if one considers the evidence of "chemoarchitecture"—the distribution of transmitter substances and enzymes in the brain.

There is now much evidence that in mammals monoamines (and particularly dopamine) are important transmitters in the striatum (but not the pallidum). The dopaminergic synapses in the striatum are thought to represent the terminals of axons arising from cells of the substantia nigra (Andén, Dahlström, Fuxe & Larrson, 1965). In both reptiles (Parent & Poirier, 1971) and birds (Fuxe & Ljunggren, 1963) monoamine containing perikarya have been found in the ventrolateral midbrain tegmentum and it is probable that the axons from these cells ascend to the striatal complex. In any event, the areas of the telencephalon that are rich in monoamine form a relatively small part of the whole. In the species so far studied, the monoamine transmitters are found only in the ventral part of the striatum (i.e. the palaeostriatum) of reptiles, and in the palaeostriatum augmentatum of birds (see Fig. 4). The nucleus accumbens also contains this transmitter. The distribution of acetylcholinesterase is similar except that it is also found in the entopeduncular nucleus (see Bertler et al., 1964; Jurio & Vogt, 1967; Jurio, 1969; Parent & Olivier, 1970). These results imply that the greater part of the striatal complex of reptiles and birds differs in its transmitter chemistry from the striatum of mammals (see Fig. 4). This information unfortunately provides no clue as to the nature of the remaining regions. This must be investigated by other means, the most obvious of which is consideration of the fibre connexions. It is not proposed to examine in detail the connexions of the archistriatum. As already noted, this region is in part related to the accessory olfactory system in reptiles, and similar conclusions have been reached in mammals (Raisman, 1972). No data are available for birds, except the interesting findings of Zeier & Karten (1971) that implicate only the ventral and caudal parts of the pigeon archistriatum as comparable with the mammalian amygdala.

THE DORSAL THALAMUS IN REPTILES AND BIRDS

For the analysis of the remaining parts of the striatal complex the most fruitful line of enquiry has been the investigation of the thalamo-telencephalic relationships. In mammals, the nuclei of the dorsal thalamus may be grouped into major subdivisions, each related to a particular part of the telencephalon. Thus, the principal nuclei of the dorsal thalamus project into the neocortex, whilst the intralaminar nuclei project to the striatum (see, for example, Powell & Cowan, 1967). Because of the enormous relative size of the "corpus striatum" of non-mammals it has been widely assumed that the dorsal thalamus of these forms must consist largely of intralaminar nuclei (see, for example, Powell & Kruger, 1960). One way of testing this assumption would be to attempt definitions of principal and intralaminar nuclei that are not completely dependent upon the relationship of these nuclei with the telencephalon. Criteria of relative size must obviously be discounted, as must those of topography, because although the reptilian and avian thalami have cytoarchitectural patterns similar enough to allow some comparison, they both differ strikingly from the mammalian form. Criteria which may be admissible are given in Table I.

TABLE I

Some characteristics of mammalian thalamic nuclei

Principal nuclei	Intralaminar nuclei
Telencephalic dependencies: degeneration occurs after small (neocortical) lesions	Telencephalic dependencies: degeneration is absent or slight unless the lesion is large and involves corpus striatum
Receive all the major sensory pathways, including those via cerebellum and superior colliculus, but excluding olfactory	Receive small number of afferents from spinal cord and cerebellum. Major source of afferents probably brain stem reticular formation

As can be seen from this Table there is no single observation that might discriminate between these two groups of nuclei. Nevertheless, the attempt is worth making. First of all it is necessary to describe briefly the cytoarchitecture of the relevant region of the reptilian and avian thalamus. (An extensive review can be found in Ariëns Kappers *et al.*, 1936; Pearson, 1972). The thalamic cytoarchitecture of birds is clearer than that of reptiles and will be described first. The most striking nucleus is the nucleus rotundus (Fig. 3, Ro), dorsal to which is the dorso-lateral complex (dla, dlp) and the nucleus superficialis parvocellularis (nucleus of the septomesencephalic tract) (nsp), both of which can be seen in Fig. 3. Caudal and medial to the nucleus rotundus is a small and clearly defined group of cells that constitute the nucleus ovoidalis (not shown in Fig. 3). These are the nuclei relevant to the present discussion. In reptiles the nucleus rotundus is also most conspicuous (see Fig. 5, Ro). It is capped by the dorsomedial and dorso-lateral nuclei (dm & dl), lateral and ventral to which are the rather ill-defined nuclei designated the dorsal and ventral parts of the lateral geniculate body (lgd, lgv) by Papez (1935). These may be seen in Fig. 5.

The afferents to these nuclei are derived from the following sources:

(a) The nucleus rotundus receives a massive input from the optic tectum in both birds (Karten & Revzin, 1966; Hart, 1969) and reptiles

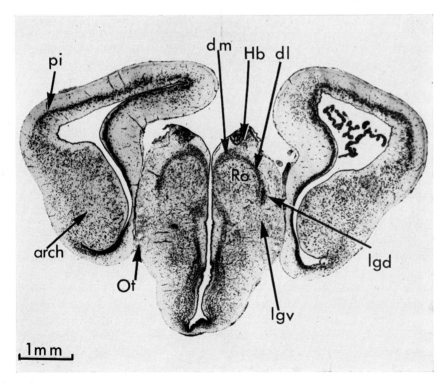

Fig. 5. Transverse section through the diencephalon of the tortoise. The section also includes the caudal telencephalon. Key to abbreviations p. 200.

(Hall & Ebner, 1970). In birds the deep tectal layers (i.e. those that receive an input from the spinal cord according to Karten (1963) and that may be considered "somatic" in the first instance*) project to thalamic regions outside the nucleus rotundus (see (e) below) but the superficial ("visual") layers may project to this nucleus only (Hart, 1969). In mammals the superficial "visual" layers of the superior colliculus project to the dorsal part of the lateral-posterior complex (ventral pulvinar)—see, for example, Casagrande et al. (1972)—i.e. to a principal nucleus that projects to the peristriate neocortex. Although the somatic input to the reptilian optic tectum is also to the deep layers in the first instance, (e.g. Ebbeson, 1969) no attempt has yet been made to investigate the thalamic projection arising in different layers of the tectum.

(b) The nucleus ovoidalis of birds receives the ascending auditory pathway (Karten, 1967). In this it resembles the mammalian medial geniculate body. The comparable nucleus of reptiles is unknown.

(c) The reptilian dorsal lateral geniculate nucleus receives fibres from the optic tract (Kosareva, 1967; Knapp & Kang, 1968 a, b; Hall & Ebner, 1970). In the owl and pigeon the rostral (anterior) part of the dorso-lateral complex receives a direct retinal input (Karten & Nauta, 1968; Nauta & Karten, 1970; Hunt & Webster, 1972). In this context it is interesting to note that Herrick (1948) states that the diencephalon of all vertebrates receives a direct retinal input. The reptilian dorsal lateral geniculate body projects to the dorsal cortex, and the avian dorsolateral complex to the Wulst (see p. 175). The system may be compared with the mammalian pathway from the retina to the striate cortex, via the dorsal lateral geniculate body.

(d) The avian nucleus superficialis parvocellularis (nucleus of the septomesencephalic tract) receives afferents from both retina and spinal cord (Karten & Revzin, 1966; Hunt & Webster, 1972). No similar findings have been reported in either reptiles or mammals.

(e) The medial lemniscus and spinothalamic tract send but few fibres into the avian dorsal thalamus. Those that can be found end in the caudal (posterior) part of the dorsolateral complex as well as in the nucleus superficialis parvocellularis (Wallenberg, 1904; Karten, 1963; Karten & Revzin, 1966). This same part of the dorsolateral complex receives fibres from the deep "somatic" layers of the optic tectum (Hart, 1969). In mammals, the thalamic region with identical inputs is the posterior group (e.g. Casagrande et al., 1972; C. M. Antonetti &

---

* The superficial tectal layers project to the deeper layers. The term "somatic" (sic) should therefore be read to mean those parts of the tectum that receive afferents from the spinal cord. Such laminae are presumably "somato-visual".

K. E. Webster, unpubl. obs.; Lund & Webster, 1967 a, b). The caudal dorso-lateral complex probably projects to the Wulst (but v.i.), whilst the mammalian posterior group appears to be related to the somato-sensory and auditory cortices. In the Tegu lizard the spinal cord projects onto a thalamic region known as the nucleus intremedius dorsalis (Ebbeson, 1969). Comparison with the findings of Hall & Ebner (1970) on the tecto-thalamic projection of turtle reveals no obvious overlap in the thalamic distribution of these two systems. The telencephalic projection, if any, of the nucleus intremedius dorsalis is unknown.

(f) The available descriptions of the thalamic input from the deep cerebellar nuclei are too incomplete to be useful in the present context (see, for example, Karten, 1964; Karten & Revzin, 1966).

Thus, on the basis of the source of their afferents, several of the reptilian and avian thalamic nuclei (including the largest single nucleus, the nucleus rotundus) bear striking resemblances to some of the mammalian principal nuclei (Table II). It is, however, necessary to consider in detail the data regarding their telencephalic relationships before attempting to arrive at firm conclusions.

THALAMO-TELENCEPHALIC RELATIONSHIPS

It is, of course, essential that the nuclei concerned must project onto the telencephalon in order to be considered as either intralaminar or principal nuclei. (The nuclei of the mid-line are related to limbic structures and will not be discussed here.) There are two possible approaches to this problem: either

(a) to remove the telencephalon (entirely or in part) and, after a survival time of some weeks, to examine the diencephalon for signs of cell-death or shrinkage; or

(b) to place lesions in the thalamus and, after a survival time usually counted in days rather than weeks, to examine the telencephalon for degenerating axons and terminals. (The appropriate survival times are determined by trial and error.) These alternatives will be considered in turn.

1. The former method is technically the easier, but is troubled by difficulties of interpretation. After using this method for mammals it is possible to divide thalamic nuclei into two groups—those that do not undergo cellular degeneration after ablation of the telencephalon in the adult animal, and those that do. The latter are said to be telencephalic "dependencies". Nuclei in which perikaryal degeneration follows cortical ablation are described as cortical "dependencies". (See Rose & Woolsey, 1943. According to these authors the concept derives from Monakow.)

The interpretation of these findings is, however, fraught with difficulties because, as Rose & Woolsey (1943) agree, to describe a region as a "dependency" does not necessarily mean that the nucleus in question projects to that part of the telencephalon upon which it "depends". Degeneration of neuronal cell bodies may occur as a transneuronal phenomenon—both retrograde and orthograde. And great care must be taken in the interpretation of negative results: some neuronal perikarya do not degenerate after axotomy. In Table I the nature of the lesion that produces degeneration is included because it is important to take account of the observation that some parts of the mammalian intralaminar nuclei (the rostral division, namely the nuclei centralis medialis, paracentralis and centralis lateralis) show slight cell changes after extensive lesions of the neocortex. This reaction shows a marked increase when the lesion involves the striatum; and the cells of the caudal intralaminar nuclei (nuclei centrum medianum and parafascicularis) degenerate only if the lesion invades the striatum (see Powell & Cowan, 1967 for a discussion).

In the pigeon, Powell & Cowan (1961) found all the thalamic nuclei listed above to be telencephalic dependencies, i.e. they form part of what would in mammals be called the dorsal thalamus. The nuclei do, however, fall into two groups: those that degenerate after localized lesions in the dorso-medial part of the hemisphere (all parts of the dorsolateral complex and the nucleus superficialis parvocellularis); and those that degenerate only after massive lesions of the telencephalon penetrating as deep as the palaeostriatum. The nuclei rotundus and ovoidalis are included in this latter group. Karten (1968) reported that the nucleus ovoidalis degenerates after lesions in the caudal neostriatum, but the lesions in fact extend beyond the cytoarchitectural boundaries of this part of the striatal complex.

In *Lacerta viridis*, Powell & Kruger (1960) found that only two thalamic nuclei degenerate, and then only after large telencephalic lesions penetrating to the palaeostriatum. These two thalamic nuclei are the nucleus rotundus and the adjacent nucleus dorsomedialis. Thus the reptilian dorsal lateral geniculate body is not a telencephalic dependency at all, and the nucleus rotundus exhibits dependency of an "intralaminar" type.

These observations may also be used to produce further hypotheses concerning the nature of the thalamic projection. It is possible that the nucleus rotundus of both birds and reptiles projects only to the palaeostriatum. On the other hand, the axons of rotundal cells may end diffusely throughout the hemisphere as a system of collaterals all of which must be destroyed before the cell bodies degenerate. If this latter explanation (i.e. the "sustaining" collateral hypothesis) is applied to the reptilian dorsal lateral geniculate nucleus, it is clear that the collaterals cannot be distributed to telencephalic structures. Kruger & Berkowitz (1960) suggest that such collaterals may be distributed to the midbrain. It is also worthy of note that in the avian diencephalon there are also nuclei that receive retinal axons but which are independent of the telencephalon—for example, the so-called lateral geniculate nucleus (Cowan, Adamson & Powell, 1961; Powell & Cowan, 1961; S. P. Hunt & K. E. Webster, unpubl. obs.). The nucleus is labelled "lg" in Fig. 3. It is assumed that such nuclei belong either to the epithalamus or to the ventral thalamus: e.g. this "lateral geniculate nucleus" may in fact be better compared with the ventral lateral geniculate nucleus of mammals. These observations, then, provide only inconclusive answers concerning the nature of the thalamic nuclei under consideration.

2. Investigations making use of the silver impregnation methods for staining degenerating axons and terminals (i.e. the Nauta & Gygax (1954) technique and its modification by Fink & Heimer (1967)) after placing lesions in the thalamus by the stereotaxic method, have yielded results which in some ways supplement and refine those described above, but which in other ways contradict them. Perhaps the most striking contradiction arises in connexion with the reptilian dorsal lateral geniculate nucleus. According to Hall & Ebner (1970) the efferents from this nucleus in turtle project to the dorsal cortex via the lateral forebrain bundle. Further, the projection is not diffuse, but is restricted to a specific part of the dorsal cortex and is topographically organized within itself. Unlike the degeneration found in the neocortex of mammals following a lesion in a principal nucleus, the degeneration in this reptile is confined to layer I of the dorsal cortex i.e. the afferents synapse only with the tips of the apical dendrites of the underlying pyramidal cells. These same authors also describe the projection of the turtle nucleus rotundus. In this case the fibres are distributed largely to the dorsal striatal area (i.e. the hypopallium of Elliot Smith), but also, more sparsely, to the ventral (palaeo-) striatum. The projection of the nucleus intremedius dorsalis was not investigated. Virtual total ablation of the thalamus never produced signs of a pathway to the "primordium neopallii". The authors also noted that the midbrain tegmentum projects only to the ventral (palaeo-) striatum.

In birds the anterior dorsolateral complex (which receives a direct retinal input) projects bilaterally to Wulst structures (Nauta & Karten, 1970; and quoted by Ebbeson & Schroeder, 1971; Hunt & Webster, 1972) The decussation occurs in the supra-optic commissure. Although reports exist of a bilateral projection from the mammalian lateral geniculate

TABLE II

Thalamo-telencephalic relationships in reptiles and birds

	Thalamic nucleus	Source of afferents	Telencephalic dependence		Target of efferents	Suggested mammalian thalamic comparison
			Large lesion	Small lesion		
Reptiles	N. rotundus	Optic tectum	Yes	No	Striatum dorsalis + Striatum ventralis	N. lateralis posterior + Intralaminar nuclei
	Dorsal lateral geniculate n.	Retina	No	No	Dorsal cortex	Dorsal lateral geniculate n.
	N. intremedialis dorsalis	Spinal cord	? No	? No	?	?
Birds	N. rotundus	Optic tectum	Yes	No	Ectostriatum	N. lateralis posterior
	Anterior dorso-lateral complex	Retina	Yes	Yes	Wulst	Dorsal lateral geniculate n.
	N. superficialis parvocellularis	Spinal cord Retina	Yes	Yes	Wulst	?
	Posterior dorso-lateral complex	Spinal cord Medial lemniscus Optic tectum	Yes	Yes	? Wulst	Posterior group
	N. ovoidalis	N. mesencephali lateralis dorsalis	Yes	? No	Neostriatum (Field L)	Medial geniculate body

body to the striate cortices (Glickstein, Miller & Smith, 1964) this has not been confirmed (see, for example, Wilson & Cragg, 1967). The nucleus superficialis parvocellularis also projects to the Wulst—chiefly to the hyperstriatum dorsale of the same side (Hunt & Webster, 1972).

The projection field of the avian posterior dorsolateral complex is unknown, but there is physiological evidence it may lie in the rostral Wulst (particularly in the intercalated nucleus) and in the caudal neostriatum (Delius & Benetto, 1972).

In pigeon the nucleus rotundus projects via the lateral forebrain bundle to the homolateral ectostriatum—more specifically to the "core" of the ectostriatum, which in turn projects to a peripheral zone known as the peri-ectostriatal belt. The Wulst also projects into the ecto-striatum (Karten, 1969). This pathway shows close parallels with the mammalian condition, in which fibres from the superior colliculus run to the lateral posterior nucleus and so to the peristriate cortex, which in turn receives axons from the striate cortex (see Fig. 6) (e.g. Diamond & Hall, 1969). Finally, Karten (1967, 1968) has shown that the avian nucleus ovoidalis and caudal neostriatum stand in a similar relation to the auditory pathway as do the mammalian medial geniculate body and auditory cortex.

Some of the parallels between the reptilian, avian and mammalian systems are illustrated in Fig. 6, and all are summarized in Table II.

Discussion of results described above

As can be seen in Table II, the avian dorsolateral complex (both anterior and posterior parts), and the nucleus superficialis parvo-cellularis apparently conform to the criteria set out in Table I to define mammalian principal nuclei. The nucleus ovoidalis is problem-atical because of the slightly contradictory reports of the nature of the lesion required to produce cell degeneration (Powell & Cowan, 1961; Karten, 1968). For the remaining nuclei the results of the two types of investigation are contradictory. A striking feature of the second group of experiments is the failure, except in the case of the reptilian nucleus rotundus, to demonstrate diffuse telencephalic projections. Extensively collateralized projections cannot therefore be used to explain the size of the lesions required to produce cell changes in many of the thalamic nuclei. It is possible that such degeneration results from a combination of both retrograde and transneuronal phenomena, and is consequent upon large telencephalic lesions that produce changes in several areas of the whole forebrain.

There remains the problem of allocating these thalamic nuclei to the "principal" or "intralaminar" group so as to categorize the regions of

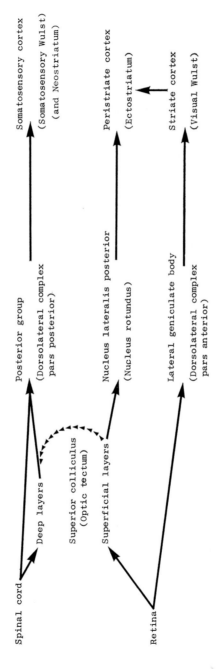

Fig. 6. Tentative scheme to compare some features of the central visual and somatosensory pathways in pigeon and rat. The brackets indicate the avian structures. The term "posterior group" in the rat we apply to a thalamic region that has the afferent/efferent relationships shown (T. R. Price, C. M. Antonetti, K. E. Webster, unpubl. obs.).

the telencephalon to which they project. The flaw in this approach is its optimism that the division into these two categories is clear in mammals. Those nuclei that are generally considered to be the intralaminar group of the mammalian thalamus may themselves be divided in two subgroups:

(a) Those that receive small inputs from the spinal cord and cerebellum, and show slight perikaryal change when large areas of neo-cortex are ablated. The cell degeneration is dramatically increased if the lesion invades the corpus striatum. This rostral group of nuclei is made up of the nuclei centralis lateralis, paracentralis, and medialis.

(b) The caudal intralaminar nuclei (the nuclei centrum medianum and parafascicularis) that receive no long ascending afferents and which degenerate *only* if the lesion invades the corpus striatum.

In their discussion of the problems involved in interpreting cell changes in the mammalian intralaminar nuclei Powell & Cowan (1967) conclude that the changes seen in the rostral intralaminar nuclei after cortical ablation are probably due to the occurrence of orthograde transneuronal degeneration. But they concede that the presence of a collateral projection to the cortex cannot be entirely eliminated. The fact that lesions in either the nucleus centrum medianum (Mehler, 1966) or the nucleus centralis lateralis (T. R. Price & K. E. Webster, unpubl. obs.) do not produce axonal degeneration in the cortex, but do so in the striatum, support the conclusion that these nuclei project only to the corpus striatum. But there are two other difficulties.

First, Jones & Powell (1971) argue that the larger part of the mammalian posterior group is best considered as part of the intralaminar system. Till now, the posterior group has been regarded as a complex of thalamic principal nuclei. But recent evidence suggests that only a relatively small part of the posterior group projects to a restricted area of cortex, whilst the remainder sends its efferents to the corpus striatum (Graybiel, 1970; Heath, 1970). In other words, although the posterior group receives a small, direct, ascending input, and appears to be a cortical dependency, it is not a principal nucleus. The cell degeneration in the greater part of the group consequent upon cortical ablation is to be explained as a transneuronal phenomenon. This clearly increases the difficulties of interpreting the evidence set out previously in the attempt to classify the avian and reptilian thalamic nuclei.

This difficulty is compounded by reports that some mammalian principal nuclei, *including tecto-receptive areas*, project to the striatum as well as the neocortex (Ebner, 1969; Leonard, 1969; Graybiel, 1970). It may be unwise to attempt to homologize the thalamic projections of mammals, birds, and reptiles: these pathways and associated nuclei

cannot be proved homologous in any more specific way than as thalamo-
telencephalic projections. But if one takes into account the results of
physiological investigations (v.i.) the comparisons suggested in Table
II and Fig. 6 are much more tenable. In brief, the physiological properties of the avian "external striatum" are closer to those of the
mammalian neocortex than to those of the mammalian corpus striatum
(see, for example, Albe-Fessard, Oswaldo-Cruz & Rocha-Miranda,
1960 a, b). If it is considered improper to take such functional considerations into account to help in establishing homologies, then the
term must be withheld, and the systems merely compared. In a recent
review of the problems encountered in applying the concept of homology
to the nervous system, Campbell & Hodos (1970) suggest that the
inclusion of such data is not only allowable, but important.

It remains possible, however, that these forebrain nuclei and pathways are unique features of the Classes Reptilia and Aves (see pp.
181–188) and that such features as they share with the mammalian
systems are the result of parallel evolution (see also pp. 191–192 and
pp. 193–195).

PHYSIOLOGICAL STUDIES OF THE THALAMO-TELENCEPHALIC PATH-
WAYS

The existence of the anatomical parallels in the afferent pathways
raises the question of possible physiological similarities.

Reptiles

There is little evidence available concerning reptiles. Kruger &
Berkowitz (1960), working with alligator, and Orego (1961) working
with turtle, described evoked potentials in the dorsal cortex following
visual and somatic stimulation. The responses were not diffusely
spread throughout the cortex, but localized, with the visual area lying
behind the somatosensory. These findings correlate well with the
observations of Hall & Ebner (1970). There has been no study of the
response properties of the dorsal striatum of reptiles.

Birds

"Evoked potential" investigations of the visual, auditory and
somatosensory pathways have been carried out, usually on the domestic
hen or pigeon (Bremer, Dow & Moruzzi, 1939; Erulkar, 1955; Adamo &
King, 1967; Harman & Phillips, 1967; Revzin, 1969, 1970; Revzin &
Karten, 1967; Biederman-Thorson, 1970 a, b; Perišič, Mihailovič &
Cuénod, 1971; Delius & Benetto, 1972; Parker & Delius, 1972). The

results of these experiments support fully the anatomical findings described above. They do, of course, provide a great deal more information than that. For example, Revzin has emphasized the duality of the avian visual pathways and compared them to the two mammalian pathways (see Fig. 6). Thus, the system from retina to Wulst via the dorsolateral complex has many properties similar to those of the mammalian geniculo-striate system (e.g. small field size, responding hyperstriatal cells arranged in columns, and the projection topographically organized) whereas the pathway to the ectostriatum via the optic tectum and nucleus rotundus is made up of units that are rarely excited by any but a moving stimulus, but which possess enormous fields (up to 180°). Cells in this second system also respond to colour, unlike those in the Wulst system. In this, the two pathways differ from the mammalian condition.

The physiological studies have however, created additional problems, in as far as evoked responses of short latency have been recorded from areas of the telencephalon which are not known to receive a specific thalamic input. Thus for the visual pathways, in addition to the potentials evoked in the Wulst and ectostriatum, others have been recorded in the caudal neostriatum. Auditory evoked responses are found not only in the caudal neostriatum (M. Rose's field L), but also in the Wulst. And both the caudal neostriatum and Wulst apparently receive a somatosensory input. To compare this multiple representation with that found in the mammalian neocortex is tempting, but it is probably unwise to do so until details of the pathways involved become available.

DESCENDING PATHWAYS FROM THE STRIATAL COMPLEX

The comparison of the reptilian and avian striatal mass with the mammalian corpus striatum would be incomplete without some mention of the efferent systems.

Reptiles

There is little or no reliable anatomical information. Johnston (1916a) working with turtle and Bagley & Richter (1924) using alligator, described an electrically excitable motor region in the dorsal cortex. In both cases this "motor" area lies in the piriform (olfactory) cortex and the significance of these results is obscure. There is no evidence of a pathway from either the dorsal or piriform cortex to the spinal cord. Bremer, Dow & Moruzzi (1939) showed that the effect does not originate from the dorsal cortex, and Bagley & Langworthy (1926), that it is mediated not through the lateral forebrain bundle, but through the medial forebrain bundle and midbrain.

Birds

Adamo (1967) investigated the descending projections from the Wulst in several species. In spite of some minor differences between species, it is broadly true that the hyperstriatum accessorium projects to all the subdivisions of the external striatum (see p. 175). To what extent these fibres might be compared to mammalian intra-cortical association fibres is uncertain. There is also a projection from the hyperstriatum accessorium to the palaeostriatum augmentum, and this might be compared to the corticostriatal projection found in mammals (e.g. Webster, 1961, 1965; Kemp & Powell, 1970). The possible inter-connexions of the remaining external striatum with each other and the rest of the brain remain to be explored.

There are, however, reports of long descending pathways from the pigeon telencephalon. Zecha (1962) described two such bundles. One, arising from "dorsal telencephalic areas" (sic), sends fibres to the diencephalon and then descends through the brain stem to end chiefly in the contralateral gracile and cuneate nuclei. The other, arising from the "caudal telencephalon" (sic) descends to the upper cervical spinal cord, giving off axons to the lateral reticular nucleus of the brain-stem as it does so. The site of origin of this tract (which closely resembles "Bagley's bundle", found as a constant "aberrant" feature of the pyramidal tract of some mammals) is now known to be the dorsal archistriatum (Zeier & Karten, 1971). The first of these tracts may explain the movements which various authors have reported after stimulation of the Wulst (e.g. Cohen & Pitts, 1967).

ONTOGENY OF THE FOREBRAIN

The descriptions of the diencephalon of the domestic hen (Kuhlenbeck, 1938) or of various reptiles (Bergquist, 1953, 1954) do not, in my view, provide any particularly reliable guide that can be used to help compare the reptilian and avian nuclear configurations either with each other or with those of mammals. In some cases, the comparisons that these studies suggest make little or no sense in the light of our knowledge of the connexions of the nuclei in the adult. The same cannot be said of the telencephalon.

Although each uses an approach that is diametrically opposed to the other, the studies of Kuhlenbeck (1937) and of Källén (1962) have both produced accounts of the ontogeny of the avian telencephalon that are highly relevant to the present discussion. In particular, Källén points out that the external striatum (see p. 175) is derived from forebrain rudiments that, in mammals, produce the cerebral cortex.

COMPARISON OF FOREBRAIN REGIONS: CONTRARY EVIDENCE

The evidence available may be summarized as follows:

The telencephalon of reptiles and especially of birds shows a marked development of the basal, "striatal" regions. The dorsal cortex of these animals is either relatively poorly developed (reptiles) or fused with the underlying "striatal" mass (birds). On the grounds of embryology and transmitter chemistry it is possible to divide the striatal eminence into two main parts: the internal striatum (palaeostriatum) and the external striatum. The internal striatum is apparently little different in mammals (in which it is known as the corpus striatum) from that found in birds and reptiles (in which forms it is known as the ventral or palaeo-striatum). The external striatum shows great variety of form and complexity, reaching a peak of differentiation in birds. It is apparently absent in mammals, but a consideration of its fibre connexions (parti-cularly its inter-relations with the thalamus) and physiology suggests that, as found in birds, it may be comparable to the mammalian neocortex, whilst, in reptiles, the external (i.e. dorsal striatum or hypopallium) plus the dorsal cortex may be compared with the mam-malian neocortex. There are, however, some observations that do not fit comfortably into this interpretation.

1. There is still controversy surrounding the precise distribution of monoamines in the avian forebrain. Bertler *et al.* (1964) found dopamine not only in the palaeostriatum augmentatum but also in the ventral part of the caudal neostriatum. As can be seen from Fig. 3, the line of demarcation between the neostriatum caudale and the palaeostriatum augmentatum is not always clear. Here, too, the neostriatum abuts on the archistriatum. It is possible that the area in question is the dorsal ("somatic") part of the archistriatum (Zeier & Karten, 1971) but this does not solve the problem. Only repeated, independent studies can resolve the discrepancy.

2. It may be argued that the use of data pertaining to the distribu-tion of monoamines and acetylcholinesterase is special pleading, because, as shown by Baker-Cohen (1968), in its content of other enzymes (e.g. succinic dehydrogenase) the palaeostriatum augmentatum is similar to a region of the external striatum (the ectostriatum in the case of birds). The distribution of transmitters and related substances *is* special

pleading—it is more likely to be related to fibre connexions than the distribution of other enzymes, and therefore to be more significant and appropriate to the way we look at the brain.

3. Another weakness of the evidence drawn from "chemoarchitecture" concerns the distribution of acetyl cholinesterase. As shown in Fig. 4, the pigeon entopeduncular nucleus is rich in this enzyme, and so is the entopeduncular nucleus of the rat (Parent & Olivier, 1960). The entopeduncular nucleus is not found in higher primates, but the globus pallidus has two segments (external and internal) instead of one, and it is generally assumed that the internal pallidal segment is comparable to the entopeduncular nucleus. This segment, however, is not rich in acetyl cholinesterase (Parent & Olivier, 1970). It remains to be seen which line of evidence will eventually prevail.

4. Baker-Cohen (1968) also argues that because the optic tectum receives a spinal (somatic) input as well as a visual one, the nucleus rotundus cannot be considered as a "visual" nucleus, but as a "general", polysensory nucleus. In birds at least the available physiological findings indicate that the nucleus rotundus *is* "visual", and this fits with what we now know of the differential connexions of the tectal laminae in both mammals and birds (see pp. 179–181 and Fig. 5), a fact that Baker-Cohen could not have been aware of at that date. It must be admitted, however, that this criticism may still be relevant in the case of reptiles, since there is no evidence of a differential tectal projection to the somatic and visual (rotundal) thalamic regions, and the nucleus rotundus projects to the palaeostriatum as well as to the dorsal striatum (see pp. 180–181 and p. 183).

4. The problems associated with the characterization of mammalian nuclei as either "principal" or "intralaminar" have been discussed already (pp. 185–188).

5. Finally there are problems relating to the recognition of types of organization that lack any comparable parallel in mammals. Under this heading one can include the bilaterality of the thalamic projection to the Wulst; and the quintofrontal tract of Wallenberg (1903, 1904) which arises in the trigeminal sensory nucleus and reaches the nucleus basalis of the telencephalon without synapsing in the diencephalon. Karten (1969) suggests that the nucleus basalis might be considered as a diencephalic cell group. There is embryological evidence that some cell groups migrate from the telencephalon into the diencephalon during development, and the nucleus basalis may be included as a group which fails to migrate. Bilateral thalamo-cortical projections have been reported in mammals (Glickstein *et al.*, 1964), but later work has not confirmed this observation (e.g. Wilson & Cragg, 1967).

CONCLUSIONS

In spite of these reservations, there is little doubt that most of the evidence supports the interpretation outlined above, i.e. it is justifiable to compare at least part of the avian striatal complex and the reptilian dorsal striatum (and general cortex) with the neocortex of mammals. If this is so, why not refer to these "striatal" regions as cortex?

First, because the "striatal" nuclei are not a cortex in the strict sense, i.e. they do not take the form of a flattened, multilaminar sheet containing pyramidal cells, some of which have apical dendrites that extend to and ramify in the outermost cortical layer. This outermost layer otherwise contains no neurons. The reptilian dorsal cortex clearly falls into this category, even though it differs in some ways from the mammalian neocortex (e.g. the fibres that reach it synapse only with the outermost parts of the apical dendrites, unlike the mammalian cortex in which such fibres synapse at several levels—see Hall & Ebner, 1970). It is also interesting to note that the region of the reptilian forebrain previously thought to represent neocortex—the "primordium neopallii" of Crosby (1917) and Elliot Smith (1919)—is now known to receive no fibres from the dorsal thalamus (Hall & Ebner, 1970). Its status is therefore unclear. In any case, as Goldby & Gamble (1957) point out, a differentiated structure in an adult animal cannot be considered a primordium of anything.

Secondly, to call these regions "cortex" would imply that they are in some way homologous with the neocortex. As Nauta & Karten (1970) point out, even if the demonstration of similar patterns of connectivity is taken as a criterion for establishing homologies in the nervous system, one may speak only of homology with respect to cell populations i.e. of the groups of cells that in fact establish synaptic contact with each other. The remaining neurons of the cytoarchitectural fields in which these populations lie cannot be included until it is established that the pattern of connectivity they make with the cell group under consideration is similar in the two animals. If this is so, it is impossible to classify the greater part of the neurons of the external striatum. It must be said, however, that even if these structures cannot be called "cortex", they should not be referred to as "striatum". This will be discussed further below.

It is not proposed to discuss here the problems that arise vis à vis the application of the concept of homology to the nervous system (see pp. 187–188). The most recent discussion of this topic is that of Campbell & Hodos (1970). It now seems a general practice to define homology as a term implying common ancestry. For the nervous system

I

there is no fossil record, and one must rely entirely on the information gleaned from comparative studies of living forms. It is perhaps unnecessary to stress that these do not form a phylogenetic series, and a great deal of extrapolation is necessary. At most, it is possible to conjecture that in the past the animal population from which the reptilian/avian and mammalian radiations sprang possessed in the basal hemisphere two groups of cells. Both of these received fibres from the dorsal thalamus that in turn received most of its sensory inflow via polysynaptic chains. The population of thalamic neurons projecting to the more ventral of the basal cell groups in the telencephalon changed relatively little during evolution. The others, however, tended to become more and more directly linked with the sensory pathways. That the turtle nucleus rotundus and some mammalian principal nuclei project onto both groups of cells in the telencephalon suggests that in the common ancestor there was, and in birds there may be intermixing of "intralaminar" and "principal" neurons in the thalamus. The tendency to establish more direct links between the thalamus and the sensory pathways reached its peak in mammals in which the more dorsal of the telencephalic cell groups expanded to form the neocortex. Reptiles and birds have moved some way towards this condition, but in this case the more dorsal of the basal telencephalic centres expanded into the lateral ventricle to form the dorsal or external "striatum". It can be imagined that neuronal proliferation in the zone labelled "X" in Fig. 1 could expand either the dorsal cortical region, or might create an elevation in the lateral ventricle dorsal to the "striatal" elevation, resulting in a form resembling that shown in Figs 2 and 4A. It is easy to see that continued proliferation of this elevation would cause it eventually to abut on the ventricular (internal) surface of the dorsal cortex. Fusion along this line of contact would incorporate the cortex into the underlying dorsal "striatum", the condition pertaining in most birds. (In some Crocodylia the striatal complex closely resembles that found in most birds i.e. the dorsal cortex apparently becomes incorporated in the striatal elevation, whilst in a few birds the condition resembles that found in Lacertilia in which the dorsal striatum and dorsal cortex remain apparently distinct, but the sulcus between dorsal and ventral striatum is not present.) This might also explain the apparent difference between reptiles and birds in the projection fields of the "dorsal lateral geniculate body" pathway—to the dorsal cortex in reptiles and *apparently* to the dorsal "striatum" in birds. Whether expansion of this sort (i.e. to form a solid, roughly spherical mass) has inherent disadvantages compared to expansion to form a sheet is not known. It seems possible, for example, that the growth of a *"solid"* accumulation of neurons will, above a certain size, present considerable problems—the larger the aggregation, the greater the difficulty of organizing the passage of afferent and efferent fibres in a way that will not interfere with the "lateral" connectivity of the outermost cells. What this critical size may be it is impossible to say. It is interesting, however, that when tested with "learning set" problems (which may be thought of as tests of an animal's power of abstraction—it learns how to learn) pigeons easily outperform laboratory rats and domestic cats, falling little short of chimpanzees (see Hodos, 1970). This underlines the desirability of abandoning the term "striatum" for these regions of the telencephalon. Whilst the word continues to be used in this context, it will remain all too easy to dismiss birds and reptiles as "animals without neocortex", possessed of no more than "phylogenetically ancient" structures, dubbed with names containing prefixes such as "archi-" and "palaeo-". These prefixes, too, are unsuitable. They imply a knowledge of the chronological events of brain evolution which we do not have. And worse, they encourage a way of thinking about the mammalian brain which is likely to be barren —viz. that the caudal brain regions resemble the organization found in reptiles, Amphibia etc. These groups of animals are successful in their own right: a monkey without neocortex is not a lizard or a bird.

ACKNOWLEDGEMENTS

I wish to thank Professor J. Z. Young, FRS, for reading the manuscript and for his useful comments and criticisms; and to thank Stephen Hunt for the running discussion of the paper whilst it was being written. This paper was written whilst the author was in receipt of a grant from the Wellcome Trust.

REFERENCES

Adamo, N. J. (1967). Connections of efferent fibres from hyperstriatal areas in chicken, raven, and African lovebird. *J. comp. Neurol.* **131**: 337–355.
Adamo, N. J. & King, R. L. (1967). Evoked responses in the chicken telencephalon to auditory, visual and tactile stimulation. *Expl. Neurol.* **17**: 498–504.
Albe-Fessard, D., Oswaldo-Cruz, E. & Rocha-Miranda, C. (1960a). Activités évoquées dans le noyau caudé du chat en réponse à des types divers d'afférence. I Étude macrophysiologique. *Electroenceph. clin. Neurophysiol.* **12**: 405–420.
Albe-Fessard, D., Oswaldo-Cruz, E. & Rocha-Miranda, C. (1960b). Activités évoquées dans le noyau caudé du chat en réponse à des types divers d'afférence. II Étude microphysiologique. *Electroenceph. clin. Neurophysiol.* **12**: 649–661.
Andén, N.-E., Dalhström, A., Fuxe, K. & Larrson, K. (1965). Further evidence for the presence of nigro-neostriatal dopamine neurons in the rat. *Am. J. Anat.* **116**: 329–333.

Ariëns Kappers, C. U. (1921). *Vergleichende Anatomie des Nervensystems*. Haarlem: Bohn, E. F.

Ariëns Kappers, C. U., Huber, G. C. & Crosby, E. C. (1936). *The comparative anatomy of the nervous system of vertebrates, including man*. (2 vols.) New York: Hafner.

Bagley, C. & Langworthy, O. R. (1926). The forebrain and midbrain of the alligator with experimental transections of the brain stem. *Arch. Neurol. Psychiat. Chicago* 16: 154–166.

Bagley, C. & Richter, C. P. (1924). Electrically excitable region of the forebrain of the alligator. *Arch. Neurol. Psychiat. Chicago* 11: 257–263.

Baker-Cohen, K. F. (1968). Comparative enzyme histochemical observations on submammalian brains. Part I Striatal structures in reptiles and birds. *Ergebn. Anat. EntwGesch.* 40 (6): 1–41.

Berquist, H. (1953). On the development of diencephalic nuclei and certain mesencephalic relations in *Lepidochelys olivacea* and other reptiles. *Acta zool., Stockh.* 34: 155–190.

Berquist, H. (1954). Ontogenesis of diencephalic nuclei in vertebrates. *Acta univ. lund.* (2) 50 (6): 1–34.

Bertler, Å., Falck, B., Gottfries, C. G., Ljunggren, L. & Rosengren, E. (1964). Some observations on adrenergic connections between mesencephalon and cerebral hemispheres. *Acta pharmacol. (Kbh).* 21: 283–289.

Biederman-Thorson, M. (1970a). Auditory evoked responses in the cerebrum (field L) and ovoid nucleus of the ring dove. *Brain Res., Amst.* 24: 235–245.

Biederman-Thorson, M. (1970b). Auditory responses of units in the ovoid nucleus and cerebrum (field L) of the ring dove. *Brain Res., Amst.* 24: 247–256.

Bremer, F., Dow, R. S. & Moruzzi, G. (1939). Physiological analysis of the general cortex in reptiles and birds. *J. Neurophysiol.* 2: 473–499.

Cairney, J. (1926). A general survey of the forebrain of *Sphenodon punctatum*. *J. comp. Neurol.* 42: 255–348.

Campbell, C. B. G. & Hodos, W. (1970). The concept of homology and the evolution of the nervous system. *Brain, Behav. Evol.* 3: 353–367.

Casagrande, V. A., Harting, J. K., Hall, W. G., Diamond, I. T. & Martin, F. G. (1972). Superior colliculus of the tree shrew: a structural and functional subdivision into superficial and deep layers. *Science, N.Y.* 177: 444–447.

Cohen, D. H. & Pitts, L. H. (1967). The hyperstriatal region of the avian forebrain and autonomic responses to electrical stimulation. *J. comp. Neurol.* 131: 323–336.

Cowan, W. M., Adamson, L. & Powell, T. P. S. (1961). An experimental study of the avian visual system. *J. Anat.* 95: 545–563.

Craigie, E. H. (1940a). The cerebral cortex in some Tinamidae. *J. comp. Neurol.* 72: 299–328.

Craigie, E. M. (1940b). The cerebral cortex in paleognathine and neognathine birds. *J. comp. Neurol.* 73: 179–234.

Crosby, E. C. (1917). The forebrain of *Alligator mississippiensis*. *J. comp. Neurol.* 27: 325–402.

Delius, J. D. & Benetto, K. (1972). Cutaneous sensory projections to the avian forebrain. *Brain Res., Amst.* 37: 205–221.

Diamond, I. T. & Hall, W. C. (1969). Evolution of neocortex. *Science, N.Y.* 164: 251–262.

Durward, A. (1930). The cell masses in the forebrain of *Sphenodon punctatum*. *J. Anat.* 65: 8–44.

Ebbeson, S. O. E. (1969). Brain stem afferents from the spinal cord in a sample of reptilian and amphibian species. *Ann. N.Y. Acad. Sci.* **167**(1): 80–101.

Ebbeson, S. O. E. & Schroeder, D. M. (1971). Connections of the nurse shark's telencephalon. *Science, N.Y.* **173**: 254–256.

Ebner, F. F. (1969). A comparison of primitive forebrain organization in metatherian and eutherian mammals. *Ann. N.Y. Acad. Sci.* **167**(1): 241–257.

Edinger, L., Wallenberg, A. & Holmes, G. M. (1903). Untersuchungen über die vergleichende Anatomie des Gehirnes. 5. Untersuchungen über das Vorderhirn der Vögel. *Abh. senkenb. naturforsch. Ges.* **20**: 343–424.

Elliot Smith, G. (1910). Some problems relating to the evolution of the brain. *Lancet* **88**: 1–16; 147–153: 221–227.

Elliot Smith, G. (1919). A preliminary note on the morphology of the corpus striatum and the origin of the neopallium. *J. Anat.* **53**: 271–291.

Erulkar, S. D. (1955). Tactile and auditory areas in the brain of the pigeon. *J. comp. Neurol.* **103**: 421–457.

Fink, R. P. & Heimer, L. (1967). Two methods for selective silver impregnation of degenerating axons and their synaptic endings in the central nervous systems. *Brain Res., Amst.* **4**: 369–374.

Fox, C. A., Hillman, D. E., Siegesmund, K. A. & Sether, L. A. (1966). The primate globus pallidus and its feline and avian homologues: a Golgi and electron microscopic study: In *Evolution of the forebrain*: 237–248. Hassler, R. & Stephan, H. (eds) Stuttgart: Georg Thieme Verlag.

Fuxe, K. & Ljunggren, L. (1963). Cellular localization of monoamines in the upper brain stem of the pigeon. *J. comp. Neurol.* **125**: 355–382.

Glickstein, M., Miller, J. & Smith O. A. (1964). Lateral geniculate nucleus and cerebral cortex. Evidence for a crossed pathway. *Science, N.Y.* **145**: 159–161.

Goldby, F. & Gamble, H. J. (1957). The reptilian cerebral hemispheres. *Biol. Rev.* **32**: 383–420.

Graybiel, A. M. (1970). Some thalamocortical projections of the pulvinar-posterior system of the thalamus of the cat. *Brain Res., Amst.* **22**: 131–136.

Hall, W. C. & Ebner, F. F. (1970). Thalamo-telencephalic projections in the turtle (*Pseudemys scripta*). *J. comp. Neurol.* **140**: 101–122.

Harman, L. A. & Phillips, R. E. (1967). Responses in the avian midbrain thalamus and forebrain evoked by click stimuli. *Expl. Neurol.* **18**: 276–286.

Hart, J. R. (1969). *Some observations on the development of the avian optic tectum.* Ph.D. Thesis, University of Wisconsin.

Heath, C. J. (1970). Distribution of axonal degeneration following lesions of the posterior group of thalamic nuclei in the cat. *Brain Res., Amst.* **21**: 435–438.

Heimer, L. (1969). The secondary olfactory connections in mammals, reptiles and sharks. *Ann. N.Y. Acad. Sci.* **167**(1): 129–146.

Herrick, C. J. (1948). *The brain of the Tiger salamander* (Ambystoma tigrinum). Chicago: Univ. Chicago Press.

Hodos, W. (1970). Evolutionary interpretation of neural and behavioural studies of living vertebrates. In *The neurosciences; Second study program*: 26–39. Schmitt, F. O. (ed.) New York: Rockefeller Univ. Press.

Huber, G. C. & Crosby, E. C. (1929). The nuclei and fiber paths of the avian diencephalon, with consideration of telencephalic and certain mesencephalic centers and connections. *J. comp. Neurol.* **48**: 1—225.

Hunt, S. P. & Webster, K. E. (1972). Thalamo-hyperstriate interrelations in the pigeon. *Brain Res., Amst.* **44**: 647–651.

Johnston, J. B. (1915). The cell masses in the forebrain of the turtle, *Cistudo carolina*. *J. comp. Neurol.* **25**: 393–468.

Johnston, J. B. (1916a). Evidence of a motor pallium in the forebrain of reptiles. *J. comp. Neurol.* **26**: 475–479.

Johnston, J. B. (1916b). The development of the dorsal ventricular ridge in turtles. *J. comp. Neurol.* **26**: 481–505.

Johnston, J. B. (1923). Further contributions to the study of the evolution of the forebrain. *J. comp. Neurol.* **35**: 337–481.

Jones, E. G. & Powell, T. P. S. (1971). An analysis of the posterior group of thalamic nuclei on the basis of its afferent connections. *J. comp. Neurol.* **143**: 185–216.

Jurio, A. V. (1969). The distribution of dopamine in the brain of a tortoise, *Geochelone chilensis* (Gray). *J. Physiol.* **204**: 503–509.

Jurio, A. V. & Vogt M. (1967). Monoamines and their metabolites in the avian brain. *J. Physiol.* **189**: 489–518.

Källén, B. (1962). Embryogenesis of brain nuclei in the chick telencephalon. *Ergebn. Anat. EntwGesch.* **36**: 62–82.

Karten, H. J. (1963). Ascending pathways from the spinal cord in the pigeon (*Columba livia*). *Int. congr. Zool.* **16(2)**: 23 (abstr.)

Karten, H. J. (1964). Projections of the cerebellar nuclei of the pigeon (*Columba livia*). *Anat. Rec.* **148**: 297–298 (abstr.)

Karten, H. J. (1967). The organization of the ascending auditory pathway in the pigeon (*Columba livia*). I. Diencephalic projections of the inferior colliculus (nucleus mesencephali lateralis pars dorsalis). *Brain Res., Amst.* **6**: 409–527.

Karten, H. J. (1968). The ascending auditory pathway in the pigeon (*Columba livia*). II. Telencephalic projections of the nucleus ovoidalis thalami. *Brain Res., Amst.* **11**: 134–153.

Karten, H. J. (1969). The organisation of the avian telencephalon and some speculations on the phylogeny of the amniote telencephalon. *Ann. N.Y. Acad. Sci.* **167(1)**: 164–179.

Karten, H. J. & Revzin, A. M. (1966). The afferent connections of the nucleus rotundus in the pigeon. *Brain Res., Amst.* **2**: 368–377.

Karten, H. J. & Hodos, W. (1967). *A stereotaxis atlas of the brain of the pigeon* (Columba livia). Baltimore: Johns Hopkins Press.

Karten, H. J. & Nauta, W. J. H. (1968). Organisation of retinothalamic projections in the pigeon and owl. *Anat. Rec.* **160**: 373 (abstr.)

Kemp, J. M. & Powell, T. P. S. (1970). The cortico-striate projection in the monkey. *Brain* **93**: 525–546.

Knapp, H. & Kang, D. S. (1968a). The visual pathways of the snapping turtle (*Chelydra serpentina*). *Brain, Behav. Evol.* **1**: 19–42.

Knapp, H. & Kang, D. S. (1968b). The retinal projections of the sidenecked turtle (*Podocnemeis unifilis*) with some notes on the possible origin of the dorsal lateral geniculate body. *Brain, Behav. Evol.* **1**: 369–404.

Kosareva, A. A. (1967). Projection optic tract fibres to visual centers in a turtle (*Emys orbicularis*). *J. comp. Neurol.* **130**: 263–276.

Kruger, L. & Berkowitz, E. C. (1960). The main afferent connections of the reptilian telencephalon as determined by degeneration and physiological methods. *J. comp. Neurol.* **115**: 125–141.

Kuhlenbeck, H. (1937). The ontogenetic development of the diencephalic centers in a bird's brain (chicken) and comparison with the reptilian and mammalian diencephalon. *J. comp. Neurol.* **66**: 23–75.

Kuhlenbeck, H. (1938). The ontogenetic development and phylogenetic significance of the cortex telencephali in the chick. *J. comp. Neurol.* **69**: 273–301.
Leonard, C. M. (1969). The prefrontal cortex of the rat. I. Cortical projections of
the mediodorsal nucleus. II. Efferent connections. *Brain Res., Amst.* **12**:
321–343.
Lund, R. D. & Webster, K. E. (1967a). Thalamic afferents from the dorsal
column nuclei. An experimental anatomical study in the rat. *J. comp.
Neurol.* **130**: 301–311.
Lund, R. D. & Webster, K. E. (1967b). Thalamic afferents from the spinal cord
and trigeminal nuclei. An experimental anatomical study in the rat. *J. comp.
Neurol.* **130**: 313–327.
Mehler, W. R. (1966). Further notes on the center median nucleus of Luys. In
The thalamus: 109–127. Purpura, D. P. & Yahr, M. D. (eds) New York:
Columbia Univ. Press.
Nauta, W. J. H. & Gygax, P. A. (1954). Silver impregnation of degenerating
axons; a modified technique. *Stain Technol.* **29**: 91–93.
Nauta, W. J. H. & Karten, H. J. (1970). A general profile of the vertebrate brain
with sidelights on the ancestry of cerebral cortex. In *Neurosciences; second
study program.* 7–26. Schmitt F. O. (ed.) New York: Rockefeller Univ.
Press.
Orego, F. (1961). The reptilian forebrain. I. The olfactory pathways and cortical
areas in the turtle. *Archs ital. Biol.* **99**: 425–445.
Papez, J. W. (1935). Thalamus of turtles and thalamic evolution. *J. comp.
Neurol.* **61**: 433–475.
Parent, A. & Olivier, A. (1970). Comparative histochemical study of the corpus
striatum. *J. Hirnforsch.* **12**: 73–81.
Parent, A. & Poirier, J. L. (1971). Occurrence and distribution of monoamine-
containing neurons in the brain of the painted turtle, *Chrysemys picta. J.
Anat.* **110**: 81–89.
Parker, D. M. & Delius, J. D. (1972). Visual evoked potentials in the forebrain of
the pigeon. *Expl Brain Res.* **14**: 198–209.
Pearson, R. (1972). *The avian brain.* London and New York: Academic Press.
Perišič, M., Mihailovič, J. & Cuénod, M. (1971). Electrophysiology of contralateral
and ipsilateral visual projections to the Wulst in pigeon. (*Columba livia*).
Int. J. Neurosci. **2**: 1–8.
Powell, T. P. S. & Kruger, L. (1960). The thalamic projection upon the telen-
cephalon in *Lacerta viridis. J. Anat.* **94**: 528–542.
Powell, T. P. S. & Cowan, W. M. (1961). The thalamic projection on the telen-
cephalon in the pigeon (*Columba livia*). *J. Anat.* **95**: 78–109.
Powell T. P. S. & Cowan, W. M. (1967). The interpretation of degenerative changes
in the intralaminar nuclei of the thalamus. *J. Neurol. Neurosurg. Psychiat.*
30: 140–153.
Raisman, G. (1972). An experimental study of the projection of the amygdala to
the accessory olfactory bulb and its relationship to the concept of a dual
olfactory system. *Expl Brain Res.* **14**: 395–408.
Revzin, A. M. (1969). A specific visual projection area in the hyperstriatum of the
pigeon (*Columba livia*). *Brain Res., Amst.* **15**: 246–249.
Revzin, A. M. (1970). Some characteristics of wide-field units in the brain of the
pigeon. *Brain, Behav. Evol.* **3**: 195–204.
Revzin, A. M. & Karten, H. (1967). Rostral projections of the optic tectum and
nucleus rotundus in pigeons. *Brain Res., Amst.* **3**: 264–275.

Riss, W., Halpern, M. & Scalia, F. (1969). The quest for clues to forebrain evolution—the study of reptiles. *Brain, Behav. Evol.* **2**: 1–50.

Rose, J. E. & Woolsey, C. N. (1943). A study of thalamo-cortical relations in the rabbit. *Bull. Johns Hopk. Hosp.* **73**: 65–128.

Rose, M. (1914). Über die cytoarchitektonische Gliederung des Vorderhirns der Vögel. *J. Psychol. Neurol. Lpz.* **21**: 278–352.

Stingelin, W. (1958). *Vergleichend morphologische Untersuchungen am Vorderhirn der Vögel auf cytologische und cytochitektonische Grundlage.* Basle: Helbing und Lichtenhahn.

Wallenberg, A. (1903). Die Ursprung des Tractus isthmostriatus der Taube. *Neurol. Zentbl.* **22**: 98–101.

Wallenberg, A. (1904). Neue Untersuchungen über den Hirnstamm der Taube. *Anat. Anz.* **24**: 357–369.

Webster, K. E. (1961). Cortico-striate interrelations in the albino rat. *J. Anat.* **95**: 532–544.

Webster, K. E. (1965). The cortico-striatal projection in the cat. *J. Anat.* **99**: 329–337.

Wilson, M. E. & Cragg B. G. (1967). Projections from the lateral geniculate nucleus in cat and monkey. *J. Anat.* **101**: 677–692.

Zecha, A. (1962). The "pyramidal tract" and other telencephalic efferents in birds. *Acta morph. neerl. scand.* **5**: 194–195. (Abstr.)

Zeier, H. & Karten, H. J. (1971). The archistriatum of the pigeon: organization of afferent and efferent connections. *Brain Res., Amst.* **31**: 313–326.

ABBREVIATIONS USED IN FIGURES

arch	archistriatum	lgd	lateral geniculate body pars dorsalis
C-P	caudate-putamen		
dc	dorsal cortex	lgv	lateral geniculate body pars ventralis
dca	dorsal corticoid area		
dl	dorsolateral nuclei	N	neostriatum
dla	dorsolateral complex pars anterior	na	nucleus accumbens
		Nc	neostriatum caudale
dlp	dorsolateral complex pars posterior	Nco	neocortex
		Nf	neostriatum frontale
dm	dorsomedial nuclei	Ni	neostriatum intermedium
ds	dorsal striatum (dorsal ventricular ridge)	nsp	nucleus superficialis parvocellularis
E	ectostriatum	On	optic nerve
ent	entopeduncular nucleus	Ot	optic tract
Gp	globus pallidus	pa	palaeostriatum augmentatum
Ha	hyperstriatum accessorium	phi	parahippocampal cortex
Hb	habenular complex	pi	piriform cortex
Hd	hyperstriatum dorsale	pne	primordium neopallii
Hi	intercalated nucleus of the hyperstriatum	pp	palaeostriatum primitivum
		Ro	nucleus rotundus
hi	hippocampal cortex	Se	septal nuclei
Hv	hyperstriatum ventrale	Sel	striatal elevation
Lfb	lateral forebrain bundle	tub	olfactory tubercle
lg	lateral geniculate body	v	vallecula
		vs	ventral striatum

DISCUSSION

YOUNG (CHAIRMAN): Thank you very much indeed Dr Webster. You have shown that birds are moderately intelligent which we always felt was the situation. There is very much to discuss here. We have the embryology, pharmacology, lots of anatomy, birds and mammals.

DIAMOND: Dr Webster's Table II shows what enormous progress has been made since the days of Herrick and Elliot Smith in understanding the relationships between birds and mammals, and I think he shows striking parallels among several vertebrate groups in the connections of various thalamic nuclei. He also raises the issue as to whether these similarities between birds and mammals imply homology. In one sense it would be almost more exciting if the similarities were due to parallel evolution than if they were homologous, since they would reflect similar behavioural requirements in apparently diverse niches as both groups evolved independently.

I was puzzled by one thing that you said in the early part. So far as I know there is evidence from anterograde degeneration studies that the principal nuclei in mammals, such as the medial geniculate nucleus, do send collaterals to the corpus striatum (Ebner, 1969*)—you do not object to that point?

WEBSTER: No. I think that it merely adds to the similarity.

DIAMOND: Yes, so the principal thalamic nuclei are *not*, according to you, projecting exclusively to neocortex?

WEBSTER: I do not think so. We accept that.

GOLDBY: Mr Chairman, I should like to say how much pleasure it has given me to hear Dr Webster today. It is, I think, almost exactly 40 years ago that in your Department, Professor Elliot Smith started me on studies of the reptile brain, and it is very satisfactory indeed to find that somebody in that same Department is still carrying on that work and that he has got so very much further than I ever did.

There was one point in what he has been telling us that interested me particularly, and which I had not known although perhaps I should have, the projection—I think you said from the dorso-lateral nucleus of the thalamus to the cortex in I think Chelonia—I do not know if I misunderstood this.

WEBSTER: No, from the dorsal lateral geniculate body, using Papez's nomenclature.

---

* Ebner, F. F. (1969). A comparison of primitive forebrain organization in metatherian and eutherian mammals. *Ann. N. Y. Acad. Sci.* **167**: 241–257.

GOLDBY: Has this been shown only in Chelonia or is it a characteristic of other reptiles?

WEBSTER: So far as I know. The study of Hall & Ebner (1970)* is the only one that I have come across in the literature.

YOUNG (CHAIRMAN): Could you say what sort of evidence this is based on?

WEBSTER: Yes—the placing of stereotaxic lesions in the thalamus followed by silver impregnation for degenerating axons and terminals (Nauta & Gygax, 1954; Fink & Heimer, 1967*).
One curious thing about this projection is that the input is only into layer 1 of the cortex, onto the very tips of the apical dendrites and nowhere else, which is very different of course from the mammalian set-up.

GOLDBY: So that although my own remark about an adult structure not being a primordium of anything may still stand, this does seem to give greater plausibility to the idea that the dorsal cortical region is at least comparable to the neopallium of mammals.

WEBSTER: Yes, it is also interesting that the "primordium neopallii" is one region that receives no thalamic input.

YOUNG (CHAIRMAN): Is it not true that the tortoises and turtles are closer to the mammalian stock than any other reptiles in general?

WEBSTER: Yes.

YOUNG (CHAIRMAN): It is mainly that there seems to be less specialization.

POWELL: I was very impressed by your caution in drawing homologies, because Dr Cowan and I first turned to the pigeon after showing that the intralaminar nuclei projected to the striatum in the mammal. We wanted to see if the concept that they were more primitive than the main nuclei was possibly true, and accepting the view at that time that most of the telencephalon of the pigeon was striatum, the working hypothesis was that if we showed that most thalamic nuclei projected upon this, then we could accept these as being homologous to the intralaminar nuclei. But when we came to write the discussion of that paper (Powell & Cowan, 1961†), I realized the difficulty in drawing homologies on the basis of comparative anatomical observations, even with experimental work as well. I would fully agree with you that however well we know the structure

---

* See list of references, pp. 195–200.

† Powell, T. P. S. & Cowan, W. M. (1961). The thalamic projection upon the telencephalon in the pigeon (*Columba livia*). *J. Anat.* **95**: 78–109.

and the connections of certain nuclei in orders as diverse as present day birds and mammals, the functional significance is the more important.

YOUNG (CHAIRMAN): May I add to that the embryological significance. I thought your last remark was very interesting.

WEBSTER: There is one snag about the significance of the embryological studies: if one examines the homologies of thalamic nuclei that such studies suggest (whether of the Kuhlenbeck or the Bergqvist/Kallén school), one finds different schemata, which are in many ways utter nonsense if compared with what exists in the adult. So in fact, in accepting the information for the telencephalon and not mentioning this difficulty with the diencephalon, I am just indulging special pleading. I suppose this merely means that we must go back and look at the diencephalon again.

Symp. zool. Soc. Lond. (1973) No. 33, 205–233

THE EVOLUTION OF THE TECTAL-PULVINAR SYSTEM IN MAMMALS: STRUCTURAL AND BEHAVIOURAL STUDIES OF THE VISUAL SYSTEM

IRVING T. DIAMOND

Departments of Psychology and Physiology, Duke University, Durham, North Carolina, U.S.A.

SYNOPSIS

Elliot Smith provided a model for the relation between comparative anatomy and comparative psychology. The first step in the inquiry is the identification of homologous structures; in this way it can be determined which structures expand and differentiate and which remain stable. For example, the pulvinar nucleus expands and differentiates conspicuously in the evolution of primates. Our studies have shown that a subdivision of this nucleus in the prosimian *Galago* relays visual impulses from the superficial layers of the superior colliculus to the temporal lobe. The lateral posterior nucleus has similar connections in lower mammals such as the hedgehog and tree shrew. The lateral posterior nucleus can thus be regarded as the extrinsic pulvinar, and the question arises whether new, intrinsic subdivisions of the pulvinar have evolved later in primate history. These phylogenetic studies lead to the conclusion that the temporal lobe began as primary visual cortex parallel to the geniculostriate system. Evidence from ablation studies suggests that these parallel projection systems play complementary roles in selective attention. If we can achieve the union between comparative anatomy and comparative psychology which Elliot Smith envisioned, perhaps we can fulfil his hope for understanding the evolution of the mind.

INTRODUCTION

". . . I am not so foolish as to imagine that the evolution of the mind can be explained by anatomical studies; but I do maintain that without a correct comprehension of the homologies of the cerebrum in different vertebrates it will be impossible to make full use of the vast knowledge of comparative psychology which is now being built up or fully to understand and interpret the varying behaviour of animals and the differences in their reactions to their environment.

This study of animal behaviour, which is being so vigorously pursued on the continent and especially in America at the present time, will provide, perhaps, the most likely channel along which the main stream of advance will be made towards the fuller comprehension of the mysteries of mental life. In this scheme of research anatomy will play the necessary, if perhaps humble, role of elucidating the differences in the structure of the nervous mechanisms of the animals examined by the comparative psychologist, and thus point the way towards an adequate explanation of their varying reactions and their different potentialities for mental processes. It will not only provide the physiologist and the psychologist with this essential information, but it will also suggest new subjects for investigation and new lines of attack." (Elliot Smith, 1910).

This remarkable passage from Elliot Smith taken from his 1910 lectures defines the method for utilizing and applying Darwin (1859) to the great question of the evolution of the mind. Expressed as a formula his two chief prescriptions are: (1) structure first, function second, and (2) identify homologies. Elliot Smith's own inquiry into the origin of the hippocampus provides an elegant illustration of the method. If the hippocampus has precursors in the medial wall of lower vertebrates, this leaves the dorsal areas as the distinctly new feature in mammalian cortex. Therefore, it is to the neocortex and the dorsal thalamus that we must look to find those behavioural traits that distinguish mammals from other vertebrates.

The passage I quoted also reveals the promise which Elliot Smith saw in the new studies in comparative psychology. I need hardly remind members of this Symposium that the "mysteries of mental life" posed a hopeless problem under the grip of Descartes' dualism. Darwin's *The Origin of Species* offered the hope of showing continuity and therefore of tracing the advance in mental organization in gradual steps. The Americans (e.g. Thorndike, 1911) recognized the implications of Darwin but, ironically, while of all the biological sciences psychology had perhaps the most to gain from *The Origin of Species*, the progress of comparative psychology has been disappointing. This assessment has been made by those with the greatest investment in Thorndike's approach (see Nissen, 1951; Warren, 1965). In my view, the reason for the failure to fulfill Elliot Smith's expectations is that psychology has failed to follow Elliot Smith's prescription.

Today I want to illustrate how we have tried to follow Elliot Smith's principles. The pulvinar nucleus is the most conspicuous subdivision of the dorsal thalamus of higher primates. As the neocortex distinguishes mammals from non-mammals so does the pulvinar distinguish higher primates from other mammals. But what is the precursor to the pulvinar in lower mammals? Le Gros Clark (1932) had the insight to see that the pulvinar was a subdivision of the lateral group and this group can be identified in lower mammals. But the specific question we asked is whether the lateral posterior nucleus in lower mammals is the homologue of one or more than one of the subdivisions of the pulvinar nucleus. As a further complication, another area in the thalamus of higher primates is designated the lateral posterior nucleus, and if we take this terminology literally, only this region is related to the thalamus of lower mammals and all of the pulvinar nucleus with its three or more divisions is new. In short, our questions were: Do new divisions of the pulvinar nucleus emerge? What old divisions expand? And what old divisions remain stable? If answers to these questions can be obtained, we can then ask what functions are served by the structural changes.

I will now turn to some data which will show that, indeed, we can trace the evolution of the pulvinar nucleus in an ascending series—

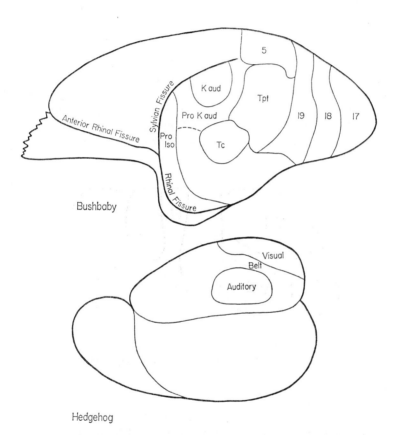

Fig. 1. Cortical development in the bushbaby as compared to the hedgehog to show the development of occipital and temporal areas. In the hedgehog the rhinal fissure divides the lateral wall into two equal halves. The boundaries in the bushbaby are based on an architectonic subdivision made by Professor F. Sanides during his visit to Duke University in September, 1972. The terminology follows Dr Sanides' analysis of the various phyletic stages of neocortex which include periallocortex, proisocortex, prokoniocortex and finally, koniocortex. For further details see Sanides (1970). Abbreviations used in this figure are as follows: K aud = primary auditory koniocortex; Pro K aud = prokoniocortex, a less granular belt surrounding the core auditory area; Tc = temporal central area; Tpt = temporal lobe, temporal parietal area; areas 17, 18, 19 and 5 are Brodmann's designations.

hedgehog, tree shrew and lemur. Further, our results have implications
for the traditional view of sensory and association cortex; namely, I
shall offer evidence that the temporal lobe of ancestral primates was
primary visual cortex and, indeed, that a good portion of the temporal
lobe in prosimians, and probably in simians as well, is primary visual
cortex.

COMPARATIVE ANATOMY OF THE TECTO-PULVINAR SYSTEM

Le Gros Clark (1929, 1959) was the first to take advantage of the
common hedgehog to represent a basal insectivore and the prototypical
eutherian mammal. There are several ways to show the primitive state
of the neocortex in this species—the most obvious one being the total
size of the neocortex in relation to the hippocampal and pyriform areas
(see Fig. 1). Dr W. C. Hall and I have continued the inquiry into the

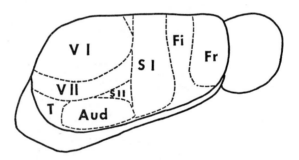

Fig. 2. Dorsal view of the hedgehog cortex showing boundaries of subdivisions
determined by architectonic and electrophysiological criteria (modified from Kaas et al.,
1970). Key to abbreviations p. 231.

hedgehog begun by Le Gros Clark and have demonstrated the paucity
of cortical subdivisions in this species as defined by evoked potentials
and by architectonics (Figs 2, 3) (Hall & Diamond, 1968a; Kaas, Hall
& Diamond, 1970). Most of the neocortex is occupied by the primary
sensory areas, but these have not attained the level of development seen
in the koniocortex of more advanced mammals. It can even be questioned whether the fourth layer of auditory or visual or somatosensory
cortex is truly granular (Sanides, 1970; Sanides & Sanides, 1972).
    The nuclei of the hedgehog's dorsal thalamus are not sharply
demarked (which may be another sign of its primitive structure), but

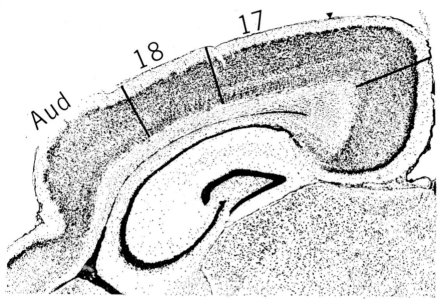

HEDGEHOG

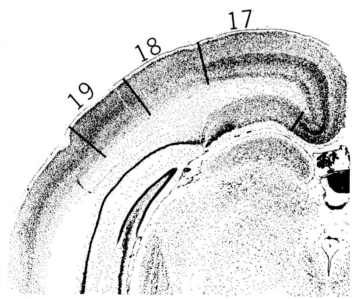

TREE SHREW

Fig. 3. Comparison of the caudal neocortex in the hedgehog and tree shrew. In both species area 17 coincides with visual area I and area 18 coincides with visual area II. In the tree shrew there are more subdivisions and each division appears more distinct. Area 17 is especially striking (taken from Harting *et al.*, 1972). Key to abbreviations p. 231.

K

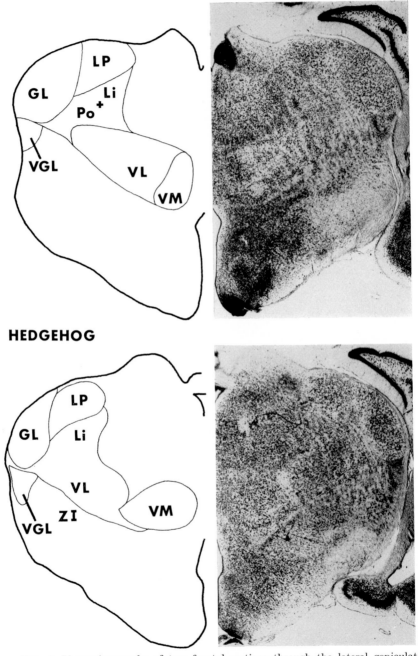

FIG. 4. Photomicrographs of two frontal sections through the lateral geniculate
nucleus and the lateral posterior nucleus of the hedgehog (taken from Harting *et al.*, 1972).
Key to abbreviations p. 231.

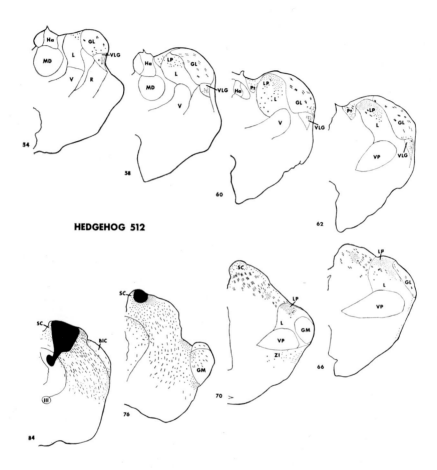

Fig. 5. Tectal lesion in hedgehog 512. Degenerated axons are shown by dashes and their terminals by dots. Note the conspicuous zone of terminal degeneration in the lateral posterior nucleus (taken from Hall & Ebner, 1970). Key to abbreviations p. 231.

both a lateral geniculate nucleus and a lateral posterior nucleus can be identified without much question (see Fig. 4). The lateral geniculate nucleus is, of course, the target of the optic tract; but the lateral posterior nucleus must also be regarded as a sensory relay to the cortex since it receives fibres from the superior colliculus. Figure 5, which is taken from a study by Hall & Ebner (1970), shows terminal degeneration in the lateral posterior nucleus after a lesion of the optic tectum. The lateral posterior nucleus, in turn, projects to a cortical area which includes visual area II or area 18 (Hall & Diamond, 1968a, b).

There is no genuine temporal lobe in the hedgehog, and this point can be made obvious from comparisons to the tree shrew and the prosimian *Galago senegalensis* (see Figs 1 & 2). In the lemur we begin to recognize features which characterize the simians: the occipital and temporal lobes have expanded greatly and are separated from the frontal lobe by a prominent sylvian fissure. Primary auditory cortex lies in this fissure but also extends onto the lateral surface (R. Ravizza, experiments in progress; F. Sanides, pers. comm.). Area 17 resembles the striate area of monkeys in several ways: the fourth layer is highly granular, while the fifth layer is sparsely populated by large pyramidal cells, and the base of layer 3 is similarly sparse. A calcarine fissure in the

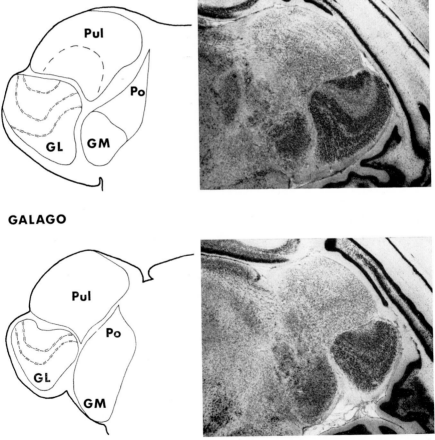

GALAGO

FIG. 6. Photomicrograph of two frontal sections through the lateral geniculate and pulvinar nucleus in the galago (taken from Harting *et al.*, 1972). Key to abbreviations p. 231.

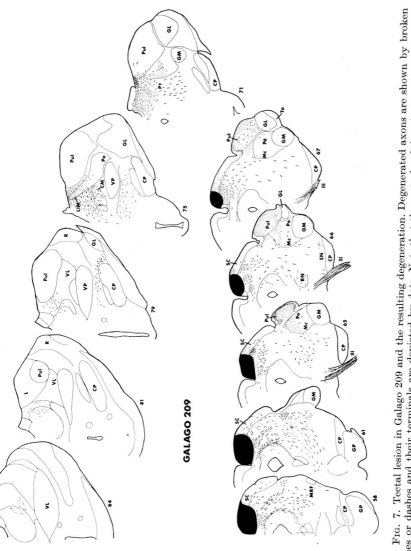

GALAGO 209

Fig. 7. Tectal lesion in Galago 209 and the resulting degeneration. Degenerated axons are shown by broken lines or dashes and their terminals are depicted by dots. Note that the rostral pulvinar is spared, but dense degeneration was located in the caudal extremity (taken from Harting et al., 1972). Key to abbreviations, p. 231.

medial wall is conspicuous and provides for a great increase in the
surface area devoted to central or foveal vision. Adjacent to area 17 are
successive belts usually termed areas 18 and 19. In the caudal
thalamus once again simian-like features are apparent. As can be seen
in Fig. 6, the lateral geniculate nucleus is large and conspicuously
laminated; several divisions of the medial geniculate are well defined,
and they adjoin the posterior group as this term was used by Rose &
Woolsey in their studies of the cat (1958). Most important for the
present purpose is the region dorsal to the lateral geniculate nucleus and

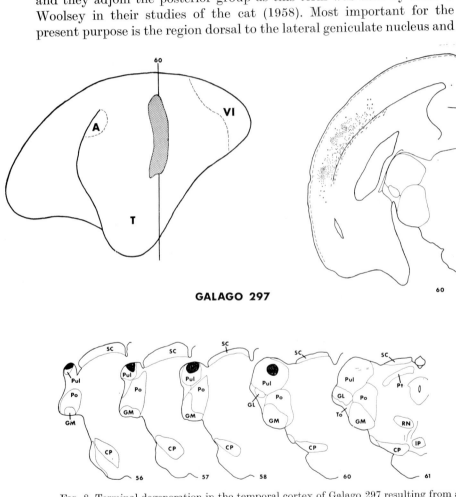

FIG. 8. Terminal degeneration in the temporal cortex of Galago 297 resulting from a
lesion in the caudal pulvinar. The extent of the cortical area receiving degenerated axons
is reconstructed on a lateral surface and is depicted by stippling. To the right is shown a
drawing of a frontal section with terminals of degenerated axons depicted by dots. Below,
the lesion in the pulvinar is shown by black (taken from Harting et al., 1972). Key to
abbreviations p. 231.

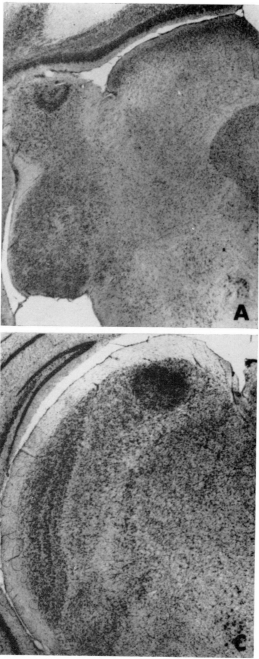

FIG. 9. Photomicrographs showing lesions of the pulvinar in the bushbaby (A) and the tree shrew (C) (taken from Harting *et al.*, 1972).

the posterior nuclei. There is little question that this dorsal area is homologous to the simian pulvinar nucleus. Now if lesions are made in the tectum, terminal degeneration as revealed by the Fink–Heimer technique (Fink & Heimer, 1967) is located in the caudal or inferior pulvinar nucleus (Glendenning, Hall & Hall, 1972; Harting, Glendenning, Diamond & Hall, 1973; Harting, Hall & Diamond, 1972). Figure 7 illustrates a representative lesion (Galago 209) of the superior colliculus and the consequent degeneration in the caudal pulvinar nucleus. When the lesion is restricted to the superficial laminae, little or no degeneration is seen in the posterior nuclei, while dense terminal degeneration can still be identified in the pulvinar nucleus. The rostral or superior pulvinar nucleus receives no input from the superior colliculus and either is intrinsic as this term was defined by Rose & Woolsey (1949) or receives fibres from some extra-thalamic source as yet unknown. If lesions are made in the caudal pulvinar nucleus, just the target of the superior colliculus, degenerated axons and their terminals can be traced to a large and well defined sector of the temporal lobe. Figure 8, taken from an ongoing study by K. Glendenning and W. Hall, shows the pulvinar lesion in Galago 297 (see also Fig. 9 for a photomicrograph of a lesion in the caudal pulvinar); the cortical site of terminal degeneration in the temporal lobe is typical of a large group of similar lesions. Thus, we reach a conclusion which breaks with the traditional view of the primate temporal lobe—in a word, a large sector of it is primary visual cortex.

In selecting the tree shrew (*Tupaia glis*) for comparison with insectivores and primates we have followed Le Gros Clark's lead once again. The tree shrew may represent the ancestor of primates (Le Gros Clark, 1959; Romer, 1967), but in any case it is especially useful for our purpose since this species appears to be intermediate in several crucial respects between the hedgehog and the bushbaby. The striate cortex has attained a conspicuous granular fourth layer (see Fig. 3), and the alternating dark and light layers give this architectonic area a most striking appearance (see Diamond, Snyder, Killackey, Jane & Hall, 1970; Kaas, Hall, Killackey & Diamond, 1972). Area 17 coincides exactly with visual area I, and even the subdivision between the binocular sector and the monocular crescent is apparent from architectonic differences (Kaas, Hall, Killackey *et al.*, 1972). Area 18 coincides with visual area II but, in contrast to the hedgehog, on the other side of this belt lies still another visual belt, area 19 or visual area III. Below area 19 several temporal regions can be identified in addition to the auditory cortex.

Corresponding to these advances in the cortex, as compared with the basal insectivore, the thalamus of the tree shrew shows advances

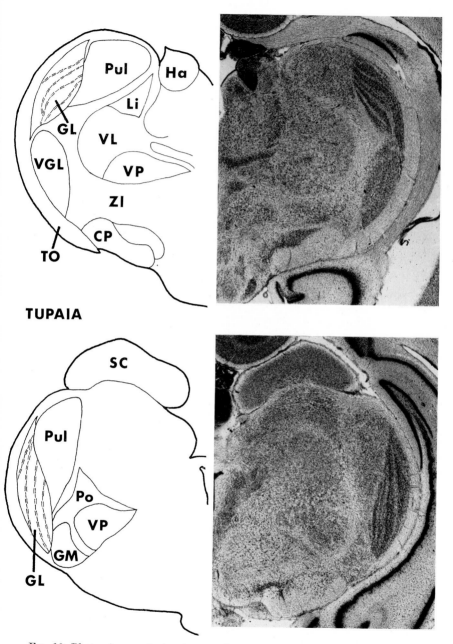

Fig. 10. Photomicrograph showing two frontal levels of the lateral geniculate and pulvinar nuclei in the tree shrew (taken from Harting *et al.*, 1972). Key to abbreviations p. 231.

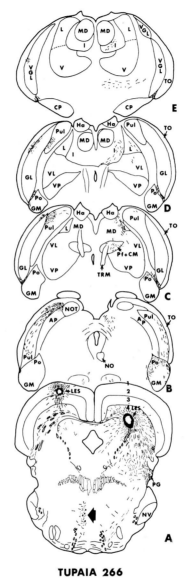

TUPAIA 266

FIG. 11. Diagram showing terminal degeneration resulting from the superficial and deep lesions shown in the following figure. On the side of the superficial lesion the terminals in the thalamus are found chiefly in the pulvinar nucleus, while on the side of the deep lesion the terminals are chiefly in the posterior nuclei. From a number of small and large lesions we concluded that all portions of the pulvinar receive fibres from the superficial layers of the superior colliculus (taken from Casagrande *et al.*, 1972). Key to abbreviations p. 231.

over that of the hedgehog (see Fig. 10). The lateral geniculate nucleus is large and clearly laminated; all six layers of the lateral geniculate nucleus can be seen in frontal sections to curve slightly and terminate at the dorsomedial extremity which represents the vertical meridian and projects to the 17–18 border in the cortex (Diamond *et al.*, 1970; Harting, Diamond & Hall, in preparation). Beyond the dorsomedial border of the lateral geniculate nucleus lies an extensive region which certainly is the caudal sector of the lateral group and therefore, could be designated either the lateral posterior nucleus or the pulvinar nucleus. The important conclusion derived from our study of connexions is that this entire nucleus receives fibres from the superficial layers of the superior colliculus; the deeper layers, in contrast, project to the posterior group (see Figs 11 and 12) (Casagrande, Harting, Hall, Diamond & Martin, 1972; Harting, Hall, Diamond & Martin, in press). Thus, the pulvinar or lateral posterior nucleus is almost certainly visual since the optic tract terminates in the superficial layers of the superior colliculus (Ramón y Cajal, 1911; Ariens Kappers, Huber & Crosby, 1936), while the deeper layers are multimodal (see, e.g. Garey, Jones & Powell, 1968;

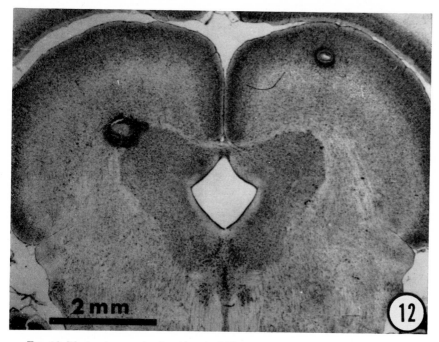

Fig. 12. Photomicrograph of section A of Fig. 11. The photograph is a mirror image of the drawing so the deep lesion is now on the reader's left (taken from Harting, Hall, Diamond *et al.*, in press).

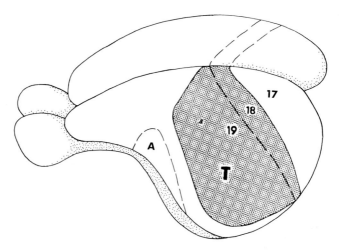

Fig. 13. View of the tree shrew cerebral hemisphere showing the target of the pulvinar nucleus as determined by anterograde degeneration (taken from Harting *et al.*, in preparation). Key to abbreviations p. 231.

Schroeder & Jane, 1971). In turn, the pulvinar nucleus relays visual impulses to an extensive cortical region which includes several architectonic subdivisions—areas 18, 19 and most of the temporal lobe below 19 (see Fig. 13). This conclusion can be derived from either retrograde studies (Diamond *et al.*, 1970) or anterograde studies (Harting, Diamond *et al.*, in preparation).

To summarize the evidence from these comparative studies, the caudal extremity of the lateral group receives fibres from the optic tectum in both the hedgehog and tree shrew and projects to cortex adjacent to area 17. They would therefore appear to be homologous. A striking difference can be seen in phyletic stages represented by these two species both in the lateral posterior nucleus and especially in the cortical target of this nucleus.

A similar pathway can be traced from the tectum to the temporal lobe in the prosimian *Galago senegalensis*. This pathway relays in a thalamic zone that is obviously the caudal or inferior division of the pulvinar nucleus. Thus, the suggestion emerges that this extrinsic division of the pulvinar nucleus is homologous in all three species. In other words, the lateral posterior nucleus is nothing but the extrinsic pulvinar nucleus or its precursor. The concept that the pulvinar nucleus or a portion of it was "extrinsic" (as this term was used by Rose & Woolsey, 1949) from the beginning of mammalian history has implications for the traditional view of the relation between this nucleus and the classical sensory relay nuclei. There is no reason to believe either

that in the beginning of mammalian history this nucleus served the so-called "higher" or associational functions or that the pulvinar nucleus represented a later development in phyletic history. However, since not all parts of the *Galago* pulvinar can be identified as a target of the superior colliculus, the possibility arises that some intrinsic divisions may have evolved between the stages represented by the tree shrew and the lemur. Whether or not the rostral or superior pulvinar nucleus of *Galago* proves to be intrinsic, we can say with confidence that it projects to cortical regions intercalated between area 17 and the target zone of the caudal subdivision of the pulvinar nucleus (studies in progress by R. Ravizza, K. Glendenning, W. Hall & F. Sanides).

Since the tree shrew has often been regarded by paleontologists (see Romer, 1967) as bearing some similarity to the primate ancestor and since the enormous temporal lobes found in primate fossils of the Eocene have captured the attention of paleontologists (see Radinsky, 1970), it is worth trying to carry our argument one step further. Our results suggest that the same factors—increased reliance on vision required by an arboreal or semi-arboreal habitat—may be responsible for the expansion of *both* occipital and temporal lobes. It should be mentioned, especially in a Symposium celebrating the works of Elliot Smith, that his close friend and colleague, Le Gros Clark, made a similar suggestion with much less evidence (1959)—a tribute to his remarkable insight.

Evidence for converging evolution in squirrels and primates (see Hall, Kaas, Killackey & Diamond, 1971; Kaas, Hall & Diamond, 1972)

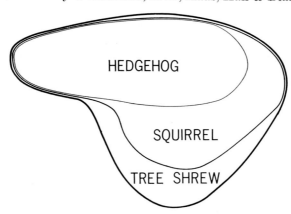

FIG. 14. Comparison of the relative development of visual cortex in the hedgehog, squirrel and tree shrew. The outline of the frontal cortex for these species has been adjusted to an approximately equal size in the figure in order to show the large amount of occipital and temporal cortex in the squirrel and tree shrew relative to the hedgehog. Similarities in the squirrel and tree shrew not present in the hedgehog are probably results of convergent evolution (taken from Kaas, Hall & Diamond, 1972).

also bears on the question of the large development of the temporal lobe in fossil primates. Figure 14 shows that in both squirrels and tree shrews there is a striking expansion of both the occipital and temporal lobes. Like the tree shrew, the pulvinar nucleus of the squirrel relays visual impulses to an extensive sector of extrastriate visual cortex, including a large portion of the temporal lobe (Kaas, Hall & Diamond, 1972). Therefore, whether or not the tree shrew actually resembles the primate ancestor, it appears likely that whatever species of insectivore first adopted an arboreal niche certain changes occurred in its visual system and these included an expansion of the geniculostriate system and an expansion of a second parallel visual system from the optic tectum to the occipital and temporal cortex via the pulvinar relay.

BEHAVIOURAL STUDIES OF THE VISUAL SYSTEM

When Elliot Smith called anatomy the "humble" partner in the study of the evolution of the mind, he was neither displaying false modesty nor denigrating anatomy. By "humble" he meant "the beginning", and no doubt he recognized the merit of Aristotle's adage that a good beginning is more than half. Certainly he recognized the elegance of comparative anatomy and the difficulty in establishing homologies. The term "humble" underlined his expectation that the next step in the inquiry would require overcoming even greater obstacles. Indeed, this has proved to be the case. But while the efforts of Thorndike and his followers did not achieve a scale of intelligence or a picture of the evolution of those abilities associated with the higher primates, this is not to say that these efforts have been to no avail. The objective tests of learning, perception and attention—those that are relatively immune to eccentricities of niche and therefore can be used to compare species on a common scale—are necessary for the second phase of the inquiry. We have tried to use such tests to gain insight into the sensory cortical areas, namely, those areas which have achieved the greatest specialization or the highest level of internal organization. The function of such areas must surely reflect the behaviour most recently evolved in the species' history. Needless to say, this second part of the inquiry lags behind, and its results are often obscure and difficult to interpret.

Before presenting some findings from ablating the striate or the extrastriate visual areas, it is useful first to suggest what we can hope to achieve from the relation between the behavioural and the anatomical aspects of the inquiry. We may anticipate that what will emerge from the anatomical side is a picture of increasing differentiation; some structures, but not all, will expand and subdivide. Thus, the number of

cortical subdivisions increases with the evolution of primates. The question to be answered by the second phase of the study is what functional subdivision corresponds to the anatomical subdivision, and especially how is the functional organization altered when the anatomical organization is altered. One possibility that is useful to discuss because it can be dismissed *a priori* is that as new structures evolve, new psychological categories such as learning or attention will emerge *de novo*. This seems to be implied by the traditional view that sensory cortex provides the basis for "sensation" in the strict sense (e.g. Hume's "impression") and that association cortex provides the substrate for the association between sensory elements (e.g. Hume's "ideas"). From a phyletic viewpoint this concept would imply that early mammals might sense colour or pitch or touch but recognize no objects. Surely even the most primitive vertebrate must perceive *objects*, recall something of past experience, shift attention, albeit all of these are performed in a crude way as compared with advanced mammals. What evolution brings about is increased refinement in the relationships between functions—"new" functions can be added only by changing the relationships between existing functions. This concept is not easy to picture, but Aristotle provided a characteristically brilliant and simple analogy to make the point. Imagine a closed figure in which the number of sides increases from 3 to 4 to 5 to 6, etc. Each additional side is not added like a layer on a cake but alters all the remaining sides and the "old" sides might be distinguished in some ways from the "new", but in a more important sense none of the sides remain exactly the same.

It is now time that concrete results illustrate the general argument. We have found in a series of experiments that the striate area and the extrastriate visual area of the tree shrew complement each other in carrying out the *same* psychological function, the function of *selective attention*. The striate area serves to focus attention and permits concentration in the face of potentially distracting cues, while the temporal visual area serves to shift attention. To support this conclusion I will select some representative results from two ablation studies (Killackey & Diamond, 1971; Killackey, Snyder & Diamond, 1971; Killackey, Wilson & Diamond, 1972). Figure 15 makes the point that without striate cortex tree shrews are quite able to discriminate between simple patterns but fail completely when the patterns are surrounded by a distracting annulus. The defect is not sensory, and the distracting influence can be overcome by patient training. Thus, if only one-half of the annulus is presented, the animals can succeed, and eventually they learn to ignore the complete annulus. The same defect can be revealed when simple patterns differ in colour. The animal's task is to select the

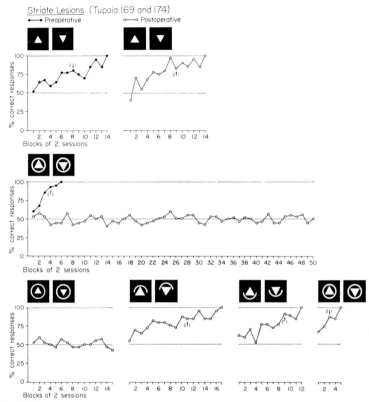

FIG. 15. Learning records for two tree shrews. Each point represents the average score for the two animals except that after one animal attained criterion level—see small arrows marking this point—the score represents the performance of the remaining case. While both animals after removal of the striate cortex readily relearned to discriminate between an upright and an inverted triangle, both failed when the same figures were embedded in an annulus. The bottom line shows various attempts at remedial training (taken from Killackey *et al.*, 1971).

horizontal orientation and ignore the hue which might be red and black stripes or blue and black stripes. This task is easy for normal animals and impossible for tree shrews deprived of the striate cortex. (For results of representative cases see Fig. 16.) Again, it is simple to show that the defect is not sensory for as soon as the hue difference is removed, the animals quickly attain a perfect performance. The hue difference, like the annulus, constitutes a distracting influence which animals deprived of area 17 cannot ignore.

Removal of the extrastriate visual area, the target of the pulvinar nucleus, produces the complementary defect. Now the animals show a strong perseverative tendency and their attention remains fixed when it is adaptive to shift. This symptom can be shown in several ways.

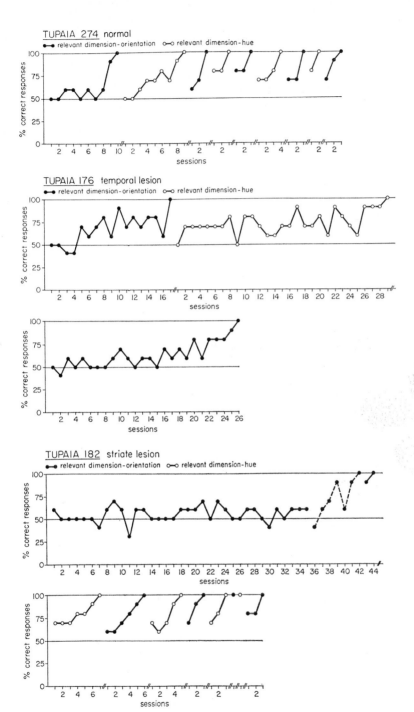

Fig. 16. Individual performance records for three tree shrews, one normal, one with removal of temporal cortex and one with removal of striate cortex. After 35 sessions, tree shrew 182 received special training in which the position of the hues was held constant for each session; this training is depicted by a broken line (taken from Killackey & Diamond, 1971).

L

Suppose normal animals are trained to choose horizontal stripes and after learning, vertical stripes are rewarded on the next session and in successive sessions the positive orientation alternates from horizontal to vertical to horizontal. At first the animals score poorly on alternate days—in particular, the days when a correct trial means choosing the vertical; but eventually the preference for horizontal wanes and the animals shift from day to day and achieve scores of about 70–80% on successive sessions (see Fig. 17, right). In contrast, the tree shrews

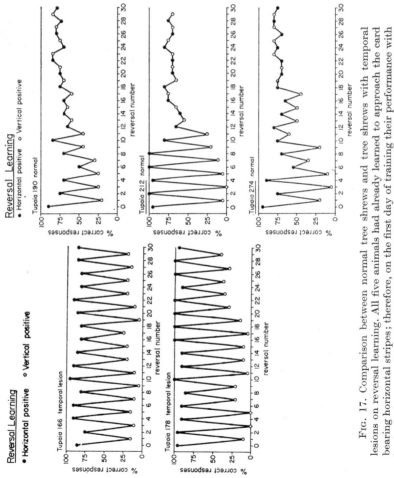

FIG. 17. Comparison between normal tree shrews and tree shrews with temporal lesions on reversal learning. All five animals had already learned to approach the card bearing horizontal stripes; therefore, on the first day of training their performance with horizontal was nearly perfect. On the second day the vertical orientation was rewarded, and of course all animals continued to select horizontal. On the third day horizontal was positive and this alternation continued for 30 sessions (taken from Killackey et al., 1972).

deprived of the extrastriate visual cortex continue to select the horizontal orientation whether this choice is rewarded or not (see Fig. 17, left). The same defect is revealed when the animals are required to shift their attention from one dimension to another. They are capable of

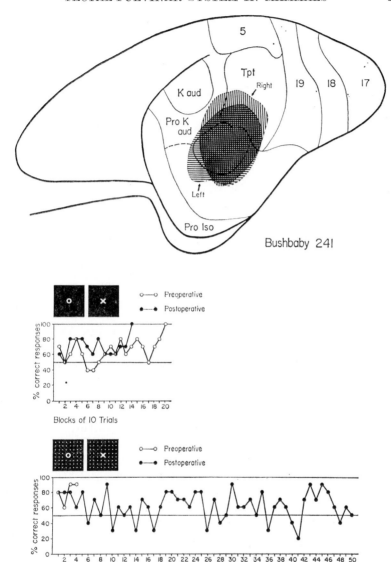

FIG. 18. The lesion and performance of bushbaby 241. On top the left and right lesions are depicted on a standard lateral view of the Galago cortex. The lesion includes some of the cortex which is the target of the extrinsic pulvinar but also includes some regions lying below this. Below is shown the performance before and after lesions on a simple discrimination of circles versus double half circles and the same discrimination except that the patterns are now surrounded by rows of white dots. After surgery the performance of the animal was excellent on the simple problem and very uneven when the pattern was surrounded by dots. Indeed, the criterion of two successive days of 90 % or better was never attained. Key to abbreviations p. 207.

learning to attend to orientation in the face of distracting differences in hue, but again, in contrast to the striate cases they are unable to shift from orientation to hue (see for example, Tupaia 176 whose performance is shown in Fig. 16).

At this point we come to the critical phase of the behavioural inquiry. If we accept for the argument's sake that selective attention in the tree shrew is subdivided "structurally" into focusing and shifting, what happens to this organization when the visual cortex grows more complex? To answer this question we have turned our efforts to the bushbaby, *Galago senegalensis*. Based on our study of connexions and architectonic subdivisions, Dr Martha Wilson, working during a leave of absence from the University of Connecticut, has ablated selectively different portions of the pulvinar projection system; in some cases her aim was to remove the cortical target of the extrinsic pulvinar, and in other cases her aim was to remove the cortical areas of the temporal lobe inferior to this target. It is too early to report her findings, except to say that preliminary results suggest a dissociation of syndromes and the subdivision is closely related to the distinction between focusing and shifting attention.

In closing, I would like to show you the results of one case studied by Mr Frank Atencio. Galago 241 (Fig. 18) received a large bilateral lesion of temporal cortex that included parts of the target of the extrinsic pulvinar nucleus as well as portions of "area 19 inferior" (as this term is used by C. and O. Vogt and by F. Sanides, pers. comm.). The chief defect seems to be an incapacity to discriminate simple patterns in the face of a potentially distracting background (in this case "dots"). The symptom appears to be similar to one produced by lesions of area 17 in the tree shrew. The results illustrate our contention that homologous structures in an ascending phyletic series will not necessarily serve the identical function. Still, the functional organization identified in the lower species will be relevant to the functional reorganization achieved by the evolution of the more advanced species.

It has been my chief purpose in this paper to honour Sir Grafton Elliot Smith, a great pioneer in the field of comparative neurology. He envisioned that one day there would be a genuine union of comparative neurology and comparative psychology. We can only wish that this vision will be fulfilled.

ACKNOWLEDGEMENTS

I would like to thank Mrs Marcia Spray for her help in preparing this manuscript and the National Institute of Mental Health for their support (MH04849).

REFERENCES

Ariens Kappers, C. U., Huber, G. C. & Crosby, E. C. (1936). *The comparative anatomy of the nervous system of vertebrates, including man.* New York: Hafner Publishing Company.

Casagrande, V. A., Harting, J. K., Hall, W. C., Diamond I. T. & Martin, G. F. (1972). Superior colliculus of the tree shrew: A structural and functional subdivision into superficial and deep layers. *Science, Wash.* **177**: 444–447.

Darwin, C. (1859). *On the origin of species.* London: John Murray.

Diamond, I. T., Snyder, M., Killackey, H., Jane J. & Hall, W. C. (1970). Thalamocortical projections in the tree shrew (*Tupaia glis*). *J. comp. Neurol.* **139**: 273–306.

Elliot Smith, G. (1910). Some problems relating to the evolution of the brain. *Lancet* **1910 (1)**: 1–6, 147–153, 221–227.

Fink, R. P. & Heimer, L. (1967). Two methods for selective silver impregnation of degenerating axons and their synaptic ends in the central nervous system. *Brain Res.* **4**: 369–374.

Garey, L. J., Jones, E. G. & Powell, T. P. S. (1968). Interrelationship of striate and extrastriate cortex with the primary relay sites of the visual pathway. *J. Neurol. Neurosurg. Psychiat.* **31**: 135–157.

Glendenning, K. K., Hall, J. A. & Hall, W. C. (1972). The connections of the pulvinar in a primate (*Galago senegalensis*). *Anat. Rec.* **172**: 316.

Hall, W. C. & Diamond I. T. (1968a). Organization and function of the visual cortex in hedgehog: I. Cortical cytoarchitecture and thalamic retrograde degeneration. *Brain, Behav. Evol.* **1**: 181–214.

Hall, W. C. & Diamond, I. T. (1968b). Organization and function of the visual cortex in hedgehog: II. An ablation study of pattern discrimination. *Brain, Behav. Evol.* **1**: 215–243.

Hall, W. C. & Ebner, F. F. (1970). Parallels in the visual afferent projections to the thalamus in the hedgehog (*Paraechinus hypomelas*) and the turtle (*Pseudemys scripta*). *Brain, Behav. Evol.* **3**: 135–154.

Hall, W. C., Kaas, J. H., Killackey, H. & Diamond, I. T. (1971). Cortical visual areas in the grey squirrel (*Sciurus carolinensis*): A correlation between cortical evoked potential maps and architectonic subdivisions. *J. Neurophysiol.* **34**: 437–452.

Harting, J. K., Diamond, I. T. & Hall, W. C. (In preparation). *Anterograde degeneration study of the cortical projections of the lateral geniculate and pulvinar nucleus in the tree shrew* (Tupaia glis).

Harting, J. K., Glendenning, K. K., Diamond, I. T. & Hall, W. G. (1973). Evolution of the primate visual system: Anterograde degeneration studies of the tecto-pulvinar system. *Am. J. phys. Anthrop.* **38**: 383–392.

Harting, J. K., Hall, W. C. & Diamond I. T. (1972). Evolution of the pulvinar. *Brain, Behav. Evol.* **6**: 424–452.

Harting, J. K., Hall, W. C., Diamond, I. T. & Martin, G. D. (in press). Anterograde degeneration study of the superior colliculus in *Tupaia glis:* Evidence for a subdivision between superficial and deep layers. *J. comp. Neurol.*

Kaas, J., Hall, W. C. & Diamond, I. T. (1970). Cortical visual areas I and II in the hedgehog: Relation between evoked potential maps and architectonic subdivisions. *J. Neurophysiol.* **33**: 595–615.

Kaas, J. H., Hall, W. C. & Diamond, I. T. (1972). Visual cortex of the grey squirrel (*Sciurus carolinensis*): Architectonic subdivisions and connections from the visual thalamus. *J. comp. Neurol.* **145**: 273–306.

Kaas, J. H., Hall, W. C., Killackey, H. & Diamond, I. T. (1972). Visual cortex of the tree shrew (*Tupaia glis*): Architectonic subdivisions and representations of the visual field. *Brain Res.* **42**: 491–496.

Killackey, H. & Diamond, I. T. (1971). Visual attention in the tree shrew: An ablation study of the striate and extrastriate visual cortex. *Science, Wash.* **171**: 696–699.

Killackey, H., Snyder, M. & Diamond I. T. (1971). Function of striate and temporal cortex in the tree shrew. *J. comp. Physiol. Psychol.* **74**: 1–29.

Killackey, H., Wilson, M. & Diamond I. T. (1972). Further studies of the striate and extrastriate visual cortex in the tree shrew. *J. comp. Physiol. Psychol.* **81**: 45–63.

Le Gros Clark, W. E. (1929). Studies on the optic thalamus of the insectivora—the anterior nuclei. *Brain* **52**: 334–358.

Le Gros Clark, W. E. (1932). The structure and connections of the thalamus. *Brain* **55**: 406–470.

Le Gros Clark, W. E. (1959). *The antecedents of man.* Edinburgh: Edinburgh University Press.

Nissen, H. W. (1951). Phylogenetic comparison. In *Handbook of experimental psychology*: 347–386. Stevens, S. S. (ed.). New York: Wiley.

Radinsky, L. B. (1970). The fossil evidence of prosimian brain evolution. In *The Primate brain* 209–224. Noback, C. R. & Montagna, W. (eds). New York: Appleton-Century-Crofts.

Ramón y Cajal, S. (1911). *Histologie du système nerveux de l'homme et des vertébrés.* Paris: A. Maloine.

Romer, A. S. (1967). Major steps in vertebrate evolution. *Science, N.Y.* **158**: 1629–1637.

Rose, J. E. & Woolsey, C. N. (1949). Organization of the mammalian thalamus and its relationships to the cerebral cortex. *Electroenceph. Clin. Neurophysiol.* **1**: 391–404.

Rose, J. E. & Woolsey, C. N. (1958). Cortical connections and functional organization of the thalamic auditory system of the cat. In *Biological and biochemical bases of behavior*: 127–150. Harlow, H. F. & Woolsey, C. N. (eds). Madison: University of Wisconsin Press.

Sanides, F. (1970). Functional architecture of motor and sensory cortices in primates in the light of a new concept of neocortex evolution. In *The Primate brain*: 137–208. Noback, C. R. & Montagna, W. (eds). New York: Appleton-Century-Crofts.

Sanides, F. & Sanides, D. (1972). The "extraverted neurons" of the mammalian cerebral cortex. *Z. Anat. EntwGesch.* **135**: 272–293.

Schroeder, D. M. & Jane, J. A. (1971) Projection to the dorsal column nuclei and spinal cord to brainstem and thalamus in the tree shrew (*Tupaia glis*). *J. comp. Neurol.* **142**: 309–350.

Thorndike, E. L. (1911). *Animal intelligence.* New York: Macmillan.

Warren, J. M. (1965). The comparative psychology of learning. *A. Rev. Psychol.* **16**: 95–118.

ABBREVIATIONS USED IN FIGURES

A, auditory area
Ap, (or Pt), pretectal area
Aud, auditory area
BIC, brachium of the inferior colliculus
CM, central median nucleus
CP, cerebral peduncle
Fi, frontal intermediate area
Fr, frontal rostral area
GL, dorsal lateral geniculate nucleus
GM, medial geniculate nucleus
GP, griseum pontis
Ha, habenular nucleus
I, intralaminar nuclei
III, oculomotor nucleus or nerve
IP, interpeduncular nucleus
L, lateral group of nuclei
LES, lesion
Li, lateral group of nuclei, intermediate division
LIM, nucleus limitans
LP, lateral posterior nucleus
Mc, magnocellular division of the medial geniculate nucleus
MD, medial dorsal nucleus
MRF, midbrain reticular formation
NO, oculomotor nucleus

NOT, nucleus of the optic tract
N V, trigeminal nerve
Pf+CM,   central median-parafascicular complex
PG, parabigeminal nucleus
Po, posterior group of nuclei
Pt, (or Ap), pretectal area
Pul, pulvinar nucleus
R, reticular nucleus
RN, red nucleus
SC, superior colliculus
S I, somatosensory area I
SN, substantia nigra
T, temporal cortex
To or TO, optic tract
TRM, tractus retroflexus of Meynert
V, ventral group of nuclei
VGL, (or VLG), ventral lateral geniculate nucleus
V I, visual area I
VL, ventral lateral nucleus
VLG, (or VGL), ventral lateral geniculate nucleus
VM, ventral medial nucleus
VP, ventral posterior nucleus
ZI, zona incerta

DISCUSSION

YOUNG (CHAIRMAN): Thank you very much indeed Professor Diamond, you have given us yet another dimension to our Symposium, and, if I may say so, whether or not one agrees with the categories that you have suggested, I think that you have given us an even more important idea to develop Aristotle, namely that by suitable means you can analyse the capacities of the mammalian brain and find particular functions related to particular parts. As a determined behaviourist, I have always felt that it was time that we found out something about mammalian brains by finding more elementary capacities, if you like, and separating them. What you have shown us is that there are very important learning capacities—learning sets for example, about which an enormous amount of nonsense has been talked—which you can destroy, and I think that is a tremendously important discovery. The learning set situation is dependent on certain parts of the brain. I do not think it has ever been suggested before, has it? And this itself shows that we can find the functional corollaries, and physiological corollaries no doubt too, of these very fundamental psychological capacities.

ZUCKERMAN: May I first join you in your congratulations? I think what we have listened to is a remarkable piece of work. You have shown how your analytical methods allow one to differentiate two of these visual areas, and your switching experiments struck me as being very convincing. Then you said that after all a shark recognizes a shark, it is not picking up isolated signals. On the other hand I take it that animals do in fact, as we do, respond to a signal as though it were the whole—that does happen, does it not?

DIAMOND: Yes.

ZUCKERMAN: Now in those experiments in which you are differentiating between vertical and horizontal stripes, did you in fact reduce the sign, the stimulus, to one horizontal stripe and one vertical stripe, or to two bits of stripes, vertical or horizontal, or was it always a series of lines ?

DIAMOND: It was always a series in the experiments reported here, but we have evidence that there would be a transfer of training from a series of stripes oriented vertically or horizontally to a single vertical or horizontal stripe.

ZUCKERMAN: If you started reducing the size in the process of transfer of the unit stimulus, so that instead of making it 2 cm or whatever it was, you had it 3 mm horizontal and 3 mm vertical, at what point would it break down and both areas fail?

DIAMOND: The question you raise is interesting and, in part, poses a puzzle for us. If the width of the black and white stripes is progressively decreased and the acuity is determined before and after removal of area 17, then there appears to be no change in threshold as a result of surgery (Ware, Casagrande & Diamond, 1972*). This finding is consistent with our observations of the tree shrews' behaviour in their home cage. After removal of visual cortex, they show no difficulties in avoiding obstacles, tracking moving stimuli, etc. However—and this is the puzzle—when the stripes are reduced in length, so they are displayed on a 1 in. circle instead of a 4 × 4 in. square, a deficit is revealed and we are at a loss to account for it.

ZUCKERMAN: It sounds like the reverse of Lashley's mass law—the mass of the stimulus and not the mass of the destruction.

DIAMOND: Since you bring up Lashley's concept of mass laws, I confess that I have hesitated to speak in public about this concept because all of us in the United States using the ablation method owe a great deal to Lashley. Yet this Symposium honouring Elliot Smith may be the time to say that the concept that the cortex acts as a whole has put us on the wrong track

* Ware, C. B., Casagrande, V. A. & Diamond, I. T. (1972). Does the acuity of the tree shrew suffer from removal of striate cortex? *Brain, Behav. Evol.* 5: 18–29.

and opened the door to a split between psychology and anatomy. Elliot Smith clearly recognized that an increase in cortical areas was a conspicuous feature in the evolution of primates. But the concept of mass action led Lashley to argue that the total number of functional areas in the cortex of man does not exceed the number in lower mammals and that it is absurd to think that evolution entails an increase in functionally distinct cortical areas (Lashley & Clark, 1946*). Lashley, of course, did find that deficits in learning and intelligence were correlated with the mass of tissue removed, and not the locus of the lesion. My own view is that while the results are undoubtedly true, they may be trivial. By the same token, the performance in a track meet would be retarded as a function of the severity of the symptom and unrelated to the cause of the syndrome, that is, to the disease itself. A rebirth of cortical architectonics is desperately needed.

YOUNG (CHAIRMAN): Well, we must not turn this into an anti-Lashley Symposium, although I must say I heartily agree with you. In some ways that man did incalculable harm, and I am beginning to think that he was not a genius, although he was an extremely nice man. Now I have said it, let the heavens fall.

ZUCKERMAN: I recall a meeting in London—it was in 1931—when Lashley first reported on his results to this country and Elliot Smith was in the chair. He had to hear Lashley demolishing all that he, Elliot Smith, had set out to construct. He listened and behaved charmingly. There was none of that controversial nonsense that we have heard so much about with Elliot Smith. He took it, and then, I hope, forgot it.

POWELL: In the monkey, as you know, the visual cortex sends very strong cortico–cortical connexions down into the temporal lobe, and there are two quite distinct sub-divisions; on the basis of these connexions the temporal lobe is very strongly related to vision. Have you any evidence for cortico–cortical connexions in these three orders of animals in which you have been studying the visual and temporal cortex—have you evidence of cortico–cortical connexions between them?

DIAMOND: Dr Powell raises an extremely important point. We have not studied cortico–cortical projections from area 17, but Tigges and his co-workers have found in *Galago* a projection from area 17 to area 18 and to a portion of the temporal lobe which we find to be the target of the extrinsic pulvinar (Tigges, Tigges & Kalaha, 1973†). While we have no study of efferent projections from area 17 to area 18 in the tree shrew, I would be most surprised if such a path did not exist. We have made this point in connexion with mapping visual area II—it seems most probable that the retinotopic organization of V II depends on cortico-cortical connexions.

* Lashley, K. S. & Clark, G. (1946). The cytoarchitecture of the cerebral cortex of *Ateles*: A critical examination of architectonic studies. *J. comp. Neurol.* **85**: 223–305.
† Tigges, J., Tigges, M. & Kalaha, C. S. (1973). Efferent connexions of area 17 in *Galago*. *J. phys. Anthrop.* **38**: 393–397.

Symp. zool. Soc. Lond. (1973) No. 33, 235–252.

THE ORGANIZATION OF THE MAJOR FUNCTIONAL AREAS OF THE CEREBRAL CORTEX

T. P. S. POWELL

Department of Human Anatomy, Oxford, England

SYNOPSIS

The possibility of dividing the neocortex of the cerebral hemisphere into several areas on the basis of their cytoarchitecture was first systematically studied in the early years of this century, particularly by Brodmann (1903, 1909) and Campbell (1903, 1905). This subject continued to be investigated in normal material by other workers, but it remained for Rose & Woolsey to test the validity of such a division in experimental material; they showed that an individual architectonic area was the projection site of a specific thalamic nucleus (Rose & Woolsey 1948a, b), and also that the boundaries of the auditory sensory area, as determined by the evoked potential method, correspond to those established with architectonic criteria (Rose & Woolsey, 1949). Later studies have shown that there are differential projections from the thalamic nuclei to the subdivisions within the sensory areas, and that there are distinct differences in the efferent connections of these subdivisions. Electrophysiological evidence has been obtained for important differences in the response properties of cells in the subdivisions of the somatic and visual areas. It has also become clear that the distribution of cortico-cortical connections—both association and commissural—is closely related to architectonic structure. The significance of recent work lies in the increasing evidence for functional differences between the cells of the constituent laminae of the sensory areas, as variations in the structure of these laminae have been one of the main criteria used in cytoarchitectural studies.

The subject of the validity of the cytoarchitectural subdivision of the neocortex has been chosen for discussion in this Symposium for two main reasons: first, because Elliot Smith was actively engaged in neuro-anatomical research at the time the results of the first systematic studies were being published and he wrote a paper (Elliot Smith, 1907) on this subject himself, and secondly, because a considerable amount of experimental work, both anatomical and physiological, during the 20 years that have passed since Le Gros Clark (1952) reviewed this question indicates that such a subdivision of the cortex on morphological criteria has functional significance. Although it was well known before the beginning of this century that certain areas of the cortex had unique structural features, such as the large pyramidal cells of Betz in the motor cortex and the clear stria of Gennari in the visual area, or the differing densities of myelinated fibres and the variation in the time of myelination, the possibility of dividing the entire neocortex of the cerebral hemisphere into distinct areas on the basis of their cyto- or myelo-architecture was first systematically studied in the early years of this century by Brodmann (1903), Campbell (1903) and Vogt & Vogt (1903). The first two of these workers noted and charted the details of the structure of the cortex in sections stained by the Nissl method to show the cell bodies; variations in the thickness of the cortex as a whole, the difference in the development and demarcation of individual laminae and the presence of specific types of cells were the major criteria used. Brodmann (1909) examined the brains of several widely different species while Campbell (1905) concentrated his attention upon those of Primates, and particularly the human. Although the Vogts initially concentrated upon the fibre systems of the cortex and thalamus, they later studied cytoarchitecture and attempted to correlate structure and function in physiological experiments. There should be little doubt now that this pioneering work formed a massive contribution to our understanding of the cerebral hemisphere, and instead of being concerned about minor differences of interpretation and of precise lines of demarcation it would be more useful in the future to follow the example set during the past few decades and to concentrate upon the correlation of the structure of the cortex with the findings made in experimental investigations. Indeed, it might be suggested that there is little to be gained from further cytoarchitectural surveys of the cortex unless these are accompanied by, and correlated with, experimental studies.

During the 20 years that followed these initial publications most attention was paid to the subject by German and Polish workers, especially the Vogts, M. Rose and von Economo and Koskinas. In general it can be stated that these workers, while confirming and extending the concept of architectonic subdivision, further divided the original areas into finer subdivisions until eventually they, and the subject, met with a certain degree of scepticism which culminated in the paper of Lashley & Clark (1946). The question was raised whether the more subtle differences that were described and utilized as criteria were not in fact due to the complex folding and curvature of the cortex by sulci of varying depth and direction.

In the 1930's, however, experimental work on thalamo-cortical connexions using the methods of Marchi for the staining of degenerating nerve fibres (Poliak, 1932) and of retrograde cellular degeneration (Le Gros Clark & Boggon, 1935; Walker, 1938) suggested that specific thalamic nuclei projected to individual architectonic areas. Although the distribution of the cellular degeneration in these investigations was always accurately described, the extent of the lesion was usually given in general terms such as the extent of involvement of the gyri or with reference to a cytoarchitectural map; rarely (Walker, 1940) was the cortical structure studied. The mapping of the sensory areas with the newly developed method of evoked potentials showed a surprising degree of correspondence with the maps of Brodmann (e.g., Adrian, 1941; Marshall, Woolsey & Bard, 1941). Furthermore, although it required a further 30 years to elapse before its significance was realized, it was clearly shown that at least one architectonic and functional area (area 17 of the monkey brain) lacked commissural connexions (Curtis, 1940). Thus experimental neurological studies were beginning to provide evidence which could be adduced in support of the significance of architectonic subdivision.

However, it remained for Rose & Woolsey (1948a) to test the validity of cytoarchitectural subdivision in experimental material, and they initiated what might be called the modern period with their statement

"It seems futile to argue whether or not so-called architectonic characteristics have functional significance if the matter can be approached experimentally."

They raised the question

"does a cortical area which receives projection fibres from a specific thalamic element possess structural characteristics which permit its morphological delimitation? In other words, can such a projection area be regarded as a cortical field?"

and they proceeded to test this hypothesis by investigating the projection of the anterior nuclei of the thalamus upon the limbic cortex with the method of retrograde cellular degeneration. They first studied the cytoarchitecture of the limbic cortex on the medial surface of the hemisphere in cats and rabbits and were able to divide it clearly into three fields, the anterior limbic region, the cingular field and retrosplenial area. Lesions of various sizes were placed in these cortical areas and the position and extent of the damage were accurately determined by careful reconstruction and by the examination of the structure of the cortex preserved at the edges. In each experiment they correlated the distribution of the retrograde cellular degeneration in the anterior nuclei of the thalamus with the degree of involvement of one or more of the architectonic fields. From a collation of the results of several experiments they showed quite conclusively that each of the components of the anterior nuclear group was related to one of the architectonic fields, and they were also able to establish homologous cortical fields in the rabbit and cat. A similar study was made by these workers of the projection of the medial nucleus of the thalamus upon the orbito-frontal cortex (Rose & Woolsey, 1948b), and again the results indicated that there was a close correlation between the structure of the cortex and the distribution of the efferent fibres from this thalamic nucleus. The experimental approach to cytoarchitecture was extended by Rose and Woolsey when they correlated the structure of the auditory sensory area of the cat not only with the findings of an experimental anatomical investigation of the projection of the medial geniculate nucleus but also with the extent of this sensory area as delimited by the evoked potential method (Rose, 1949; Rose & Woolsey, 1949). They found a close correspondence between the total area responsive to electrical stimulation of the cochlear nerve and the whole auditory region as defined on morphological grounds; a remarkably close correlation of the extent of the first auditory field as defined on the basis of its cytoarchitecture and as delimited by the evoked potential method and the experimental anatomical data showing it to be the site of projection of the anterior two-thirds of the medial geniculate nucleus; and a good correlation between structure and function for the anterior portion of the second auditory area. These studies of Rose & Woolsey laid the foundation for the later investigations concerned with the correlation of structure and function, and in addition Rose (1949) introduced a new concept of an architectonic field. He emphasized that the structural features of a cortical "field" are not necessarily, or usually, constant throughout but that they may vary, and

> "a number of closely related variants, often unfortunately referred to as fields or subfields, must be grouped together for functional considerations and that it is a 'natural' series of such variants which constitutes a cortical field".

Depending upon whether the change in such variants is abrupt or gradual the borders of the field will be sharp or blurred; a border determined on the character of the morphological changes should be considered preliminary until checked by other methods. He also suggested that, if cytoarchitectural features reflect the total connexions of that area,

> "one cortical sector may prove to be in regard to one set of connections an independent field while it may share with another sector connections from another set and thus be a part of a morphologically different definable field".

As a consequence, there may be more than one map of the cortex depending upon which set of afferent or efferent connexions have been used for the experimental analysis.

Subsequent studies on thalamo-cortical projections have confirmed the hypothesis of Rose & Woolsey. Thus lesions within the lateral geniculate nucleus of the monkey (Wilson & Cragg, 1967; Hubel & Wiesel, 1969) have resulted in fibre degeneration in area 17 only of the visual cortex and stimulation of, and lesions within, the ventral posterior nucleus have resulted in evoked potentials or degeneration of fibres in the first and second somatic areas (Guillery, Adrian, Woolsey & Rose, 1966; Jones & Powell, 1969a, 1970a). Furthermore, the hypothesis has been extended as it has been shown the subdivisions of the sensory areas defined on a cytoarchitectural basis have differential projections from *within* a specific thalamic nucleus. In a study of the retrograde cellular degeneration in the ventral posterior nucleus of the monkey following selective damage to one or other of the subdivisions, areas 3, 1 and 2, of the somatic sensory area Le Gros Clark & Powell (1953) concluded that area 3 was receiving the major projection and areas 1 and 2 mainly collateral axonal branches; after a lesion restricted to area 3 there was a diffuse but severe cell loss throughout the antero-posterior extent of the nucleus whereas damage to area 2 resulted in cell shrinkage only; a lesion of area 1 caused cell loss in the caudal part only of the nucleus. These conclusions from cellular degeneration studies have been confirmed by more recent experiments using axonal degeneration techniques (Jones & Powell, 1970a); following a lesion within the ventral posterior nucleus of the thalamus of the monkey severe fibre and terminal degeneration is found in area 3, but in areas 1 and 2 there is strikingly less degeneration and the fragments of the degenerating fibres in these areas are appreciably finer; in additon, the change in the density and nature of the degeneration within the cortex changes abruptly at the boundary between areas 3 and 1 (Fig. 1). Similar observations have been made in studies of the projection of the lateral geniculate nucleus upon the visual cortex in the cat; from an analysis

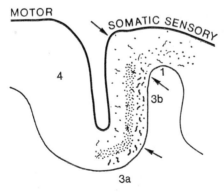

Fig. 1. Semi-schematic figure of a sagittal section of the pre- and post-central gyri of the monkey, stained with the Fink-Heimer method to show the differences in the density and nature of the fibre and terminal degeneration in areas 3 and 1, five days after a large lesion of the ventral posterior nucleus of the thalamus.

of the cellular degeneration in this thalamic nucleus after varying
extents of damage to areas 17 and 18 it was possible to infer that cells
of different sizes were projecting predominantly to one or the other of
these areas (Garey & Powell, 1967). As in the somatic sensory cortex,
the pattern of fibre degeneration in areas 17 and 18 due to damage of
the lateral geniculate nucleus is distinctly different, the fibre fragments
in area 18 being much coarser than in area 17 (Garey & Powell, 1971;
Rossignol & Colonnier, 1971). Indeed, while it is rather difficult to
define precisely the boundary between these two areas of the visual
cortex in Nissl-stained sections the margin is clearly indicated in
sections of such experimental brains stained by the Nauta method as
the change from the coarse fibre fragments to the finer granularity is
abrupt and striking (Fig. 2). The earlier observations of Rose & Woolsey

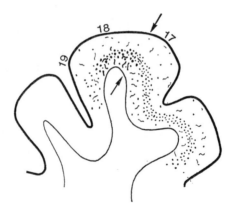

Fig. 2. Semi-schematic representation of the differences in the fibre and terminal
degeneration in areas 17 and 18 of the visual cortex of the cat, five days following lesion
of the lateral geniculate nucleus.

(1949) on the thalamic projection to the auditory cortex have been
extended, and it has been found that the caudal third of the medial
geniculate nucleus, which does not degenerate after removal of the first
and second auditory areas, is related to the insulo-temporal region
(Diamond, Chow & Neff, 1958). As well as being related to different
cortical areas these two parts of the nucleus receive different groups of
ascending afferent fibres (Morest, 1965), and these observations may be
considered to vindicate the well-known statement of Elliot Smith (1910)
that the key to an understanding of the cortex lies in an intensive study
of the thalamus.

One of the first areas of the cortex to be identified on the basis of its
structure was area 17 of man by Gennari due to the marked development

of the stria which bears his name. The stria is really an accentuation of the outer band of Baillarger, and it is of interest, therefore, that recent light and electron microscopic studies of the distribution of degenerating terminals of thalamo-cortical axons in the monkey have shown that whereas there is a dense continuous band of degeneration in the deep part of layer III and in layer IV in the somatic sensory cortex, the degeneration in area 17 is split by the stria into a dense band deep to it in layer IV and a little immediately superficial to it in layer III (Hubel & Wiesel, 1969; Garey & Powell, 1971), and in the motor cortex the degeneration is split by the inner band of Baillarger in layer V. These experimental findings are thus pointing to a relationship between the site of termination of afferent fibres from the thalamus and the degree of development of the intrinsic fibre connexions.

Although most of the experimental anatomical evidence which has been adduced in support of the validity of cytoarchitectural subdivision has been derived from studies on thalamo-cortical projections, a number of recent investigations on cortico-cortical connexions have shown that these fibre systems also respect architectonic boundaries. It was probably the experiments of Hubel & Wiesel (1965) on the visual areas of the cat which stimulated the present interest in association connexions of the sensory areas of the cortex. These authors found that the response properties of the cells in areas 17, 18 and 19 were different and suggested that "visual messages are transmitted from visual I to visual II and III for further processing". That such connexions are not only present but are well organized was shown by Hubel & Wiesel using axonal degeneration techniques; following a small lesion in area 17 they traced degenerating fibres to distinct foci in areas 18, 19 and the lateral suprasylvian area. Subsequent experimental studies on the ipsilateral connexions of other sensory areas (e.g., Jones & Powell, 1969b) have shown that the limits of the areas containing fragments of these degenerating association fibres coincide precisely with those based on cytoarchitecture and this correlation is certainly compelling when, as in the case of the motor cortex of area 4, there is no dispute either with regard to the morphological criteria used or to the boundaries. Damage to the first somatic sensory area results in degeneration clearly limited to area 5 of the parietal cortex, to the second somatic sensory region, to area 4 and the supplementary motor area. Other investigations have shown that the cortical connections of the various areas of the frontal and temporal lobes can also be correlated with cytoarchitecture (Pandya & Kuypers, 1969; Jones & Powell, 1970b; Heath & Jones, 1971). In the convergence of the sensory pathways in the cortex of the monkey and cat there is a step by step progression in *both* the frontal and parieto-temporal lobes and each step corresponds with one of the areas described by Brodmann (1909) or with those of the closely similar maps of von Bonin & Bailey (1947). It has already been mentioned that the first somatic sensory area sends fibres to area 5 of the parietal lobe and to area 4 of the frontal lobe; from area 5 in turn fibres pass to area 7 and to area 6, and from area 4 to the first somatic sensory area and to area 6. The further links in these sequential progressions similarly connect well recognized architectonic areas.

At least one of the principles of organization of the commissural cortical connexions would also appear to be based upon the morphological structure of the cortex in that the fibres arising in a particular area also terminate principally in the corresponding area of the opposite hemisphere. The corresponding architectonic areas may not be the sole site of termination, however, as damage of the boundary of areas 17 and 18 in the cat (Hubel & Wiesel, 1965; Wilson, 1968; Garey, Jones & Powell, 1968) or monkey (Zeki, 1970) may result in degeneration in the other subdivisions of the visual cortex of the other side, and a lesion of the first somatic sensory area results in degeneration in the second as well as the first somatic sensory area of the opposite hemisphere. The absence of commissural connexions between area 17 of the two hemispheres was first shown by the method of evoked potentials several years ago (Curtis, 1940), and recent anatomical observations on these commissural connexions of the visual cortex have provided a remarkably clear example of the significance of even relatively minor structural features. In the monkey brain there are no commissural fibres either arising or terminating within area 17, but such fibres do end in a narrow strip, only a few millimetres wide, along the boundary of area 17 with peristriate cortex, in the representation of the vertical meridian of the retina. At this boundary of area 17 large cells have been described in layer III (Bonin & Bailey, 1947), and in the sections showing axonal degeneration the coincidence of the narrow band of degeneration and the large cells is very striking (Fig. 3). As most of the available evidence suggests that commissural fibres only terminate in areas which are also giving rise to such connexions, the suggestion that the large cells are the origin of the commissural fibres (D. Whitteridge, pers. comm.) seems reasonable. It should therefore be of interest to determine whether the other sites in the peristriate cortex in which commissural fibres terminate are also characterized by the presence of large cells. A similar question arises in the case of the commissural connexions of the somatic sensory area where the regions of cortex related to the hand and foot do not send or receive commissural fibres (Jones & Powell, 1968; 1969c; Pandya & Vignolo, 1968): do these regions differ from adjacent parts of this

MONKEY

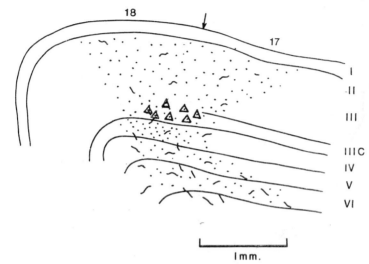

FIG. 3. The distribution of the commissural fibre and terminal degeneration at the boundary of areas 17 and 18 of the visual cortex of the monkey, and the coincidence of this degeneration with the extent of the large pyramidal cells in the deep part of layer III.

sensory area in their intrinsic structure? In view of the density of the commissural connexions it would be reasonable to expect that the absence of such connexions would be reflected in the structure.

For several reasons these findings on the cortical connexions provide one of the most powerful arguments in support of the validity of cyto-architecture: in most of these experiments it has been possible to define accurately the site and extent of the lesion, and in many cases the damage has been very slight, has not involved white matter and has unquestionably been restricted to an architectonic area; the Nauta method for demonstrating axonal degeneration is extremely sensitive and precise and the limits of the degeneration are often remarkably sharp and clear.

There is also increasing evidence to indicate that the structural subdivisions within the sensory areas of the cortex differ in their efferent connexions to subcortical structures. Thus, while it is now agreed that area 17 as well as area 18 of the visual cortex sends fibres back to the lateral geniculate nucleus (Guillery, 1967; Garey et al., 1968; Holländer, 1972) there is no doubt that area 18 contributes more heavily to this projection (Holländer, 1970). There are clear-cut differences in the distribution of the fibres from the various subdivisions of the auditory cortex to the several components of the medial geniculate nucleus

(Diamond, Jones & Powell, 1969) and, at least in the case of the first area and the insulo-temporal region, these appear to be the reciprocal of the thalamo-cortical projection. Electrophysiological studies indicate a similar degree of differentiation in certain of the efferent connexions to sub-cortical sites from the somatic sensory cortex, as the fibres to the dorsal column nuclei appear to arise solely from area 3a (Gordon & Miller, 1969).

The evidence which has been obtained from experimental anatomical studies indicating that areas which differ in their structure also have different afferent and efferent connexions strongly suggests that they would be functionally distinct. It was mentioned earlier that in their combined anatomical and functional study of the auditory region of the cortex of the cat Rose & Woolsey (1949) first showed the significance of such a correlation of structure and function, and further direct evidence that this is so has now been obtained from microelectrode recording of individual neurons in several sensory areas, and more recent work also shows that the cells of different laminae of the sensory areas may respond to distinct types of physiological stimuli. Substantial evidence of functional differences between the cells of the subdivisions of a sensory area was presented by Powell & Mountcastle (1959a, b) from their combined histological and physiological study of the somatic sensory cortex of the monkey. This investigation was an extension of earlier work of these authors: on the one hand, the findings of Le Gros Clark & Powell (1953) that the three subdivisions of this sensory area have differential afferent connexions from the ventral posterior nucleus had suggested that these subdivisions may have important functional differences, and on the other Mountcastle (1957) had found, in the corresponding area in the cat, that all neurons responded to one, but never more than one, of three types of mechanical stimulation of the skin or deep tissues and also that there was a functional organization of the cortex into columns such

> "that the elementary pattern of organization in the cerebral cortex is a vertically oriented column or cylinder of cells capable of input-output functions of considerable complexity, independent of horizontal intragriseal spread of activity".

In the microelectrode studies of areas 3, 1 and 2 the type of stimulus and the site of the receptive field were determined for a large sample of individual neurons, the tracks of the majority of the microelectrode penetrations were found histologically and from reconstructions it was possible to identify for most cells the depth within the cortex and the site in terms of architectonic subdivisions—the cytoarchitecture of the postcentral gyrus being studied for each experimental brain. As well as fully confirming the hypothesis of a functional columnar organization of the cortex a correlation of the response properties of the cells with their sites in the cortex showed a differential distribution of neurons related to the skin and those excited by deformation of deep tissues in the three architectonic areas. Two-thirds of the cells in area 3 were activated by cutaneous stimuli whereas 90% of the cells in area 2 responded to stimulation of deep tissues—fascia or joints—whilst the proportion of cells in area 1 responding to these stimuli was inter-mediate—30% to skin. This change in response properties of the cells as one moved posteriorly across the postcentral gyrus was a gradual one, with no abrupt transitions, and this pattern is in accord with the histo-logical observation of a gradient of morphological change from the bottom of the central sulcus to the posterior margin of area 2. These findings of Powell & Mountcastle have been confirmed in recent studies of the details of the topographical representation of the body surface in the same region by Werner & Whitsel (1968).

One observation of Powell & Mountcastle (1959b), the significance of which has only been realized in the last few years, was that in the narrow band of cortex immediately behind the depth of the central sulcus there was an abrupt increase in the proportion of cells responding to stimulation of deep tissues, and this represented a distinct change in the trend found in the cortex of the posterior wall of the sulcus. This narrow band of cortex, area 3a, forms a region of transition between somatic sensory cortex proper posteriorly and the motor cortex of area 4 anteriorly. The impression of Powell & Mountcastle that area 3a "does in fact receive a heavy projection from deep tissues" has been confirmed in the more systematic studies of this transition region in recent years. A comparison of the area of responses in the cortex to stimulation of group 1 muscle afferent fibres in the cat with the architectonic maps of the sensori-motor region of Hassler & Muhs-Clement (1964) led Oscars-son & Rosen (1966) to suggest that the projection field corresponds to area 3a, and this was confirmed and extended by Landgren & Silfvenius (1969). A combined electrophysiological and histological study in the baboon (Phillips, Powell & Wiesendanger, 1971) provided direct evidence of a similar projection in the primate. It should be mentioned, however, that these conclusions that area 3a is a specific cortical projection area for muscle afferents has recently been questioned in a preliminary report by Burchfiel & Duffy (1972) who have found responses to muscles as well as joints in area 2.

In the first somatic sensory area, and particularly in area 2, cells are found which respond to movement of joints; and they usually respond to movement of one joint only, and often to only a limited degree of movement. In area 5, however, to which this sensory area is known to project, there are cells which although responding to joint movement do so in response to movement of more than one joint and these may be in different limbs (Duffy & Burchfiel, 1971).

The combined electrophysiological and histological study of the visual cortex of the cat by Hubel & Wiesel (1965) has provided strong evidence that the subdivisions of this region, areas 17, 18 and 19, described on anatomical cellular and myelinated fibre criteria by Otsuka & Hassler (1962) also have functional significance. These workers showed that there are three well organized representations of the contralateral half of the visual field, visual I, visual II and visual III, and that the limits of these areas coincide with those of areas 17, 18 and 19 as determined in sections stained for cells or myelinated fibres. Furthermore, as has been mentioned, Hubel & Wiesel also showed that following a small lesion in area 17 degenerating fibres could be traced to areas 18 and 19, and they concluded that the limits of these areas as defined by four independent methods coincided. In an earlier study of area 17 of the cat Hubel & Wiesel (1962) had found two types of neurons with distinct response properties, the simple cell responding to a line stimulus precisely placed and oriented and a complex cell which responds to an appropriately oriented stimulus regardless of where in the receptive field it is placed. In visual areas II and III (areas 18 and 19) they found further types of neurons, the lower order hypercomplex cell which is activated by a correctly oriented line appropriately terminated at one or both ends, and a higher order hypercomplex cell which also requires a stimulus of critical length but which would respond when this was oriented in either of two orientations, 90° apart. The proportions of these types of neurons were found to be markedly different in the three areas: in visual I (area 17) approximately two-thirds of the cells were simple and one-third complex, whereas in visual II (area 18) the majority were complex and 5–10% were lower order hypercomplex, and in visual III (area 19) there were approximately equal numbers of complex and lower order hypercomplex with a small number of higher order hypercomplex. Hubel & Wiesel therefore suggested that there was sequential processing of information between the simple cells of area 17 and the hypercomplex cells of areas 18 and 19, and considered that the fibre connexions which they had shown to pass between these areas provided an anatomical basis for such mechanisms.

It would probably be generally agreed that these combined anatomical and functional studies in the somatic and visual sensory areas have provided the most compelling evidence in support of the validity of cytoarchitectural subdivisions in showing that the cells within the subdivisions of these areas as defined on morphological criteria differ in their responses to physiological stimuli. Two recent investigations on these same sensory areas of the monkey have shown further that the constituent cells of different laminae of the cortex have clear differences in response properties. The importance of these findings lies in the fact that variations in the development of the laminae have been one of the major anatomical criteria used in architectonic studies, and also because they suggest close similarities in the mechanisms for the processing of information within these two sensory areas. In a study of the visual cortex of the monkey Hubel & Wiesel (1968) found that the majority of the cells in layer IV and in the deep part of layer III are simple and respond to stimulation of one eye only, but in the more superficial and deep layers the cells are complex or hypercomplex and are binocularly activated. As the thalamo-cortical afferent fibres are known to terminate in precisely those layers in which simple cells are found (Hubel & Wiesel, 1969; Garey & Powell, 1971), Hubel & Wiesel consider that there are synaptic connexions from these cells to the complex and hyper-complex neurons in the more superficial and deep laminae. In the somatic sensory area Whitsel, Roppolo & Werner (1972) have also found differences in the responses of cells in different laminae to move-ment of a stimulus across the skin, and these authors also postulate a sequential intracortical processing of information with "a resorting of afferent activity in the transition from lamina IV to more superficial and deeper cortical laminae". In layer III, particularly of area 1, the cells showed different response patterns to stimuli moving across their receptive fields in different directions whereas such directionally selective responses were not a feature of cells in layer IV.

In view of the demonstration of functional differences between the cells of various laminae, and the fact that some of these layers tend to be composed predominantly of one cell type, the present attempts of neurophysiologists to correlate specific functional characteristics with a particular morphological cell type—pyramidal or stellate—are of particular importance (Kelly & Van Essen, 1972). As the cell somata and dendrites of these two types of cells can be readily identified with both the light and electron microscopes it is equally important that the synaptic connexions of these types of cells be determined, as this would provide a structural basis for the information processing that clearly occurs within the sensory areas.

If it is accepted that the experimental approach to cytoarchitecture has established the validity of the subdivisions by showing that they also differ in their connexions—afferent and efferent, both cortical and subcortical—and have functional differences, and in addition that one of the major structural characteristics of these areas—the varying degree of development of the individual laminae—is functionally distinct, it means that the several cortical areas differ *both* in the nature of their afferent connexions *and* in the processing of information within them. It would appear, therefore, that further investigations on the detailed structure and response properties of individual cells in areas in the parieto-temporal (Duffy & Burchfiel, 1971; Gross, Rocha-Miranda & Bender, 1972) and frontal lobes, to which the sensory areas are now known to project (Jones & Powell, 1970b), should help us to understand how these parts of the cortex in turn process information from the cells which respond to relatively simple stimuli in the sensory areas to the integration of a total pattern within and between individual sensory systems. In addition, it should be clear that experimental neurological and psychological studies on the cortex should take into serious account the limits of the cytoarchitectonic areas, and the observations and lesions should be correlated with structure and described in terms of these areas. Finally, there is probably little to be gained from further descriptions of anatomical detail in normal material unless these are correlated with experimental findings.

REFERENCES

Adrian, E. D. (1941). Afferent discharges to the cerebral cortex from peripheral sense organs. *J. Physiol., Lond.* **100**: 159–191.

Bonin, G. von & Bailey, P. (1946). *The neocortex of* Macaca mulatta. Urbana, Ill.: University of Illinois Press.

Brodmann, K. (1903). Beiträge zur histologischen Lokalisation der Grosshirnrinde. 1 Mitteilung: die Regio rolandica. *J. Psychol. Neurol., Lpz.* **2**: 79–107.

Brodmann, K. (1909). *Vergleichende Lokalisationslehre der Grosshirnrinde in ihren Prinzipien dargestellt auf Grund des Zellenbaues.* Leipzig: J. A. Barth.

Burchfiel, J. L. & Duffy, F H. (1972). Muscle afferent input to single cells in primate somatosensory cortex. *Brain Res., Amst.* **45**: 241–246.

Campbell, A. W. (1903). Histological studies on cerebral localisation. *Proc. R. Soc.* **72**: 488–492.

Campbell, A. W. (1905). *Histological studies on the localisation of cerebral function.* Cambridge: Cambridge University Press.

Curtis, H. J. (1940). Intercortical connections of corpus callosum as indicated by evoked potentials. *J. Neurophysiol.* **3**: 407–413.

Diamond, I. T., Chow, K. L. & Neff, W. D. (1958). Degeneration of caudal medial geniculate body following cortical lesion ventral to auditory area A II in cat. *J. comp.. Neurol.* **109**: 349–362.

Diamond, I. T., Jones, E. G. & Powell T. P. S. (1969). The projection of the auditory cortex upon the diencephalon and brain stem in the cat. *Brain Res., Amst.* **15**: 305–340.

Duffy, F. H. & Burchfiel, J. L. (1971). Somatosensory system: organizational hierarchy from single units in monkey area 5. *Science, Wash.* **172**: 273–275.

Elliot Smith, G. (1907). A new topographical survey of the human cerebral cortex, being an account of the distribution of the anatomically distinct cortical areas and their relationship to the cerebral sulci. *J. Anat. Physiol., Lond.* **41**: 237–249.

Elliot Smith, G. (1910). Some problems relating to the evolution of the brain. *Lancet* **1910(1)**: 1–6.

Garey, L. J. & Powell, T. P. S. (1967). The projection of the lateral geniculate nucleus upon the cortex in the cat. *Proc. R. Soc.* (B) **169**: 107–126.

Garey, L. J. & Powell, T. P. S. (1971). An experimental study of the termination of the lateral geniculo-cortical pathway in the cat and monkey. *Proc. R. Soc.* (B) **179**: 41–63.

Garey, L. J., Jones, E. G. & Powell, T. P. S. (1968). Interrelationships of striate and extrastriate cortex with the primary relay sites of the visual pathway. *J. Neurol. Neurosurg. Psychiat.* **31**: 135–157.

Gordon, G. & Miller R. (1969). Identification of cortical cells projecting to the dorsal column nuclei of the cat. *Q. Jl exp. Physiol.* **54**: 85–98.

Gross, C. G., Rocha-Miranda, C. E. & Bender, D. B. (1972). Visual properties of neurons in inferotemporal cortex of the macaque. *J. Neurophysiol.* **35**: 96–111.

Guillery, R. W. (1967). Patterns of fiber degeneration in the dorsal lateral geniculate nucleus of the cat following lesions in the visual cortex. *J. comp. Neurol.* **130**: 197–222.

Guillery, R. W., Adrian, H. O., Woolsey, C. N. & Rose, J. E. (1966). Activation of somatosensory areas I and II of the cat's cerebral cortex by focal stimulation of the ventrobasal complex. In *The thalamus.* Purpura, D. P. & Yahr, M. D. (eds). New York: Colombia University Press.

Hassler, R. & Muhs-Clement, K. (1964). Architektonischer Aufbau des senso-motorischen und parietalen Cortex der Katze. *J. Hirnforsch.* **54**: 377–420.

Heath, C. J. & Jones, E. G. (1971). The anatomical organization of the supra-sylvian gyrus of the cat. *Ergebn. Anat. EntwGesch.* **45**: 1–64.

Holländer, H. (1970). The projection from the visual cortex to the lateral geniculate body (LGB). An experimental study with silver impregnation methods in the cat. *Expl Brain Res.* **10**: 219–235.

Holländer, H. (1972). Autoradiographic evidence for a projection from the striate cortex to the dorsal part of the lateral geniculate nucleus in the cat. *Brain Res., Amst.* **41**: 464–466.

Hubel, D. H. & Wiesel, T. N. (1962). Receptive fields, binocular interaction and functional architecture in the cat's visual cortex. *J. Physiol., Lond.* **160**: 106–154.

Hubel, D. H. & Wiesel, T. N. (1965). Receptive fields and functional architecture in two nonstriate visual areas (18 and 19) of the cat. *J. Neurophysiol.* **28**: 229–289.

Hubel, D. H. & Wiesel, T. N. (1968). Receptive fields and functional architecture of monkey striate cortex. *J. Physiol. Lond.*, **195**: 215–243.

Hubel, D. H. & Wiesel, T. N. (1969). Anatomical demonstration of columns in the monkey striate cortex. *Nature, Lond.*, **221**: 747–750.

Jones, E. G. & Powell, T. P. S. (1968). The commissural connexions of the somatic sensory cortex in the cat. *J. Anat.* **103**: 433–455.

Jones, E. G. & Powell, T. P. S. (1969a). The cortical projection of the ventro-posterior nucleus of the thalamus in the cat. *Brain Res., Amst.* **13**: 298–318.

Jones, E. G. & Powell, T. P. S. (1969b). Connexions of the somatic sensory cortex of the rhesus monkey. I. Ipsilateral cortical connexions. *Brain* 92: 477–502.

Jones, E. G. & Powell, T. P. S. (1969c). Connexions of the somatic sensory cortex of the rhesus monkey. II. Contralateral cortical connexions. *Brain* 92: 717–730.

Jones, E. G. & Powell, T. P. S. (1970a). Connexions of the somatic sensory cortex of the rhesus monkey. III. Thalamic connexions. *Brain* 93: 37–56.

Jones, E. G. & Powell, T. P. S. (1970b). An anatomical study of converging sensory pathways within the cerebral cortex of the monkey. *Brain* 93: 793–820.

Kelly, J. & Van Essen, D. (1972). Physiological and morphological identification of neurons in the visual cortex of the cat. *J. Cell Biol.* 55 Abstr. *12th A. Mtg Am. Soc. Cell Biol.* 133a.

Landgren, S. & Silfvenius, H. (1969). Cortical projection of Group 1 muscle afferents from the cat's hindlimb. *J. Physiol., Lond.* 200: 353–372.

Lashley, K. S. & Clark, G. (1946). The cytoarchitecture of the cerebral cortex of *Ateles*: a critical examination of architectonic studies. *J. comp. Neurol.* 85: 223–305.

Le Gros Clark, W. E. (1952). A note on cortical cyto-architectonics. *Brain* 75: 96–108.

Le Gros Clark, W. E. & Boggon, R. H. (1935). The thalamic connections of the parietal and frontal lobes of the brain in the monkey. *Phil. Trans. R. Soc.* (B.) 224: 313–359.

Le Gros Clark, W. E. & Powell, T. P. S. (1953). On the thalamocortical connexions of the general sensory cortex of *Macaca. Proc. R. Soc.* (B). 141: 467–487.

Marshall, W. H., Woolsey, C. N. & Bard, P. (1941). Observations on cortical somatic sensory mechanisms of cat and monkey. *J. Neurophysiol.* 4: 1–24.

Morest, D. K. (1965). The lateral tegmental system of the midbrain and the medial geniculate body: study with Golgi and Nauta methods in cat. *J. Anat.* 99: 611–634.

Mountcastle, V. B. (1957). Modality and topographic properties of single neurons of cat's somatic sensory cortex. *J. Neurophysiol.* 20: 408–434.

Oscarsson, O. & Rosen, I. (1966). Short-latency projections to the cat's cerebral cortex from skin and muscle afferents in the contralateral forelimb. *J. Physiol., Lond.* 182: 164–184.

Otsuka, R. & Hassler, R. (1962). Uber Aufbau und Gliederung der corticalen Sehsphäre bei der Katze. *Arch. Psychiat. Nervenkr.* 203: 212–234.

Pandya, D. N. & Kuypers, H. G. J. M. (1969). Cortico-cortical connections in the rhesus monkey. *Brain Res., Amst.* 13: 13–36.

Pandya, D. N. & Vignolo, L. A. (1968). Interhemispheric neocortical projections of somatosensory areas I and II in the rhesus monkey. *Brain Res., Amst.* 7 300–303.

Phillips, C. G., Powell, T. P. S. & Wiesendanger, M. (1971). Projection from low-threshold muscle afferents of hand and forearm to area 3a of baboon's cortex. *J. Physiol., Lond.* 217: 419–446.

Poliak, S. (1932). *The main afferent fiber systems of the cerebral cortex in Primates.* Berkeley, California: University of California Press.

Powell, T. P. S. & Mountcastle, V. B. (1959a). The cytoarchitecture of the post-central gyrus of the monkey *Macaca mulatta. Bull. Johns Hopk. Hosp.* 105: 108–131.

Powell, T. P. S. & Mountcastle, V. B. (1959b). Some aspects of the functional organization of the cortex of the postcentral gyrus of the monkey: a correlation of findings obtained in a single unit analysis with cytoarchitecture. *Bull. Johns Hopk. Hosp.* **105**: 133–162.

Rose, J. E. (1949). The cellular structure of the auditory region of the cat. *J, comp. Neurol.* **91**: 409–440.

Rose, J. E. & Woolsey, C. N. (1948a). Structure and relations of limbic cortex and anterior thalamic nuclei in rabbit and cat. *J. comp. Neurol.* **89**: 279–348.

Rose, J. E. & Woolsey, C. N. (1948b). The orbitofrontal cortex and its connections with the mediodorsal nucleus in the rabbit, sheep and cat. *Assn. Res. Nerv. Ment. Dis.* **27**: 210–232.

Rose, J. E. & Woolsey, C. N. (1949). The relations of thalamic connections, cellular structure and evocable electrical activity in the auditory region of the cat. *J. comp. Neurol.* **91**: 441–466.

Rossignol, S. & Colonnier, M. (1971). A light microscope study of degeneration patterns in cat cortex after lesions of the lateral geniculate nucleus. *Vision Res.* Supplement No. 3: 329–338.

Vogt, O. & Vogt, C. (1963). Zur anatomischen Gliederung des Cortex cerebri. *J. Psychol. Neurol.* **2**: 160–180.

Walker, A. E. (1938). *The Primate thalamus.* Chicago: Chicago University Press.

Walker, A. E. (1940). A cytoarchitectural study of the prefrontal area of the macaque monkey. *J. comp. Neurol.* **73**: 59–86.

Werner, G. & Whitsel, B. L. (1968). Topology of the body representation in somatosensory area 1 of Primates. *J. Neurophysiol.* **31**: 856–869.

Whitsel, B. L., Roppolo. J. R. & Werner, G. (1972). Cortical information processing of stimulus motion on Primate skin. *J. Neurophysiol.* **35**: 691–717.

Wilson, M. E. (1968). Cortico-cortical connexions of the cat visual areas. *J. Anat.* **102**: 375–386.

Wilson, M. E. & Cragg, B. G. (1967). Projections from the lateral geniculate nucleus in the cat and monkey. *J. Anat.* **101**: 677–692.

Zeki, S. M. (1970). Interhemispheric connections of prestriate cortex of monkey. *Brain Res., Amst.* **19**: 63–75.

DISCUSSION

YOUNG (CHAIRMAN): Thank you very much indeed, Dr Powell, for this very clear and complete view of the cerebral cortex. Is it ever going to be possible for the techniques of physiology to show us the pattern of the more complex responses you are seeking? I still feel there is a big future in investigating the pattern of interaction in the brain and that this may be better done with the rather old-fashioned method of microscopy revealing individual units with the precision which people like you can read. I do not know how far you can go. Would you care to comment on that?

POWELL: I would fully agree with that. During the last few years all of us have concentrated on refining our approach and have tended to forget, as you said, that we can often learn a lot by studying the total populations.

This is where I think cytoarchitecture may help in studying the whole population for the whole area, and I think that a combination of cyto-architecture with experimental psychology can contribute a lot in regard to the function of a total area rather than the units within that area.

WEBSTER: I fully agree with most of what Dr Powell has said, but I would like his views on one particular point. He commented that the columnar pattern of degeneration in the cortex following a thalamic lesion is in fact not cylindrical. It has a funnel shape that narrows from layer one down to the junction of layers three and four, where it becomes roughly cylin-drical. That is exactly what we have seen in our studies of mammalian thalamo-cortical projections. My reaction to this is that you cannot make one cortical map. You have to make a map for each layer. Would you like to comment?

POWELL: Rose (1949)* made a similar point in discussing the redefinition of cytoarchitecture, when he said that previously cytoarchitecture had been based on the normal structure of each area, but one might have different maps on the basis of different connexions—afferent and efferent—and upon certain functions. I think that is quite possible, but to me the significant thing is that during the last 20 years we have seen the establish-ment of a concept on an experimental basis and, as Professor Diamond said earlier, having established that a certain area in different species has the same connexions, he could then go on to other animals from that baseline. The work of several independent people using different tech-niques has shown this, and I hope therefore that people will accept it and use it for their descriptions of degeneration, lesions, etc.

DIAMOND: Elliot Smith said that anatomy plays a humble role in the science he envisioned, but he was of course only referring to the fact that psy-chology would face enormous obstacles, and I am sure he appreciated more than anyone how difficult it is to establish through comparative studies homologies and how challenging it is to make architectonic sub-divisions. I was struck in Dr Powell's presentation by the difference between the geniculo-striate system in the primate and the cat. Dr Powell has shown in a beautiful series of studies with Garey and Jones, see, for example Garey & Powell (1967),* that the lateral geniculate nucleus projects to cortical areas 17 and 18 in the cat, while in primates, of course, the lateral geniculate nucleus projections are entirely restricted to area 17. So it is perhaps worth mentioning how difficult it is to establish homologous cortical areas on the basis of any criterion.

---

* See list of references, pp. 248–251.

PRIMATE SYSTEMATICS AND THE TARSIUS PROBLEM

CHAIRMAN: N. A. BARNICOT

CHAIRMAN'S INTRODUCTION

We have heard an illuminating survey of Sir Grafton Elliot Smith's scientific career by Lord Zuckerman and considered his association with work on hominid evolution in this Symposium so far. Professor Day told us about his own recent work on the lower limb bones of *Homo erectus* and on some of the new hominid material from East Africa and Mr Leakey showed us a new and exciting skull from the Lake Rudolf area. We recalled that Elliot Smith was associated with the discovery of Peking man through his former student Davidson Black and with the first of the australopithecine finds through another former student, Raymond Dart. We then went on to consider another of his major interests, the evolution of the mammalian brain which was the first problem he tackled, and the localization of functions in the cerebral cortex.

We pass on to yet another facet of his wide interests, the question of primate phylogeny in general. We have a series of four papers, the first of them by Dr Oxnard. I feel sure that Dr Oxnard will be dealing with work involving multivariate statistical analysis and we have already had a taste of these methods in Professor Day's and Professor Ashton's communications. One of the big changes that has come about in evolutionary studies on man and other Primates since Elliot Smith's time has been the increasing use of sophisticated statistical analysis. It is interesting to recall that when Elliot Smith was at University College, Karl Pearson, in another region of that institution, was doing much to lay the foundations of the statistical treatment of biological material. I have asked Lord Zuckerman whether he thinks that there was any contact between these two great figures and he thinks that there was not. I have certainly failed to find any mention of Pearson in Elliot Smith's writings; but it is worth remembering today that Pearson's coefficient to racial likeness was one of the earliest examples of the multivariate approach to classification problems. We should also remember that the application of later and more powerful multivariate methods to problems of primate phylogeny was pioneered by the Birmingham School of which Dr Oxnard was formerly a member.

The second paper in this set is by Dr Martin who is going to discuss and illustrate the kind of logical problems that arise when one tries to construct phylogenetic trees using morphological data. The next paper is by Professor Goodman who is going to deal with the phylogenetic information that can be obtained from protein sequences. It is only in the last 20 years that it has become possible to do these detailed analyses of proteins and to appreciate the relevance of such data to gene divergence. Nevertheless immunological comparisons of primate blood proteins, motivated by an interest in phylogeny, were started by Nuttall early in this century, and Elliot Smith mentioned this work and seems to have regarded it as particularly important and decisive. This is an interesting example of his intuition, for at that time, 20 years or so before the birth of molecular genetics, he could not possibly have known why protein differences were so significant. He also mentioned some of the early work on the ABO blood groups of apes.

Dr Martin will be discussing phylogenetic trees and I feel sure that Professor Goodman will show us examples of trees constructed by computer methods. It occurred to me that our preoccupation with trees may be yet another example of the influence of our arboreal ancestry to which Elliot Smith and Wood Jones pointed so eloquently.

In the last of this set of four papers, Professor Garnham is going to consider the evolutionary implications of the distribution of malaria parasites in Primates. Darwin, of course, mentioned parasitological evidence but I do not know whether Elliot Smith ever wrote about it. I think it is a line that should interest anyone who is keen on biochemical evolution because I imagine that these specialized parasites are very sensitive to the biochemical environment in which they live; their adjustments to the various branches of the primate evolutionary tree are a reflection of the subtle biochemical changes involved in the divergence of their hosts.

Symp. zool. Soc. Lond. (1973) No. 33, 255–299.

SOME LOCOMOTOR ADAPTATIONS AMONG LOWER PRIMATES: IMPLICATIONS FOR PRIMATE EVOLUTION

CHARLES E. OXNARD

Department of Anatomy, The University of Chicago, Chicago, Illinois, U.S.A.

SYNOPSIS

The problem of the early origins of the primates has long been associated with the concept of adaptation to the arboreal environment. Recent studies have particularly suggested that an arboreal locomotor mode of "vertical clinging and leaping" as exhibited by many extant prosimians such as the Indri and the tarsier, may well have been the chief locomotor specialization of Eocene primates such as *Notharctus*. Examination of individual anatomical features does not deny the idea that there exists a single morphological adaptation to "vertical clinging and leaping". But examination of mensurational structural parameters, taken together in a manner that allows for their intercorrelations, provides a different picture. Thus separate multivariate studies of detailed measurements of the lower and upper limb girdles suggests that more than one morphological mode characterizes those anatomical parts in species thought to be vertical clingers and leapers. Further statistical examination of the major dimensions of each of the entire limbs confirms and extends this conclusion. Finally, morphometric investigations of the overall form of these animals demonstrates that intermediate and incipiently adapted species exist. On this basis, a second look at the behavioural literature indicates, if only tentatively, that there may indeed be locomotor differences among these species that preclude the concept of a single group of "vertical clingers and leapers". Because presumed morphological adaptations and putative behaviours of prosimians have been implicated in the evolution of higher primates, further study, both behavioural and morphological, of both extant and fossil species, appears to be indicated.

INTRODUCTION

The early evolution of the Order Primates has long been associated with the assumption that the basic primate specialization is to an arboreal environment. Grafton Elliot Smith contributed largely to such ideas through the suggestion of arboreal habitats in primate ancestors that lived prior to the Cenozoic era. The ancestral primates were presumed to have become arboreal before adaptation to terrestrial habits had changed the generalized flexibility of their limbs to morphologies more stringently associated with quadrupedal cursorial locomotion. Presumably this had indeed also occurred in a number of other mammalian orders, but possibly nowhere to the extent seen in the primates. Since those early studies, although much work has been done relating to the morphology of living prosimians and presumed early primate ancestors, most has been aimed at other aspects (e.g. teeth, jaws and cranium) of these particular species.

Rather more recently however, attention has again been directed towards the locomotion and habitat of living prosimians, with appropriate study, inference and speculation resulting for various fossil species. Perhaps the most explicit of these investigations is that of Napier & Walker (1967) who, in their attempt to characterize the locomotor patterns of many different primates, provide an account of a new "natural locomotor group among the Primates". All the living members of this locomotor mode are prosimians and they include: *Tarsius, Indri, Propithecus, Avahi, Lepilemur, Galago* and *Euoticus* (*Hapalemur simus* is also tentatively included in this grouping by Napier & Walker, but not upon any behavioural evidence). In addition to being primarily arboreal, the animals concerned are said to have a vertical clinging posture at rest, and a leaping mode of progression during which the hindlimbs, used together, provide the propulsive force. Napier & Walker suggest that the special interest of this locomotor mode of vertical clinging and leaping is that it appears to constitute the only known locomotor adaptation of Eocene primates such as *Notharctus, Necrolemur* and *Smilodectes*. They speculate that it may be regarded as the earliest locomotor specialization of primates and preadaptive to the various arboreal adaptations of all the living higher primates. This view has been fairly generally accepted not only by primate morphologists such as Pilbeam (1972) and Simons (1972) who have adopted this grouping of vertical clinging and leaping and applied it to many fossil species, but also by students of the behaviour of prosimian forms. Thus Jolly (1972) retains Napier & Walker's vertical clingers and leapers but nevertheless cautions that while the morphological adaptations have been relatively extensive and are fairly detailed and refined, the behavioural basis is much less extensive; "Napier's (or any other) classification explicitly depends on the degree to which a primate uses various patterns, (and) this leaves the behavioral side of the definitions still hanging in midair". In fairness to Napier & Walker (1967) it must be recognized that they were clearly aware of the preliminary nature of their ideas. Thus they state that their grouping of these particular prosimians as vertical clingers and leapers is "a 'working plan' which will undoubtedly be subject to considerable future amendment as the behaviour and morphology of primates becomes better understood and the grades between apparently 'discrete' locomotor groups are revealed".

Perhaps only one author has taken Napier & Walker directly to task about the reality of vertical clinging and leaping. Thus Cartmill

(1972) suggests that "vertical clinging and leaping" is an artificial category. His arguments include, first, a disagreement as to the adaptive advantages of vertical clinging and leaping, a line of discussion that, on both sides, can be little more than speculative in our present state of knowledge. Secondly, and more importantly, he summarizes the morphological evidence and shows that most of the supposedly diagnostic features, when taken one by one, do not separate extant vertical clinging and leaping forms from those species that he believes (and Napier & Walker concur) to be more generally quadrupedal (e.g., *Microcebus*, *Phaner*, *Cheirogaleus* and *Lemur*). Thirdly, he reminds us of the specific differences between *Tarsius* and *Galago* on the one hand with special coxal and tarsal modifications for leaping, and the indriids on the other with only a generally lengthened lower limb.

These last differences have, of course, long been known. Morton (1924) for instance, understood the importance of the elongation of the tarsus in the leaping of *Galago* and *Tarsius*, in contrast to the elongation of the legs in the leaping of *Propithecus* and other indriids. Volkov (1903, 1904) was also aware of the two different pedal morphologies and he understood, if in a less detailed manner, their relationship to movement in the leaping of these forms. Mivart (1867, 1873), though apparently not as overtly aware of the functional implications, had a clear view of the morphological differences between these two groups of species.

INDIVIDUAL ANATOMICAL FEATURES

The reality of the doubt that can be introduced into a discussion of the morphological adaptation to "vertical clinging and leaping" as a single locomotor mode can be seen by attempting to view individual anatomical features in the two following ways.

1. It is possible to define the various diagnostic structural characters rather broadly so that they include all the known vertical clingers and leapers. These definitions (Table I) must, therefore, be so broad that a large number of forms that are not vertical clingers and leapers (*sensu* Napier & Walker) are of necessity also included. The Table shows, for instance, that although no extant primate other than a vertical clinger and leaper shares every trait, nevertheless a number of other living genera share many of them. Thus if we suppose that *Lemur* (which is missing only a single feature) were known only as a fossil form incomplete enough that the absent feature (mean lower limb index) could not be assessed, then that species might be grouped, presumably incorrectly, with the vertical clingers and leapers. Such exceptions are

M

TABLE I

Anatomical features characterizing all "vertical clingers"

Anatomical feature	Some other primate genera possessing it			
	Lemur	*Daubentonia*	*Callimico*	*Loris*
Intermembral Index < 72	•	•	•	
Brachial Index > 99	•	•	•	•
Hand Length Index > 27	•	•	•	
Phalangeal Length Index > 54	•	No data	•	•
Lower Limb Length Index > 125		•		•
Hallucial Length Index > 70	•	•		•

Also included are: long ilium and short ischium, straight femoral shaft, small to medium body size.

These all characterize many Primates.

Comment: *Lemur* is missing only a single feature; if *Lemur* were only known as a fossil with lower limb missing, then it might well be thought to be a "vertical clinger".

indeed so numerous that a fossil for which it is only possible to procure a partial list may be said to be a vertical clinger and leaper at only a rather poor level of likelihood.

2. It is also possible to define the various individual features so tightly that they are confined to the known vertical clingers and leapers (*sensu* Napier & Walker). When this procedure is carried out (Table II), although by definition every anatomical feature is found in one or other vertical clinger and leaper and only in such forms, many of the features are absent from a number of individual vertical clinging and leaping species and specimens. Because it is so easy to find vertical clingers that lack several of these traits, it is also likely that fragmentary evidence from a fossil that did practise vertical clinging and leaping, may show no definitive evidence of such behaviour. For example, if we suppose that *Propithecus* were known only as a fossil, and that the remains were incomplete enough that the lower end of the femur were missing and an intermembral index could not be estimated, Table II shows that there would be very little to indicate the remarkable locomotion that we know the Sifaka employs.

It seems clear that the problem described here can be clarified by further studies. First, more detailed descriptions of the behaviours of the extant forms, both those thought to be vertical clinging and leaping

TABLE II

Anatomical features confined to "vertical clingers"

Anatomical feature	Lepilemur	Hapalemur simus	Galago	Tarsius	Propithecus	Avahi	Indri
				"Vertical clingers" in which feature may be absent			
Intermembral Index <68	•	•			•	•	•
Brachial Index >117 (Prosimii)	•	•	•		•	•	•
Hand Length Index >33 (ex Aye-aye)	•	•	•		•	•	•
Phalangeal Index >64	•	•	•		•	•	•
Cylindrical Femoral Head*	•	•	•		•	•	•
Elongated Tarsus†				•			
Anteriorly Projected Patellar Femoral Grooves	No data				○‡		○‡

* Somewhat similar in *Cheirogaleus* and *Microcebus*.
† Somewhat similar in *Cheirogaleus* and *Phaner*.
‡ One femoral groove only.

Comment: *Propithecus* is missing many of these features; if *Propithecus* were only known as a fossil and the lower end of the femur and intermembral index were missing, then it might well *not* be recognized as a "vertical clinger".

species, and those believed to be otherwise, may help to elucidate the behavioural reality of a single locomotor mode of vertical clinging and leaping. Part of the behavioural problem is, of course, that it is not easy to discover precisely what happens during leaping in these animals. Hall-Craggs (1965) has analysed the leap of *Galago* in appropriate detail but no other species has been so investigated. Secondly, analyses of primate morphology that are able to encompass many individual anatomical features taken together (rather than one by one) may be able to confirm or deny the reality of a single type of morphological adaptation to vertical clinging and leaping. For although there are known tarsal and coxal features in which *Propithecus*, *Indri* and *Avahi* differ from *Galago* and *Tarsius*, these specific differences could be either minor morphological variations resulting from parallel evolutionary pathways toward a single adaptation or evidence of two distinct adaptive modes. Thirdly, in such a series of studies, both behavioural and morphological, quantification and analysis may suggest species that are intermediately or even incipiently adapted toward such behavioural and morphological modes. It is against a detailed background such as this that it may be possible to be somewhat more definitive about the functional status of different early primates.

MULTIVARIATE MORPHOMETRIC INVESTIGATIONS

Some basis for morphological study is already partly available in the data that have been collected by several groups of workers over a period of many years. For instance, one approach is to investigate the detailed structure of specific anatomical regions critical in understanding the function of vertical clinging and leaping. A detailed study of the foot is obviously important here, as is a similar study of the pelvis. Other regions are not precluded. Even in the forelimb, a study of the hand for instance, or the shoulder girdle, for example, may provide information bearing upon these questions, although it is likely that the form of these upper limb elements may be less related to vertical clinging and leaping than the structure of the hindlimb components. A series of features of such skeletal parts can be defined and quantified, and the analysis of their dimensions in combination (using methods that make allowance for their intercorrelations) is capable of providing an overall view (or total morphological pattern) that gives evidence bearing upon these questions. A set of measurements of the hip and shoulder girdles is already available through the anatomical laboratories at Birmingham and Chicago and involving collaboration with Professor Lord Zuckerman, Professor E. H. Ashton, Dr R. M. Flinn and Mr T. F.

Spence; the analysis of these data with the question of vertical clinging and leaping in mind has here been attempted.

Although such regional studies are able to provide considerable information, they suffer from a particular defect. That is, through being studies of localized regions of the skeleton, some at least of the information that they present may be distorted or hidden because they are the developmental products of a localized region of the embryo. In other words, some of the intercorrelation that may exist among the dimensions characterizing such localized structures, may be the resultant of common morphogenetic fields or growth gradients occurring during the process of ontogeny. This may partially obscure such adaptive, phylogenetic or taxonomic information as may be inherent in the data.

One way of circumventing this problem to some or other degree is to utilize, as the object of study, a considerably larger portion of the anatomy of the animal. Thus, in addition to a set of dimensions of the hip and shoulder girdles, it is possible to examine a series of overall measurements of (say) the lower limb. It would seem self-evident that whatever else such a series of measurements is related to, it surely reflects the propensities of the impulsively operating lower limb in any group of animals thought to be vertical clingers and leapers. Again one can also analyse a suite of dimensions relating to the upper limbs; although presumably less important in vertical clinging and leaping, the upper limb does indeed participate and should reflect, at least in part, concomitant adaptations.

Yet another way of examining this morphological problem is to attempt to synthesize the information in a full range of measurements of the whole animal. Of course this is a summary of the morphological correlates of its entire life processes. However, if the mode of locomotion differs sufficiently from that of more orthodox species (such as is postulated for vertical clinging and leaping as compared with more generalized quadrupedalism) then information about the eccentric locomotor mode should nevertheless shine through. A study of this last type is also able to provide a final test of the reality of a single morphological adaptation to vertical clinging and leaping. For although animals displaying this adaptation may differ in terms of the detailed anatomy of particular parts (differences that may be related to small variations in the parallel pathways actually undergone during evolution) their similarities over all should be apparent if truly a single adaptive mode exists.

These latter studies (analyses of dimensions of each limb, and of overall bodily dimensions) have here been undertaken on the basis of the data from the extensive and careful studies of Professor A. H. Schultz. Through his kind cooperation, and again through collaboration between the laboratories in Birmingham and Chicago, the data have been made available so that analyses enabling them to be viewed as a whole have been carried out in an attempt to assess the morphological reality of "vertical clinging and leaping".

THE PELVIC GIRDLE

In the case of the pelvic girdle, examinations of a series of nine dimensions of the innominate bone have been carried out and are reported earlier in this symposium. In that report the data are primarily used to investigate the structure of the pelvis in the higher primates, especially the great apes and man, with a study of the Sterkfontein innominate bone in mind. But, as presented in that communication, there is considerable evidence that in other groups of primates the form of this bony complex is related, to a not inconsiderable degree, to the functions of the part in locomotion. Re-analysis of a subset of these data has been carried out here in order to reveal relationships in the structure of the pelvis (a) within groups of animals that appear to be relatively generalized quadrupedal forms (e.g., among New World monkeys, species such as *Saimiri*, *Callicebus*, *Aotus* and *Leontocebus*, and among the Prosimii, genera such as *Tupaia* and *Lemur*, Napier & Walker, 1967), (b) within the group of animals that are believed by Napier & Walker (1967) to be unequivocal vertical clinging and leaping forms (e.g. *Avahi*, *Galago*, *Indri*, *Propithecus* and *Tarsius*) and (c) including species which, to an intermediate or incipient degree (Stern & Oxnard, 1973), may utilize vertical clinging and leaping within their locomotor patterns more than the first of the above but decidedly less than the second (e.g., *Microcebus* and *Lemur catta*, possibly *Cheirogaleus*, and even *Phaner*).

These data consist of a series of measurements designed to provide information about relative positions of muscular attachments to the bone and about the relative disposition and orientation of joint surfaces. The features are defined in Zuckerman, Ashton, Flinn, Oxnard & Spence (this volume, pp. 71–165) and comprise dimensions such as the relative caudal extent of the ischium, a measure possibly related to the length of the lever arm of the hamstring muscle block and the position of the iliac blade relative to the acetabulum, a measure possibly related to the position of the lesser gluteal muscles. Other dimensions include such features as the dorso-ventral positions of the auricular facet and the acetabular cavity, together with their relative cranio–caudal separations. In all a total of nine such features were examined in 41 primate genera for most of which reasonable samples were available.

The relationships of the various species to one another as defined by these nine dimensions were obtained through the computation of Mahalanobis generalized distances. As the matrix of these distances actually exists within the nine-dimensional space of the overall analysis, the presentation has been simplified by displaying those connections between genera that are of shortest length as they exist within the bivariate plot (e.g. Fig. 1) of the first two canonical axes (containing some 90% of the total information). In such a method of display, the relative arrangement of the various genera is provided and the distances between connected genera are correct representations of their propinquity to one another. The distances of unconnected genera from one another are incorrect because of the impossibility of a multi-dimensional representation. However, the use of the first two canonical variates as the reference plane for such a view ensures that only a small amount of information is missing in the pictorial display. Of course, no information is lost in the full matrix of generalized distances.

Figure 1 shows that there is a central tightly-knit grouping of animals that include *Lemur*, *Chiropotes*, *Callicebus*, *Lepilemur*, *Aotus*, *Leontocebus*, *Callithrix* and *Saimiri*. All of these may be defined, in terms of their overall locomotor patterns, as generalized quadrupedal primates capable of running, jumping and climbing in the trees but in no way specialized for particular activities as are some other primate species. Figure 1 also shows that the prehensile and semi-prehensile tailed cebids and the slow-clinging lorisines, each in their own way uniquely different from the previous more generalized quadrupedal group of primates, are also located in their own special places within the nine-dimensional space of the analysis. But the figure throws into clear relief the marked difference between *Propithecus* and *Indri* on the one hand and *Galago* and *Euoticus* on the other (*Avahi* and *Tarsius* were not available for this investigation). Therefore, at least in the structure of the pelvis, the galagine vertical clingers and leapers differ completely from the indriid vertical clingers and leapers, and both are closer to the more generalized quadrupedal species than either is to the other.

It is also of interest that, of these various species that form the major bulk of the more quadrupedal forms, one genus, *Cheirogaleus*, lies at that edge closest to the galagines (although still very far from them), and two genera (*Lemur* and *Hapalemur*) are at the opposite edge closest to the indriids again nonetheless distant from this last very outlying group.

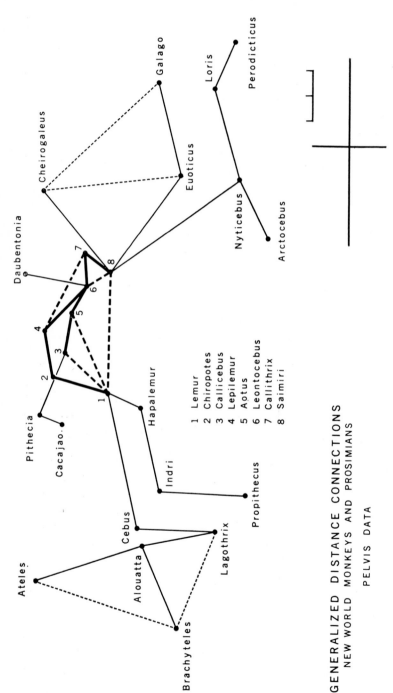

GENERALIZED DISTANCE CONNECTIONS
NEW WORLD MONKEYS AND PROSIMIANS

PELVIS DATA

1 Lemur
2 Chiropotes
3 Callicebus
4 Lepilemur
5 Aotus
6 Leontocebus
7 Callithrix
8 Saimiri

Fig. 1. Analysis of nine dimensions of the pelvis. The connections between various genera are those minimum generalized distances (in distance units, D, as shown by the marker), plotted within the set of general directions provided by the graph of the first two canonical axes. In cases where the minimum distances are almost the same as the next nearest distances, these latter links have been shown as dotted lines. Because the generalized distances (multi-dimensional) are compressed into the (two-dimensional) canonical plot, the only distances on the graph that are correct are the solid lines.

THE SHOULDER GIRDLE

In the case of the shoulder girdle, examination of a total of 17 dimensions of the scapula, clavicle and humerus (upper end) in a total of 39 genera of primates has revealed that its morphology is very closely related to the functions that are imposed upon the girdle during the movements of which the various primate species are capable (Oxnard, 1967, 1973). Some of these 17 dimensions had already been chosen as being related possibly to locomotion through including measures of such features as the angulation of insertion of the trapezius muscle and the caudal prolongation of insertion of the serratus magnus muscle (Ashton & Oxnard, 1964). Others of these dimensions are less likely to be related directly to the biomechanics of locomotion and are more measures of the overall shape of the bone, e.g., one dimension specifying the shape of the acromion and another the form of the superior scapular border (Ashton, Oxnard & Spence, 1965). These features have all been defined in a detailed manner by Ashton, Flinn, Oxnard & Spence (1971). Previously, these data have been primarily used to investigate the shoulder structure in the Anthropoidea, the results referring to the Prosimii being mainly corroborative of certain morphological and functional trends that seem to exist in parallel in the two suborders. But re-analysis of these data has been carried out here in order to reveal the relationships in the structure of the shoulder within the groups of species representing relatively generalized quadrupedal animals and those characterized by Napier & Walker (1967) as vertical clinging and leaping forms. The information has been displayed in the same way as for the study of the pelvic girdle: the closest generic relationships as given by the Mahalanobis generalized distances are plotted within the plane of the first two canonical axes. The latter contain 85% of the information and therefore, although the only distances in the diagram that are accurate are those of genera joined by their minimum connections, other distances are not grossly incorrect.

Figure 2 demonstrates these results and shows again that the quadrupedal forms lie in a central close-knit position and that they comprise the genera *Cebus*, *Callicebus*, *Callithrix*, *Saimiri*, *Aotus*, *Hapalemur*, *Lemur*, *Lepilemur* and *Leontocebus*. As in the hip, so in the shoulder, the markedly modified prehensile-tailed cebids and the slow-clinging lorisines lie in unique outlying positions from the generalized quadrupedal species. In addition however, though to a less marked extent than in the hip, those genera that are included as vertical clingers and leapers lie at two separate peripheral locations almost diametrically opposite to one another: indriids versus galagines. We cannot say where *Avahi*, *Indri*

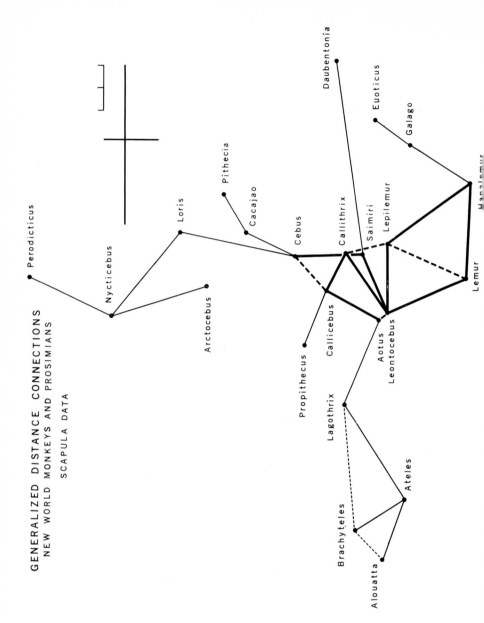

Fig. 2. Analysis of 17 dimensions of the scapula. The connections between various genera are those minimum generalized distances (in distance units, D, as shown by the marker) plotted within the set of general directions provided by the graph of the first two canonical axes. In cases where the minimum distances are almost the same as the next nearest distances, these latter links have also been shown as dotted lines. Because the generalized distances (multi-dimensional) are being compressed into the (two-dimensional) canonical plot, the only distances on the graph that are correct are the solid lines.

or *Tarsius* might have fallen if they had been included in the analysis, although the implication must be that each would have fallen near its appropriate fellow. This study of the shoulder girdle, though a little weak due to unavoidably missing data, nevertheless confirms the general picture provided by the hip analysis.

THE LOWER LIMB

Similar analyses have also been carried out on a suite of measurements related to the overall dimensions of the lower limb. Here the measures are more broadly representative of the structure of the lower limb and its attachment to the trunk. They have not been chosen as having particular biomechanical importance, but presumably characterize the relative shape of this part. They comprise seven dimensions: the relative lower limb length, the crural index, intermembral index, relative foot breadth, foot length relative to lower limb length, and relative hip breadth and have been taken on a total of 57 primate genera. Although consisting of a limited number of variables, these data contain information about a rather larger number of measurements of primate lower limbs, and include also trunk length and simple upper limb length. However, they have been compounded into ratios by Professor A. H. Schultz and the original measurements are no longer available. The analysis that is carried out here comprises once more a determination of Mahalanobis generalized distances and these are plotted within the space of the first three multiple discriminant functions. It is necessary in this case to use the first three functions because the effective dimensionality of the data is truly higher than in the previous examples, three functions containing more than 80% of the information. Accordingly, a three-dimensional model of the generalized distance connections may be constructed and Fig. 3 is a photograph of that model.

Figure 3 shows, to a degree more marked than any of the previous examples, that there is a grouping of genera that comprises *Aotus*, *Saimiri*, *Leontocebus*, *Cebus*, *Tupaia*, *Lemur* and *Microcebus*. These genera, which may well be thought of as relatively generalized arboreal running, leaping, climbing quadrupedal forms, all fall together as a central nucleus. Emanating from this nucleus is a series of arms somewhat resembling a star, and each arm includes separate clusters of genera that resemble one another in different ways. One such cluster includes the slow-clinging lorisines (unlabelled) which have unique patterns of movement. Another includes *Ateles*, *Lagothrix* and *Alouatta*, the prehensile-tailed cebids, again with patterns of movement that are

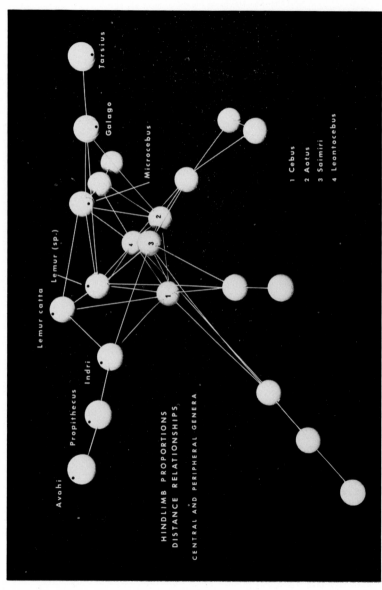

Fig. 3. Analysis of the seven dimensions of the hindlimb. The connections between the various genera are those generalized distances (in distance units, D, between neighbouring genera) forced into a three-dimensional model. The general scale of the model is 25 distance units but because of the perspective effect, a marker cannot be provided. The distances between genera that are not shown by connections are not necessarily correct, although it is remarkable how closely the three-dimensional model does represent the actual multi-dimensional generalized distances.

unique among these primates (also unlabelled). The vertical clinging and leaping genera, however, are separated into two widely divergent arms; one of these contains the genera *Propithecus*, *Avahi* and *Indri* and although a long distance from the central group, this cluster finds its closest link with the generalized arboreal quadrupedal forms through *Lemur catta* and next *Lemur* species. The second arm of the vertical clinging and leaping forms is placed in an almost directly opposite direction from the first and contains the genera *Galago* and *Tarsius*, and again, though a long way from the central group of generalized quadrupedal species (the perspective of the model appears to shorten this distance somewhat), these two genera find their nearest linkage through the genus *Microcebus*. (It is important to realize that, although a measure of tarsal length is included in these dimensions, the separations achieved do not depend upon this feature alone; indeed, they do not even depend largely upon this character; the distinctions that emerge from these multivariate studies depend upon *all* the original parameters, never simply a select few.)

THE UPPER LIMB

Although a study of the hindlimb relates more closely to the functional elements of vertical clinging and leaping, the results of a study of a variety of forelimb dimensions are also presented. Here the measures included relative upper limb length, brachial index, intermembral index, relative hand length, relative hand breadth, relative thumb length, relative shoulder breadth, relative chest circumference and chest index. Again ratios comprising more than nine measurements are used and again include trunk length and in one ratio lower limb length, because the original direct measurements are not available. And once more, this total of nine variables representing each specimen is studied by means of multiple discriminant function analysis and Mahalanobis generalized distances. A similar attempt is made to characterize the various genera by means of the shortest generalized distance links among them displayed within the space of the first three functions (containing more than 80% of information). Again, therefore, a three-dimensional model may be built and a photograph is shown in Fig. 4.

This model displays several interesting features. As in all the other studies there is a tangle of forms that all fall very close together; and as before, it turns out that these are the genera *Saimiri*, *Cebus*, *Aotus*, *Leontocebus*, *Tupaia*, *Microcebus* and *Lemur* forming the group of arboreal, running, leaping and climbing quadrupedal animals. As with the study of hindlimb dimensions, the group of prehensile–tailed cebids

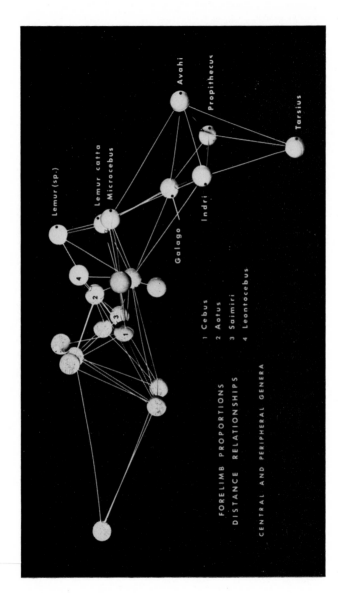

Fig. 4. Analysis of nine dimensions of the forelimb. The connections between various genera are those generalized distances (in distance units, D, between neighbouring genera) forced into a three-dimensional model. The general scale of the model is 30 distance units from side to side but because of the perspective effect, a marker cannot be provided. The distances between genera that are not shown by connections are not necessarily correct, although it is remarkable how closely the three-dimensional model does represent the actual multi-dimensional generalized distances. In this case perusal of some of the generalized distances that cannot be represented in a three-dimensional model show that the true relationships among the vertical clingers require further study for their elucidation.

and the group of slow-moving lorisines (both unlabelled) occupy their own independent parts on the model. With regard to the various vertical clinging and leaping genera, however, the immediate impression is that *Propithecus*, *Indri*, *Avahi*, *Tarsius* and *Galago* all fall together as a "tail" emanating from the more quadrupedal species. This is the first analysis in these investigations where it appears that there is any special affinity between the indriids on the one hand and tarsioids and galagines on the other.

It is also of interest that the genera at the "root of the tail" where it joins the general "corpus" of quadrupedal forms, are precisely *Microcebus* and *Lemur catta*, two animals representative of those species also so identified in the study of the other anatomical regions.

This particular analysis, if carried no further, suggests that there really is a morphological entity, in terms of the structure of the forelimb at least, that parallels the putative behaviour of vertical clinging and leaping as defined by Napier & Walker (1967). If this is truly so, however, it is not shown by the hindlimb as a whole, nor by the studies of the two limb girdles. *Nor, as we shall see later, is this a correct interpretation of the forelimb data just presented; some information is contained in higher dimensions not considered by this technique.*

THE MULTI-DIMENSIONAL PROBLEM

These various studies of the different anatomical elements have been obtained through the use of multivariate statistical methods applied to mensurational data organized in the form of measurements or ratios, and with logarithmic and other transformations (e.g., regression adjustment) superimposed when necessary in order (a) to render more equal the variances and covariances of the different groups and (b) to take some account of the differing absolute sizes of the individual animals and species. That part of the information that is represented by the generalized distances (connected genera in the diagrams) contains information from the full multi-dimensionality of the data space; but, as only the shortest distance connections are shown, these do not (and cannot) contain all the information. The other elements of these analyses are the canonical axes and multiple discriminant functions. Here, though most of the information is contained within the early axes and functions, some is confined within later axes and functions, and this latter part of the result is not displayed. In the case of the scapular and pelvic analyses, examination of the minimum spanning tree and single linkage cluster analysis of the total generalized distance matrices shows that the missing information does not materially change the picture;

these additional analyses merely confirm and slightly augment what has already been shown by the above generalized distances plotted against the first two canonical axes. But in the case of the more complicated data of overall dimensions of the fore- and hind limbs, examination of the totality of the generalized distances and the later discriminant functions not included in the figures shows that significant information is indeed being lost.

Accordingly we have undertaken further studies utilizing a technique recently described by Andrews (1972) for displaying, in graphical form, information contained within a data space of dimensionality greater than three.

Although we are unaccustomed to imagining multi-dimensional objects, mathematicians quite regularly examine plots and functions that are multi- or even infinite-dimensional. On this basis Andrews (1972) suggests the possibility of examining multi-dimensional data by embedding them within a high-dimensional space of functions and then visualizing them by plotting the functions in the usual manner. Such a procedure does not, of course, actually allow us to see the multi-dimensional structure of our data, but it does allow us to view the groups and clusters that may be contained within the data, and to locate intermediate members and outliers that may be present. Thus, although the geometric representation that can be seen in bivariate plots or three-dimensional models of canonical variates or discriminant functions is hidden, the structure, in terms of the propinquity of groups one to another, can be most clearly revealed without any loss of information.

This can be achieved by mapping each dimension (x) of a k-dimensional figure $[x = (x_1, x_2, x_3, \ldots x_k)]$ into a function of the form

$$f_x(t) = x_1/\sqrt{2} + x_2 \cdot \sin t + x_3 \cdot \cos t + x_4 \cdot \sin 2t + x_5 \cdot \cos 2t + \ldots$$

the function then being plotted over the interval $-\pi < t < +\pi$. In the use of this method, the dimensions $x_1 \ldots x_k$ must be uncorrelated so that it is not possible to use original data. But such a requirement means that the method is tailor-made for examining canonical variates or discriminant functions or any other product of a multivariate transformation as long as these are uncorrelated. Thus for our purposes, the position of each genus of animal within a given analysis (e.g., shoulder, pelvis, upper or lower limb) may be plotted by using its mean score on each canonical or discriminant axis as the x_1's, x_2's, x_3's $\ldots x_k$'s in the above equation. Such a representation preserves means, distances and variances. It yields one-dimensional projections that may show clusterings, intermediates and outliers. The continuum of such one-dimensional representations or projections for each group is plotted as a curve from $-\pi$ to $+\pi$ so as to give the full relationship between the groups (Fig. 5).

This method of display can also be easily understood by examining the way in which alterations in only a few "dimensions" may affect the picture that it provides. Thus the first frame of Fig. 6 shows the effect, upon the multi-dimensional representation of a hypothetical group of animals in a hypothetical analysis, of changing only the value for the first canonical axis (i.e., of x_1 in the plot of the function). As can also be readily judged from the mathematical form of the function, such an

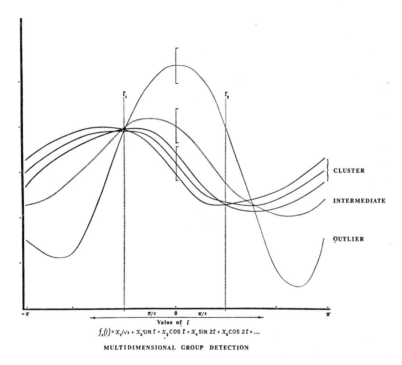

$$f_x(t) = x_1/\sqrt{2} + x_2 \sin t + x_3 \cos t + x_4 \sin 2t + x_5 \cos 2t + \ldots$$

MULTIDIMENSIONAL GROUP DETECTION

FIG. 5. Explanation of multi-dimensional group display by the method of Andrews (1972). A theoretical set of functions is plotted for a number of dimensions x_1 to x_k (representing canonical variates for the mean positions of five groups of objects) for values of t from $-\pi$ to $+\pi$. A cluster of means that are similar is shown by a similar set of neighbouring wavy lines on the plot. A mean which is widely outlying from the cluster is shown by a curved line widely different on the plot. An intermediate mean is shown lying in an intermediate position. The three markers show that standard deviations are conserved in this method of display. The members of the cluster fall together within a standard deviation of one another; the outlier and the intermediate are appropriately arranged in relation to the standard deviation markers. Particular values of t may be of especial interest. Thus t_1 shows that combination of the original variables in which all of the means are indistinguishable. t_2 shows a particular combination of the original variables which completely distinguishes the three means.

N

alteration does not change the shape of the curve but moves it vertically up or down the plot. The second frame of Fig. 6 shows that a change in a second canonical variate (x_2 in the plot) has the effect of moving one

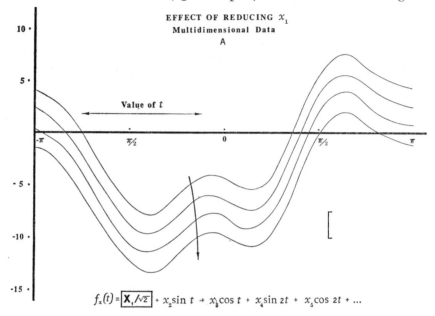

$$f_x(t) = \boxed{X_1/\sqrt{2}} + x_2\sin t + x_3\cos t + x_4\sin 2t + x_5\cos 2t + \dots$$

Fig. 6a

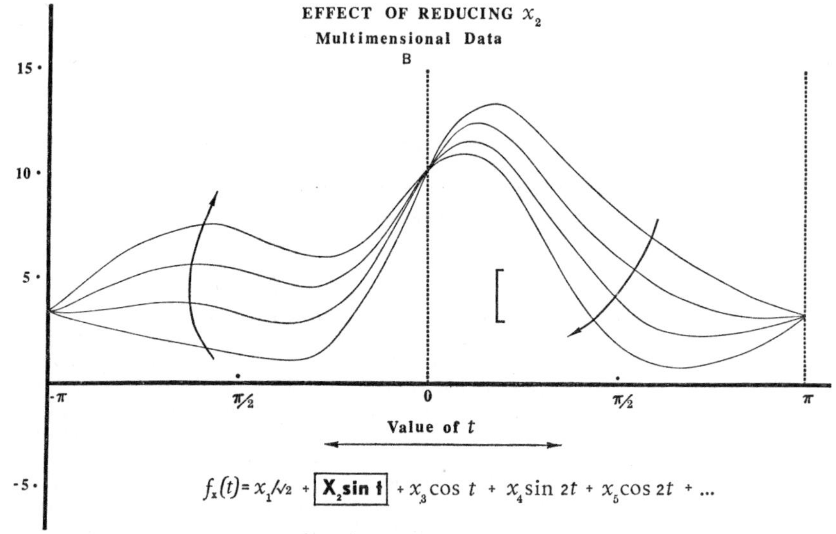

$$f_x(t) = x_1/\sqrt{2} + \boxed{X_2\sin t} + x_3\cos t + x_4\sin 2t + x_5\cos 2t + \dots$$

Fig. 6b

half of the function in one vertical direction on the page and the other half in the opposite direction. The third frame of Fig. 6 shows how a change in a third canonical variate (x_3 of the function) has again a predictable effect upon the plot, moving the first quarter in one direction, the middle two quarters in the reverse direction, and the last quarter in the obverse direction again. In a similar manner, changes in the various other canonical variates affect the position and shape of the plots, and

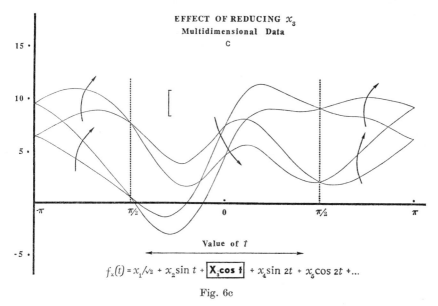

Fig. 6c

FIG. 6. This figure shows three more hypothetical sets of multivariate data represented by x_1's to x_k's plotted in the manner of Andrews (1972) from $-\pi$ to $+\pi$. Figure 6a (see facing page) shows the effects on such a plot of changing values in x_1 (the first canonical variate) alone. Figures 6b (see facing page) and 6c show similarly the effect of changing values of x_2 and x_3 (second and third canonical variates) alone. Similar perturbations are related to higher canonical variates.

it is in this way that a series of similar many-dimensional objects will appear as similarly curved plots, while a different many-dimensional object will appear as a distinctly different curved line (Fig. 5).

This technique has been applied to the analyses of pelvic and scapular dimensions already described in this study. But in those investigations the effective dimensionality of the data is truly expressed in the first two canonical axes so that little that is new is revealed. However, in the case of the more complicated studies of the hind- and fore-limbs, it is clear that considerable significant information remains in the higher dimensions; this new technique is able to display it.

Thus when applied to the multivariate analysis of the hindlimb dimensions, it becomes very obvious indeed (a) that the grouping of central genera comprising generalized, arboreal, running, leaping, climbing quadrupeds, truly exists, (b) that the genera *Propithecus*, *Indri* and *Avahi* genuinely form a cluster outlying from the first and (c) that the genera *Tarsius* and *Galago* (hardly definable as a group because there are only two) nevertheless share a pattern that differs widely both from the group of generalized forms and from the cluster of Madagascan species (Fig. 7).

This confirmation of the picture provided by the generalized distance analysis and the multiple discriminant functions is dramatically focused by the consideration of those species which the previous analysis tentatively suggests are intermediate. Thus Fig. 7 demonstrates unequivocally that *Lemur* species and especially *Lemur catta* depart from the generalized pattern in ways that fit the peculiar curvatures demonstrated by the plot for the group, *Propithecus*, *Indri* and *Avahi*. In like manner, Fig. 7 shows that *Microcebus* departs from the structure that

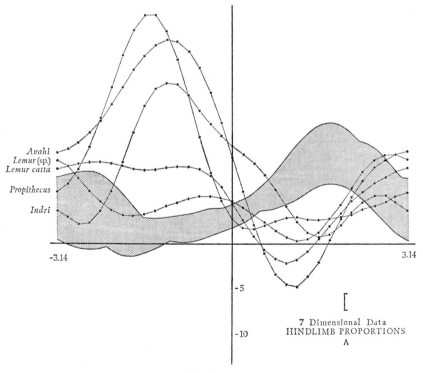

Fig. 7a

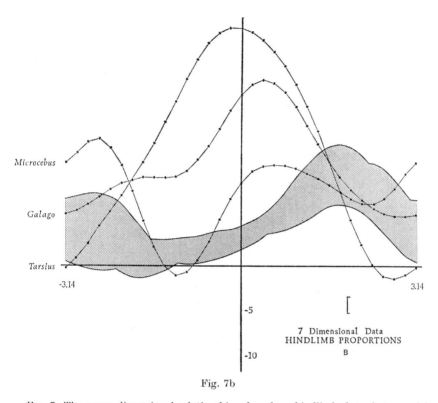

Fig. 7b

Fig. 7. The seven-dimensional relationships, based on hindlimb data, between (a) generalized running, leaping and climbing quadrupedal genera and (b) the various vertical clinging and leaping genera. The functions (derived as explained in the text) are plotted for values of t along the horizontal axis from $-\pi$ to $+\pi$ so that the multidimensional position of a single generic mean can be represented by a wavy line. The vertical distance of a point on such a line from a line representing another given genus is related to the statistical separation, in all seven dimensions, of the two genera. An indication of the degree of this is provided by the marker which is a single standard deviation unit in vertical extent. The first plot (Fig. 7a) shows curves typical of the various indriid genera as compared with the envelope (shaded area) formed by the plots of the generalized quadrupedal genera. *Lemur* (sp.) and *Lemur catta* can be seen to deviate from the generalized quadrupedal pattern in the direction of the indriid mode but to a minor extent at each value of t. Thus these two genera are genuinely intermediate between the indriids and the quadrupedal forms in the full seven-dimensional space of the analysis. The second plot (Fig. 7b) shows curves typical of *Galago-Tarsius* as compared with the envelope (shaded area) formed by the plots of the generalized quadrupedal genera. *Microcebus* can be seen to deviate from the generalized quadrupedal pattern in the direction of the *Galago-Tarsius* mode but to a minor extent at each value of t. Thus this genus is genuinely intermediate between *Galago-Tarsius* and the quadrupedal forms in the full seven-dimensional space of the analysis. The special way in which the indriid genera deviate from the envelope of quadrupedal forms differs completely from the manner in which the *Galago-Tarsius* spectrum deviates from the envelope. A similar remark applies to the positions of the intermediate genera.

defines the generalized series of genera and precisely in those ways that move it towards but not identical to the form of the curvatures of the *Tarsius-Galago* spectrum. These demonstrations do not mean that the picture provided by the previous more standard analyses is wrong. Rather do they mean that the complexity of the previous analyses was too great to demonstrate using a three-dimensional model; information of considerable importance to these separations is contained in multivariate axes higher than three and has been neatly demonstrated by this multi-dimensional display.

It is of especial interest to study the result of applying this same technique to the investigation of the overall proportions of the forelimb. Here, it will be remembered, the generalized distance analysis and multiple discriminant functions, though separating all vertical clinging and leaping forms from the generalized quadrupedal species, do not effect any marked separation among the more specialized animals. In this sense, then, it superficially appears to support the idea of vertical clinging and leaping as a single morphological entity. That this is truly spurious, as was hinted earlier, is clearly seen when the method of Andrews is applied. Thus Fig. 8 shows the results and they demonstrate

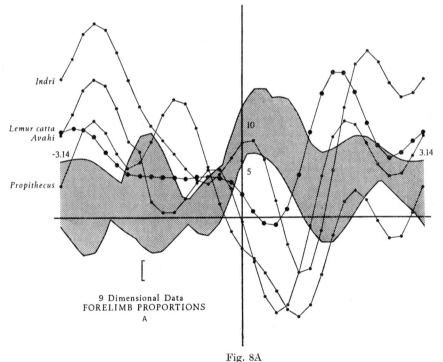

9 Dimensional Data
FORELIMB PROPORTIONS
A

Fig. 8A

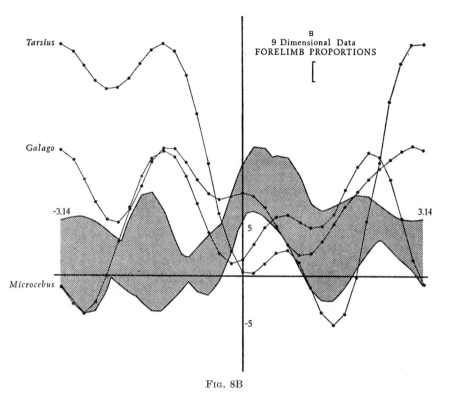

FIG. 8B

FIG. 8. The nine-dimensional relationships, based on forelimb data, between (A) the generalized running, leaping and climbing quadrupedal genera and (B) the various vertical clinging and leaping genera. The functions (derived as explained in the text) are plotted for values of t along the horizontal axis from $-\pi$ to $+\pi$ so that the multi-dimensional position of a single generic mean can be represented by a wavy line. The vertical distance of a point on such a line from a line representing another given genus is related to the statistical separation, in all nine dimensions, of the two genera. An indication of the degree of this is provided by the marker which is a single standard deviation unit in vertical extent. The first plot (Fig. 8a) shows curves typical of the various indriid genera as compared with the envelope (shaded area) formed by the plots of the generalized quadrupedal genera. *Lemur catta* can be seen to deviate from the generalized quadrupedal pattern in the direction of the indriid mode but to a minor extent at each value of t. Thus this genus is genuinely intermediate between the indriids and the quadrupedal forms in the full nine-dimensional space of the analysis. The second plot (Fig. 8b) shows curves typical of *Galago-Tarsius* as compared with the envelope (shaded area) formed by the plots of the generalized quadrupedal genera. *Microcebus* can be seen to deviate from the generalized quadrupedal pattern in the direction of the *Galago-Tarsus* mode but to a minor extent at each value of t. Thus this genus is genuinely intermediate between *Galago-Tarsius* and the quadrupedal forms in the full nine-dimensional space of the analysis. The special way in which the indriid genera differ from the envelope of quadrupedal forms differs completely from the manner in which the *Galago-Tarsius* spectrum deviates from the envelope. A similar remark applies to the positions of the intermediate genera.

clearly that when higher dimensional information is taken into account, there can easily be discerned, in addition to the close-knit group of generalized, arboreal, running, leaping and climbing quadrupedal species, two additional groups. One of these comprises *Indri, Propithecus* and *Avahi* and it is noteworthy that *Lemur catta* tends toward the particular pattern displayed by these genera; the other of these groups contains the genera *Tarsius* and *Galago*, and again as a special case, *Microcebus* tends towards the particular pattern displayed by them. It is thus in the case of the study of the forelimb dimensions that Andrews' technique is of especial importance, revealing information that is almost completely hidden if one has only a partial view of the analysis obtained by studying the earlier discriminant axes alone.

OVERALL BODILY FORM

As the final test of these concepts, an analysis is available carried out on the entire range of dimensions taken by Professor Schultz on these species. These comprise, in addition to the dimensions of the fore- and hind-limbs already studied, a series of measurements of the trunk, of the overall proportions of the head (for example width relative to height), and of detailed proportions of particular structures of the head (for instance relating to the ears and nose)—a total of 23 dimensions. When these are examined for this particular group of species by means of generalized distances and multiple discriminant analysis, it is immediately obvious that so many individual discriminant axes are providing information of importance about the relative grouping of the animals, that any two- or three-dimensional representation is not very helpful. In this case, the dimensionality of the data is so high that it is worth passing directly to the plots of functions provided by substituting, in the sine-cosine equation, the values of each discriminant function for the x's from the first ($x = x_1$) to the last ($x = x_{23}$). The result is shown in Fig. 9. Even here, where it may well be thought that other facets of the life processes of these individuals may interfere with the locomotor adaptation, a separation of the various generalized arboreal, running, leaping and climbing quadrupedal forms into a single rather closely-knit group is evident, although not figured, as is also the distinction from them, each in its separate pattern, of the indriid genera on the one hand, and the complex of tarsioid and galagine genera on the other. Here also, the genus *Lemur*, especially *Lemur catta*, tends towards the indriid pattern although not so extreme; the genus *Microcebus* evinces similarities with the *Galago-Tarsius* spectrum, again being rather less extreme than *Galago*, the less modified of the two.

Of course, the presence of these trends in the total bodily form and proportions of these species does not necessarily mean that locomotion is an over-riding factor in their adaptation. While this is certainly one explanation, it is also always possible that adaptations to other functional modes may be influencing the results; however, if this is so, then the other adaptations must in general be related to morphological changes that are in the same direction as the locomotor ones or that in some other way help to introduce similar separations among these particular primate genera.

CONCLUSIONS FROM MORPHOLOGY

This information suggesting that the morphological adaptations to locomotion of these various prosimian genera do not indicate an adaptation to a unique locomotor mode but rather to at least two types of locomotion, is derived from a number of quite separate and independent investigations and is therefore very strong. These contrasting modes are represented among living forms by the indriids on the one hand, and the galagine-tarsioids on the other.

In addition, the morphological data suggest, also strongly, that a number of other prosimian forms, hitherto thought to be primarily adapted for a rather generalized arboreal, quadrupedal mode of movement, are in fact morphologically adapted to a considerable extent along the way towards each of the two possible extant functional patterns. *Lemur catta* is strongly implicated in relation to the indriid mode and *Microcebus* similarly for the galagine–tarsioid spectrum.

Of course, a number of prosimian genera have not been examined in most of these studies because of the limitation of material. It would be of considerable interest to obtain full data from genera such as *Hapalemur, Lepilemur, Cheirogaleus, Phaner* and more tupaioid genera to see to what extent they too might be placed in intermediate positions within the matrix of results.

The morphological information that has here been uncovered suggests that, even if the behavioural information that is currently available points to a single category of vertical clinging and leaping, we should be searching most diligently for indications of behavioural differences among the many species. It would appear that these are most likely to be found among the most extreme genera *Propithecus, Indri* and *Avahi* on the one hand, and *Tarsius* and *Galago* on the other. In addition, the morphological data suggest that as behaviourists, we might also pay special attention to the incipiently modified forms, for example *Lemur catta* on the one hand and *Microcebus* on the other.

CHARLES E. OXNARD

Pending new field studies yet to be undertaken, we may ask if the current literature supplies any indications at all, overt or subtle, that the behavioural situation might possibly parallel the morphological complexities.

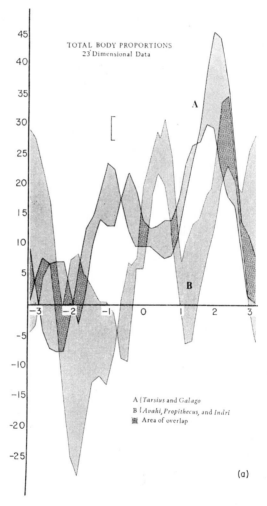

Fig. 9. Using all of the conventions of the previous figures, this figure shows the multi-dimensional relationships between various genera for the 23-dimensional study of the total body proportions of these primates. Figure 9a shows the marked difference between the two types of vertical clinging and leaping form. The generalized quadrupedal forms (not figured) fall in a simple belt straight across the figure and thus differ from both. Figure 9b shows how closely *Lemur catta* and *Lemur mongoz* approach the position of the indriids. Figure 9c shows how closely *Microcebus* approaches *Tarsius* and *Galago*. All of these comparisons cut completely across taxonomic boundaries.

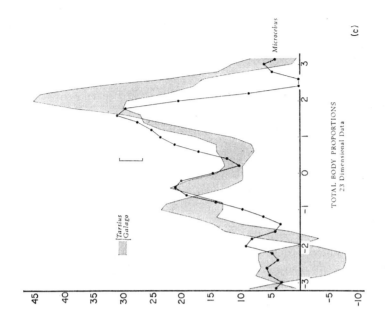

(c)

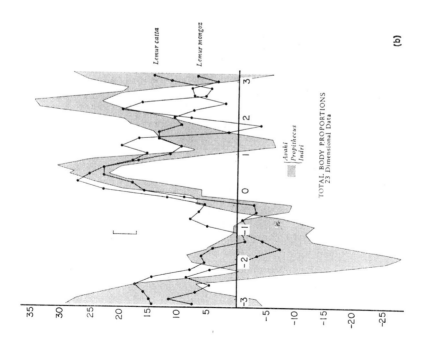

(b)

TENTATIVE BEHAVIOURAL CORRELATES

A review of the behaviour of these species has been presented by Stern & Oxnard (1973). There can be little doubt, as shown by the extensive study of Petter (1962) in conjunction with the earlier observations of Rand (1935) and Attenborough (1961) together with the more recent work of Charles-Dominique & Hladik (1971) that vertical clinging and leaping is a major mode of locomotion both among the indriid genera and in *Lepilemur*. There is, moreover, no question but that they are behaviourally distinct from the more quadrupedal Madagascan lemurs even though a number of workers have pointed to facets of the locomotion of indriids and *Lepilemur* in which they display a wide variety of methods of progression in different circumstances (e.g., Attenborough, 1961; Petter, 1962; and especially Jolly, 1966).

The inclusion of *Tarsius* and *Galago* in the group of vertical clingers and leapers by Napier & Walker (1967) is not based on as detailed wildlife observations as exist for the Madagascan primates. Early studies of Hoogstraal (1947), Schmidt (1947), Wharton (1948) and Davis (1962) indicate the frequent utilization of vertical postures and movement by *Tarsius* in its natural habitat. Observations of captive animals (Le Gros Clark, 1924; Harrison, 1963; Napier & Walker, 1967; Grand & Lorenz, 1968) confirm and extend the above studies. Napier & Walker suggest that *Galago*, while as specialized morphologically as the tarsiers, is rather more generalized behaviourally. It is only recently that actual wildlife observations depicting the locomotor repertoire of these creatures have been published. Charles-Dominique (1971) provides excellent descriptions of the behaviour of *Galago alleni*, *G. demidovii*, and *Euoticus elegantulus* in Gabon. Of these, *G. demidovii* is the only species that, under the conditions observed by Charles-Dominique, shows a marked preference for vertical supports. The remaining species do not show such a preference, *G. alleni* being seen more frequently upon horizontal branches, and *Euoticus elegantulus* showing a marked choice of horizontal or oblique branches of large diameter. Observations of other species of *Galago* are scarce but do not controvert the above reports.

However, although the foregoing suggest that *Tarsius* and *Galago* are relatively similar to the indriids and *Lepilemur*, other information points to distinct differences. Thus the anatomical specializations of the tail of *Tarsius* for supporting the animal while it clings to a vertical stem though certainly a strong indication of this primate's natural behaviour, also draws attention to differences from the indriids and *Lepilemur*. Thus while *Tarsius* rests in the manner indicated and *Galago* also rests in a clinging position, the indriids and *Lepilemur* most frequently rest by sitting in the fork of two branches using the more horizontal to bear the weight and grasping the more vertical with the feet (Sprankel, 1965; Grand & Lorenz, 1968).

Perhaps of more importance is the fact that photographs of captive tarsiers leaping to vertical supports show that the animals are curled and upright throughout the flight, rather than horizontal and stretched out as is reported for indriids (Walker, 1948; Wharton, 1948). Similar reports exist for galagines, the observations of Charles-Dominique (1971) stating that although the body of *G. demidovii* is held horizontally at the beginning of a long leap, it rapidly becomes upright. *G. alleni* and *Euoticus elegantulus* hold the body entirely upright and somewhat flexed during the flight phase of their leaps.

Differences also seem to exist in relation to additional locomotor abilities displayed by these animals. Thus *Tarsius* is not, of course, constrained to vertical progression, at least in captivity, but such typical quadrupedal walking and climbing behaviours as it engages in (Hill, Porter & Southwick, 1952; Grand & Lorenz, 1968) do not encompass the variety, range, power and agility of the additional movements reported for the indriids. In the galagines, also, although a number of other movements have been noted (e.g., "bush-babies can climb in any position that the substrate requires, for example upside-down on a horizontal branch, upwards and downwards on vertical branches" Sauer & Sauer, 1963), there is no evidence of the extremely acrobatic behaviours observed in the indriids. For this latter group a wide variety of other acrobatic behaviours such as are involved in climbing vertical lianes and feeding and are characterized by movements such as suspension by all four limbs, by fore- or hind-limbs alone, and even on occasion by a degree of movement using pendulum-like arm suspension have all been reported as part of the wide variety of other movements carried out by indriids (e.g., Attenborough, 1961; Jolly, 1966; Petter, 1962).

There is thus some tentative evidence that the locomotor pattern that characterizes the genera *Propithecus*, *Indri* and *Avahi* may differ in terms of the nature of the primary leaping habit, the form of branch resting, and extent and kind of additional locomotor behaviours, from the overall type of movement that is found in *Tarsius*, *Galago* and *Euoticus*.

Napier & Walker (1967) also classify two other genera as vertical clingers and leapers. First, they include *Lepilemur*, and a number of already quoted sources, and especially Charles-Dominique & Hladik (1971), confirm that this is so. Secondly, Napier & Walker include *Hapalemur* (*simus*) as a tentative vertical clinger and leaper, apparently mainly on the basis that this species inhabits regions in which most arboreal supports are vertical, together with their observation that a captive specimen spent almost all of its time sitting in an upright position. Except for this suggestion, *Hapalemur* had been described up to that time as practising the same relatively unspecialized locomotion as the majority of other arboreal quadrupedal forms. More recently, however, Petter & Peyrieras (1970) have published detailed observations on this genus and consider that *Hapalemur* (*griseus*) is intermediate between lemurids and indriids in terms of locomotion. These observations provide a fairly substantial confirmation of Napier & Walker's perspicacity in evaluating this genus. In our own morphological analyses, we have not consistently had specimens representing either of these genera. But on those occasions when some information has been available, these genera have fallen in positions intermediate towards the indriids, or in locations unique to themselves. Clearly they differ from members of a generalized quadrupedal mode.

There are also a number of genera which are considered unequivocally by most workers to be generalized quadrupedal animals not particularly specialized for vertical clinging and leaping in their locomotions, e.g. Napier groups the mouse lemur, the dwarf lemur and the forked lemur under this category, and Cartmill refers specifically to quadrupeds: *Microcebus*, *Phaner* and by implication *Cheirogaleus* and *Lemur*. Yet although most of these prosimians have been thought of as primarily quadrupedal animals a number of recent reports are providing tentative contrary evidence. Thus occasional progression by leaps from vertical trunk to vertical trunk with the body upright is actually fairly common among all small primates. It has been especially reported, however, for *Microcebus* (*M. murinus* by Charles-Dominique & Martin, 1970 and *M. coquereli* by Petter, Schilling & Pariente, 1971). In these descriptions, it seems as though special stress is laid upon the upright position of the trunk during leaps (as in *Galago* and *Tarsius*). Again, Napier & Walker (1967) claim that *Lemur catta* may show features in its locomotor pattern that are intermediate between more quadrupedal primates and those that most frequently employ vertical clinging and leaping. Jolly (1966) does not support Napier & Walker's claim that *Lemur catta* adopts vertical clinging postures in the field, but she has observed that *L. catta* does often leap on vertical trunks when moving through the forest. Rand (1935) reports episodes of vertical clinging and leaping in *Lemur fulvus*, though he notes that this was not the standard manner of progression.

To summarize the behavioural data of vertical clinging and leaping, it can be said that there are excellent wildlife observations indicating that this is an important and frequent means of rest and progression among the lemuroids: *Propithecus, Indri* and *Avahi* and *Lepilemur*, among the galagines: *Galago* and *Euoticus* and in the single extant tarsioid: *Tarsius*. However, with the hindsight of the morphological study just presented, a search of the literature suggests the possibility, and it remains tentative only, that among these various genera there may exist two different ways of vertically clinging and leaping and associated behaviours. That there is such a division in terms of morphology has been shown beyond reasonable doubt; that this has behavioural associations remains to be fully confirmed or possibly denied. Further, and again with the hindsight of this investigation, it does seem as though for each of these two major ways of vertically clinging and leaping, there may well be a number of extant species that practise these behaviours to an incipient or intermediate degree. It would seem that morphologically this is established. Whether or not a degree of behavioural intermediacy is also true is hinted at, but by no means certain. If anything, this behavioural possibility is even more tentative than the one above. Finally, the most tentative suggestion of all, is that both the behavioural information and the morphological data *suggest* that there may even exist a third mode. For the poor behavioural descriptions and the inadequacy of the morphological information that can be presented at this time for *Lepilemur* and perhaps even *Hapalemur*, do not rule out the possibility that these genera participate in yet a third kind of vertical clinging and leaping.

Many small primates can vertically cling and leap; it may well be that the reports here cited are nothing more than special selection of such published items. However, the strength of the morphological information suggests that careful studies designed to discover, in a comparative manner, the differences in life styles of these various species, together with elucidation of the precise biomechanical details that are involved in them, may provide a series of most interesting hypotheses of significance in primate evolution.

VERTICAL CLINGING AND LEAPING IN EOCENE PRIMATES

On the basis of their anatomico-behavioural suggestions Napier & Walker (1967) drew the conclusion that the skeletal anatomy of Eocene primates points to these genera all having been vertical clingers and leapers. These genera include: *Notharctus* (on the basis of five diagnostic characters), *Necrolemur* (five characters), *Smilodectes* (four characters), *Hemiacodon* (four characters), *Nannopithex* (three characters), *Tetonius* (two characters), *Microsyops, Omomyis, Teilhardina, Pseudoloris,*

Aelolopithecus, *Parapithecus*, *Amphipithecus*, and *Microchoerus* (all one character each), representing four separate families, Adapidae, Microsyopidae, Omomyidae and Tarsiidae (after Simons, 1963). These numbers of diagnostic characters are limited and, knowing as we do the difficulties of diagnosis from single characters, there may well be considerable doubt about a number of the suggestions.

Indeed it is the rejection of most of these traits, and the imprecision of others, as necessarily diagnostic of vertical clinging and leaping, that allows Cartmill (1970) to question 10 of the above genera as being vertical clinging and leaping species. He concurs that there is some evidence of such a locomotor mode in *Notharctus*, *Smilodectes* and perhaps *Nannopithex*, although he allows far fewer characters to be of value as detectors in even these genera. Szalay (1972) provides similar criticisms of these ideas and extends them by suggesting that some of the groups in Napier & Walker's listing may not be primates at all (McKenna, 1966).

Simons (1972) proceeds in almost the opposite direction. He accepts the idea of vertical clinging and leaping as a major mode and plumps for many of these genera to be in this category. Thus he suggests the possibility that it was in Eocene times that a nearly universal pattern of vertical clinging and leaping may have evolved, though he also notes that this view has not been without challenge. He adds to the list of Napier & Walker by suggesting the possibility (a) that the mode of locomotion of some of the sub-fossil lemurs, *Megaladapis* and *Archaeoindris*, may have been of a koala-like pattern representing a modified form of vertical clinging and leaping (citing Walker, 1967), and (b) that the earlier form *Plesiadapis* of the Paleocene may have had features that were preadaptive to the vertical clinging and leaping of subsequent Eocene ancestors. Napier (1970) had previously suggested that Paleocene primates, while being tree and possibly forest floor dwellers had probably not embarked upon the specializations to arboreal life that characterize the Eocene forms. In the opinion of Szalay (1972) *Plesiadapis* shows no indication whatever of having been adapted to vertical clinging and leaping.

The evidence of this paper shows that morphologically there are at least two kinds of vertical clinging and leaping in extant primates and that yet a third mode may be identifiable when the adaptations of *Hapalemur* and *Lepilemur* can be more closely examined. We also know that there are at least two groups of incipiently adapted living species. Any review of Eocene primate remains therefore must not only attempt to discover whether there is any indication that some kind of vertical clinging and leaping was practised, but must certainly direct itself to which type, if feasible to what degree, and even as a logical possibility, to the option that yet other kinds of vertical clinging and leaping may have existed.

For instance the adaptations of the genera *Notharctus* and *Smilodectes* seem to indicate a degree of vertical clinging and leaping in their locomotor repertoire. At the same time there is no real indication of a resemblance to *Galago* or *Tarsius* (Stern & Oxnard, 1973); are these species modified in the manner of the modern indriids? In like manner the morphological adaptations of the genera *Hemiacodon*, *Nannopithex* and *Necrolemur* seem to indicate some degree of vertical clinging and leaping among their locomotor abilities. Again, however, there is little indication of a resemblance to indriids (Stern & Oxnard, 1973). Are these species morphologically modified in the manner of the modern *Galago* and *Tarsius*? Finally, both of these extinct sets of species may be compared with the incipient forms that this study indicates may exist.

It has been possible to make estimates for *Notharctus* of three of the parameters that were used in the analysis of the dimensions of the forelimb presented earlier in this paper. This is not extensive enough data to allow us to interpolate this fossil within the matrix of living species. But it does allow us to eliminate some of the extant forms as possible models along the lines of which *Notharctus* may have been adapted. Thus, if we take the known values for the three parameters for *Notharctus*, and add them in turn to the values of the remaining dimensions for different living genera, then we can gauge which living genera are reasonable models. Thus, if we pretend that in its missing dimensions *Notharctus* resembled either an indriid morphological type or a *Galago-Tarsius* morphological mode, then multiple discriminant function analysis indicates that this hybrid creature lies in a part of the eigenvector space not occupied by any other primate, some 30 standard deviation units distant. In like manner, a second study shows that, if the missing dimensions were similar to those now found in the incipiently adapted species such as *Lemur catta* or *Microcebus*, then again, the "new" creature lies in an unlikely part of the eigenvector space some 23 standard deviation units away from any primate. These separations are of an order much greater than we usually find in the studies of primate genera. They indicate that whatever the missing dimensions were like, it is rather unlikely that they were similar to those of any of the extant highly specialized leaping prosimians or either of the incipiently modified leaping prosimians.

When however, generalized quadrupedal primates such as *Saimiri* are used for this examination (Fig. 10), then we find (a) that different "*Notharctus*-extant species composites" fall close to the positions of the actual living genera, and (b) that, in particular, they lie within the envelope of generalized quadrupedal forms. Even if the missing dimensions are taken as being similar to those in such specialized (although still basically quadrupedal) primates as *Daubentonia* and *Cacajao*, the "*Notharctus-Daubentonia* composite" on the one hand, and the "*Notharctus-Cacajao* combination" on the other, differ scarcely at all from the actual patterns for the respective living species themselves. Equivalent findings exist for many other extant prosimian and New World primates and they indicate that there is no evidence to suggest

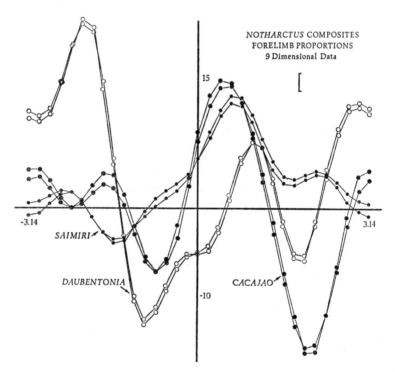

FIG. 10. Using again the previous conventions, this figure shows the multi-dimensional relationships between (a) the position for *Saimiri* as compared with that for the *Notharctus-Saimiri* composite, (b) the position for *Cacajao* as compared with the *Notharctus-Cacajao* composite and (c) the position for *Daubentonia* as compared with the *Notharctus-Daubentonia* composite. In each case the composite differs very little from the respective extant species although these latter are widely different from one another. This indicates that the missing dimensions of *Notharctus* could well have been similar to those of each of these three relatively quadrupedal primates. A similar study using *Notharctus* and any of the vertical clinging and leaping genera produces composites that lie 23 to 30 standard deviation units away from any living primate; such a distance is so great as to suggest that it is rather unlikely that *Notharctus* could indeed have had its missing dimensions so composed.

that the missing dimensions of *Notharctus* are any different from each of these *non*-vertical clinging and leaping genera.

These results suggest, very tentatively, that *Notharctus* has a morphology different from extant vertical clinging and leaping forms. They do not exclude the possibility that *Notharctus* may have had a morphological adaptation to some other, as yet unknown, type of vertical clinging and leaping; nor do they exclude a morphological adaptation of *Notharctus* similar to those of almost any of the extant quadrupedal prosimian or New World genera. This last suggestion is not at variance with Gregory's (1920) original view that *Notharctus* may have been a "primitive quadruped". Of course, these conclusions, far from being the last word, are merely a tentative beginning that throws doubt upon the idea that *Notharctus* is modified along the same lines as the overt or incipient extant vertical clinging and leaping species. A study of the actual fossil remnants of *Notharctus*, attempting to obtain more of the parameters used in the above investigations, is now required for definitive functional diagnosis of this extinct genus.

The more general findings of this study mean that we may take a new look at the question of the basic arboreal specialization of the early primates. Certainly vertical clinging and leaping cannot now be the basic specialization of this group. It seems clear that the evidence presented here forces us to accept the likelihood of a multiple partitioning of behaviours associated with leaping in Eocene and perhaps earlier times. Elements of this multiple partition may have been associated with plant feeding, or with an insectivorous diet; environmental locomotor restrictions or predator pressures may have played their parts; yet other possibilities exist; this wider aspect of the problem has scarcely been touched upon. From this arises the possibility that one or more of the specialized products of different cells of the partitioning process, may have given rise to various later higher primates. However, the possibility also exists that such later evolutionary products may have arisen from one or more of the incipiently adapted forms that must have existed prior to or during the partitioning procedure. And finally it is far from ruled out that the later higher primates may have arisen from early forms that remained generalized and were not implicated at all in these leaping adaptations. Only further study, both behavioural and morphological, of both living and extinct species, especially quantitative and multivariable, may be able to provide less speculative answers to such questions.

One final comment is of interest in indicating the degree to which our knowledge is dependent upon the investigations of workers in the past. The present results are based upon attempts to view large numbers

292 CHARLES E. OXNARD

of characters when taken as a totality and eliminating redundant information. It is well worth recording that, on the basis of detailed, extensive and careful studies of the foot alone, not only was Morton (1924) very clear in his understanding of the differences between *Galago* and *Tarsius* on the one hand and the indriids on the other, but also he understood the morphological and functional implication of the fact that the foot of *Microcebus* among the cheirogaleines may be allied in functional terms to those of *Tarsius* and *Galago*, and that the foot and lower limb of *Lepilemur* among the lemurines resembles somewhat that of the indriids. Volkov (1903, 1904) also understood, although with less apparent recognition of the functional implications, the associations between different structures in the foot to arboreal posture and movement; his data suggest some adaptive relationship between certain *Lemur* species and *Avahi* and *Indri*. Finally, it must be recorded that as early as 1867 and 1873, Mivart, who provides in his written work few indications at all of functional implications, was nevertheless well aware of the marked distinction between the structure of the foot in *Galago* and *Tarsius* as compared with that in the indriids. He especially noted that the structure of the lower limb, the foot, and in particular the tarsus placed *Lemur*, and more so *Lepilemur*, intermediate towards *Indri*, and that the structure of the foot of *Cheirogaleus* and *Microcebus*, again including especially the tarsus, placed these forms toward *Galago* in morphological resemblance. It is of interest to turn once more to the works of Grafton Elliot Smith (1927) and Frederick Wood Jones (1926). There it is on record that *Lemur* has adaptations and habits of hopping "that mark it off from more primitive prosimians" and that *Microcebus* possesses "adaptations of body form associated with the aerial phase" of its locomotion.

I am especially indebted to my colleague Professor Jack T. Stern, Jr for collaboration, discussion and criticism at all stages of this study. This study includes a review of some of the findings of our previous joint publication (Stern & Oxnard, 1973). Professors Ronald Singer and Edouard Boné have also contributed through discussion and criticism. Mr Gene Albrecht provided some computational analyses.

Some of the data that have been analysed are those undergoing more extensive study with Professor E. H. Ashton, Dr R. M. Flinn and Mr T. F. Spence. Results have also been made available from other collaborative studies through Professor Lord Zuckerman and the aforementioned. The data on overall bodily proportions were kindly made available by Professor A. H. Schultz.

The investigations are supported by NSF Grant GS 30508 and by grants from the Louis Block Bequest Fund, the Dr W. C. and C. A. Abbott Memorial Fund, and the Wenner-Gren Foundation for Anthropological Research.

REFERENCES

Andrews, D. F. (1972). Plots of high-dimensional data. *Biometrics* **28**: 125–136.

Ashton, E. H. & Oxnard, C. E. (1964). Functional adaptations in the primate shoulder girdle. *Proc. zool. Soc. Lond.* **142**: 1–28.

Ashton, E. H., Oxnard, C. E. & Spence, T. F. (1965). Scapular shape and primate classification. *Proc. zool. Soc. Lond.* **145**: 125–142.

Ashton, E. H., Flinn, R. M., Oxnard, C. E. & Spence, T. F. (1971). The functional and classificatory significance of combined metrical features of the primate shoulder girdle. *J. Zool., Lond.* **163**: 319–350.

Attenborough, D. (1961). *Zoo quest in Madagascar.* London: Butterworth.

Cartmill, M. (1970). *The orbits of arboreal mammals: a reassessment of the arboreal theory of primate evolution.* Univ. of Chicago: Ph.D. Dissertation.

Cartmill, M. (1972). Arboreal adaptations and the origin of the order Primates. In *The functional and evolutionary biology of primates.* Tuttle, R. H. (ed.) Chicago: Aldine-Atherton.

Charles-Dominique, P. (1971). Éco-éthologie des prosimiens du Gabon. *Biologia gabon.* **7**: 121–228.

Charles-Dominique, P. & Hladik, C. (1971). Le *Lepilemur* du sud de Madagascar: écologie, alimentation et vie sociale. *Terre Vie* **25**: 3–66.

Charles-Dominique, P. & Martin, R. D. (1970). Evolution of lorises and lemurs. *Nature, Lond.* **227**: 257–260.

Davis, D. D. (1962). Mammals of the lowland rain-forest of north Borneo. *Bull. natn. Mus. St. Singapore* No. 31: 1–129.

Grand, T. I. & Lorenz, R. (1968). Functional analysis of the hip joint in *Tarsius bancanus* (Horsfield, 1821) and *Tarsius syrichta* (Linnaeus, 1758). *Folia primatol.* **9**: 161–181.

Gregory, W. K. (1920). On the structure and relations of *Notharctus*, an American Eocene Primate. *Mem. Am. Mus. nat. Hist.* (n.s.) **3**: 49–243.

Hall-Craggs, E. C. B. (1965). An analysis of the jump of the Lesser galago (*Galago senegalensis*). *J. Zool., Lond.* **147**: 20–29.

Harrison, B. (1963). Trying to breed *Tarsius. Malay. Nat. J.* **17**: 218–231.

Hill, W. C. O., Porter, A. & Southwick, M. (1952). The natural history, endoparasites and pseudo-parasites of the tarsiers (*Tarsius carbonarius*) recently living in the Society's menagerie. *Proc. zool. Soc. Lond.* **122**: 79–119.

Hoogstraal, H. (1947). The inside story of the tarsier. *Chicago nat. Hist. Mus. Bull.* **18**(11): 7–8 & **18**(12): 4–5.

Jolly, A. (1966). *Lemur behavior. A Madagascar field study.* Chicago: Univ. of Chicago Press.

Jolly, A. (1972). *The evolution of primate behavior.* New York: Macmillan.

Le Gros Clark, W. E. (1924). Notes on the living tarsier (*Tarsius spectrum*). *Proc. zool. Soc. Lond.* **1924**: 217–223.

McKenna, M. C. (1966). Paleontology and the origin of the primates. *Folia primatol.* **4**: 1–25.

Mivart, St. G. (1867). Additional notes on the osteology of the *Lemuridae. Proc. zool. Soc. Lond.* **1867**: 960–975.

Mivart, St. G. (1873). On *Lepilemur* and *Cheirogaleus*, and on the zoological rank of the *Lemuroidea. Proc. zool. Soc. Lond.* **1873**: 484–510.

Morton, D. J. (1924). Evolution of the human foot. II. *Am. J. Phys. Anthrop.* **7**: 1–52.

Napier J. R. (1970). *The roots of mankind.* Washington, D.C.: Smithsonian Inst.

Napier, J. R. & Walker, A. C. (1967). Vertical clinging and leaping—a newly recognized category of locomotor behaviour of primates. *Folia primatol.* **6**: 204–219.

Oxnard, C. E. (1967). The functional morphology of the primate shoulder as revealed by comparative anatomical, osteometric and discriminant function techniques. *Am. J. Phys. Anthrop.* **26**: 219–240.

Oxnard, C. E. (1973). *Form and pattern in human evolution: some mathematical, physical and engineering approaches.* Chicago: Univ. of Chicago Press.

Petter, J.-J. (1962). Recherches sur l'écologie et l'éthologie des lémuriens malagaches. *Mém. Mus. natn. Hist. nat. Paris* (A. Zool.) **27**: 1–146.

Petter, J.-J. & Peyrieras, A. (1970). Observations éco-éthologiques sur les lémuriens malagaches du genre *Hapalemur. Terre Vie* **24**: 356–382.

Petter, J.-J., Schilling, A. & Pariente, G. (1971). Observations éco-éthologiques sur deux lémuriens malagaches nocturnes: *Phaner furcifer* et *Microcebus coquereli. Terre Vie* **25**: 287–327.

Pilbeam, D. (1972). *The ascent of man: an introduction to human evolution.* New York: Macmillan.

Rand, A. L. (1935). On the habits of some Madagascar mammals. *J. Mammal.* **16**: 89–104.

Sauer, E. G. F. & Sauer, E. M. (1963). The South West African bush-baby of the *Galago senegalensis* group. *Jl S.W. Afr. scient. Soc.* **16**: 5–36.

Schmidt, K. P. (1947). Pangolins, tarsiers, and flying lemurs of Philippines, *Chicago nat. Hist. Mus. Bull.* **18**(7): 1–3.

Simons, E. L. (1963). A critical reappraisal of tertiary primates. In *Evolutionary and genetic biology of primates* **I**: 66–129. Buettner-Janusch, J. (ed.) New York and London: Academic Press.

Simons, E. L. (1972). *Primate evolution: an introduction to man's place in nature.* New York: Macmillan.

Smith, G. E. (1927). *The evolution of man.* London: Oxford University Press.

Sprankel, H. (1965). Untersuchungen an Tarsius. I. Morphologie des Schwanzes nebst ethologischen Bemerkungen. *Folia primatol.* **3**: 153–188.

Stern, J. T. Jr & Oxnard, C. E. (1973). *Primate locomotion: some links with evolution and morphology.* Basel: Karger.

Szalay, F. S. (1972). Paleobiology of the earliest primates. In *The functional and evolutionary biology of primates.* Tuttle, R. H. (ed.) Chicago: Aldine-Atherton.

Volkov, M. Th. (1903). Variations squelettiques du pied chez les primates et dans les races humaines. *Bull. Mém. Soc. Anthrop. Paris* **4**: 632–708.

Volkov, M. Th. (1904). Variations squelettiques du pied chez les primates et dans les races humaines. *Bull. Mém. Soc. Anthrop. Paris* **5**: 1–50; 201–331 & 720–725.

Walker, A. C. (1967). Patterns of extinction among the subfossil Madagascan lemuroids. In *Pleistocene extinctions: the search for a cause*. Martin, P. S. & Wright, Jr, J. H. E. (eds) New Haven: Yale Univ. Press.

Walker, E. P. (1948). The curious but practical peculiarities of the tarsier—a long jump champion—shown by high speed photography. *Illus. Lond. News* **123**: 738–739.

Wharton, G. H. (1948). Seeking Mindanao's strangest creatures. *Natn. geogr. Mag.* **94**: 389–408.

Wood Jones, F. (1926). *Arboreal man*. London: Edward Arnold.

DISCUSSION

BARNICOT (CHAIRMAN): Thank you very much for your interesting paper, Professor Oxnard. May I just ask you one thing? When you distinguish the *Indri* group from the *Galago-Tarsius* group in your analysis it immediately comes to mind that the members of these two groups differ greatly in body size. Does this contribute to the differentiation of the two groups?

OXNARD: Yes, I think that this is rather important and it brings a whole series of questions towards us that I have not touched on in my presentation. These all relate to the biological milieu within which this set of morphological adaptations occurs. One of these is certainly the matter of size difference; the indriids are rather large animals, the *Galago-Tarsius* group include only small species. Of the various intermediate forms, those which are intermediate towards the indriid cluster tend to be intermediate in size, those which are intermediate towards the *Galago-Tarsius* mode are especially small. However, with these size differences go other facets of the life processes of these animals; insectivorous and vegetarian diets are surely related at least in part to the question of size. Size may relate to the differences between predation upon food species as contrasted with escape from predation. There may even be an association with substrate size (in the sense of the differing sizes of branches and trunks) upon which the vertical clinging and leaping is being performed, and this will in turn bear some relationship to body size. But for these particular genera and the various habitats that they occupy, our information is scanty and certainly scarcely good enough to be able to give weight to these factors, and of course others not mentioned here, in order to understand size relationships.

As regards the way in which actual size differences may have affected our raw data and final results, we have been under certain constraints. For instance, with the data of Professor Schultz we must operate upon indices because the raw measurements are unfortunately no longer available. And although within physical anthropology the index is a hallowed method for eliminating size differences among specimens, we all know that it is at best only a crude technique. Hence in this part of the study we accept that there may still be some size related effect built into the results. However, there are a variety of other ways of allowing for size that can be applied and in the pelvic and scapular studies, for instance, we have been able to investigate the problem using logarithmic transformations, regression adjustments, and principal components manipulations; we thus feel relatively happy that, though all aspects of size have not been covered in our analyses, the eventual results and discussion have not been heavily biased by such effects.

GOODMAN: Where would the Slow loris, the Potto and the general group of lorisines fall on your diagrams?

OXNARD: These species were included simply because they are prosimians. They fall, in each analysis, in their own unique part of the multi-dimensional space well away from all other forms. Their closest morphological links to other species are always through the central group of generalized quadrupedal species. This is so whether one looks at data from limb girdles, from whole limbs, or from total bodily proportions.

I included lorisines exactly because they are curiously different from all of the other species examined. The prehensile-tailed cebids and the Pitheciinae (*Cacajao* and *Pithecia*) were included for similar reasons. The prehensile-tailed cebids have obvious locomotor peculiarities and although *Cacajao* and *Pithecia* are much more quadrupedal, they also show some unusual locomotor features. Correlating with this is the finding that these groups also are unique outliers, each being placed at an edge of the set of all species. But the lorisines, the group that you asked about, are very clearly unique.

DAY: I should like to pursue this size question just a little further, if I may, because it did strike me particularly on the first slide relating to the pelvis, and also the second relating to the shoulder, and then again in Schultz's data of total body proportion, that in all of these we had a distinct size run from left to right. All the large ones were on the left and all the smaller ones were on the right, with a few minor exceptions which I would accept.

In a canonical analysis the first variate very often has a very large size component, but you say that you took great steps in the pelvic data to eliminate the size component, and yet it comes out at the end that you have a size run. This I find hard to understand, and I would appreciate your comments.

OXNARD: The answer here is very clear. *First,* it is a misunderstanding of the nature of these analytical techniques to think that a coherent size pattern is displayed running from right to left on the diagrams. These are multi-dimensional analyses and the relationships that you mention do not lie in a single plane of the analysed data space. If we pretend that things are in three dimensions, there are both big and small species that lie in front of and behind the plane of the diagrams. The canonical axes are only used to supply direction vectors for the minimum spanning tree of connections. The actual distances are in many dimensions. There is not, in these data, any consistent size related axis.

[*Note added after discussion:* For instance, a glance at the diagram relating to the analysis on the hindlimb dimensions appears to suggest that at the left hand end we have the prehensile-tailed cebids (large bodied animals) and at the right hand end *Galago* and *Tarsius* (small bodied animals). But this appears so only because we are projecting a large generalized distance ("*D*") out of its own higher dimension into the plane of the first two canonical axes. This is why non-connected genera in these diagrams cannot be correctly placed in the canonical two-space.

Furthermore, if we redo the analysis without the prehensile-tailed cebids (one group only of the larger species), then we find the set of canonical axes are entirely reframed so that they lie at about 45 degrees in the euclidean space in which they are projected in the diagram as compared with their original positions.

Again, if we look at the analysis on the forelimb dimensions we see that the indriids (in this context, big animals) and *Tarsius* and *Galago* (small animals) all lie together at the right hand end of the separations. Here examination of higher dimensions indicates clearly that the separation of these two species is not at all in the same "direction" as that *apparently* separating out other large forms (prehensile-tailed cebids). It is a misunderstanding of the high-dimensionality situation to believe this.]

Secondly, I should reiterate that we have tried a number of different methods in attempts to allow for such size related aspects as must presumably be present in the data. The first is simply through the use of indices and though crude, it is clear that this does allow for a great deal of the size related effect. The second is the method of regression adjustment; although not simple, this is now a highly standard way of allowing for size differences in the data before performing these analyses (logarithmic transformation is also useful here and has been used). The third and rather more speculative method is to perform a principal components analysis on the data and pretend that size falls out as the first principal component. It is well established that with raw measurements, this certainly does occur; whether the method is also so dependable when used upon indices and angles is open to question and we are investigating this. However, I have indeed used the method and have found that the differences between canonical analyses before and after subtracting the first principal component from these data is small. That is: whatever is being subtracted in that first principal component has little effect on the results. If it truly represents a size effect then that effect is minimal presumably because the major effect of size is already removed in the use of indices. It may not reflect a size effect but if it does not, certainly no other principal component does.

ZUCKERMAN: I should like to raise just one simple point to elaborate what was said. Obviously Professor Day was taking too simple a view of that picture. With these multi-dimensional representations one has to be very careful which way one interprets what the eye sees on the graph.

There is another very simple fact: all these quadrupedal groups were in one band, and there is just as much variation between those quadrupedal creatures as there is between the creatures that are outlying.

OXNARD: I agree.

NAPIER: It is very exciting, of course, to see an idea that one has had, and developed up to a point, being taken to such an elegant conclusion as we have been treated to today. This may sound rather like a case of sour grapes, but not a bit of it, I do not intend it that way; but we have always felt that there were at least two groups within the vertical clingers and leapers. We may have been influenced by size. That is to say, one felt that the *Avahi*, the *Indri* and *Propithecus* obviously were likely to have rather a different method of moving about than the smaller types. While we always felt that a separation existed, we never took the matter further.

I think this is a very good example of where simple methods based on observation and measurements, and less sophisticated statistical methods, can only take one so far. The methods that we have shown this morning take us much further towards a fuller understanding.

So far as *Lemur catta* is concerned, I have always been interested in this lemur as the "odd man out". If one looked at the dimensions, there was *Lemur catta*, particularly in its intermembral index, right in the middle between the quadrupeds, between the New World monkey quadrupeds mainly, the rest of the genus *Lemur*, and the group which we call vertical clingers and leapers. It has the lowest intermembral index of the whole of the *Lemur* group. One really could not get anywhere until, as Alison Jolly says in her book—you have quoted it, and many other people, of course, have made similar statements—we look at the *behaviour* of the animal. Walker and I quite independently observed such vertical clinging and leaping under experimental conditions, if the animal is prevented from coming to the ground.

I set up an experiment some years ago to show this, and there is no doubt that *Lemur catta* is as good a vertical clinger as any of the recognized ones. But I think, apart from the intermembral index which does not really say all that much, the behaviour was the thing that really gave it away and confirmed for us the feeling that it belonged in that vertical clinging group; although morphologically it was a little different from the others.

Ashton: If I may, Mr Chairman, I should like to refer back to the question of interpretation of canonical and related analyses, and especially to some of the difficulties that may be experienced in the interpretation of the relatively complex analyses that can be carried out with modern large-scale computers.

Any misunderstandings that exist may possibly stem from the fact that in the very early days when canonical analysis was carried out— one of the pioneer studies being on tooth dimensions, carried out in collaboration with Dr Yates and his colleagues at Rothamsted during the mid-1950's—we were dealing with relatively small numbers of groups of animals—in fact about six genera only—and also with relatively small numbers of dimensions, not more than three or four in many cases, upon each tooth.* Possibly because of this, possibly by a stroke of good fortune, virtually all the information in the canonical analysis was contained in the first two axes. However, this does not obtain in more complicated analyses, and one therefore has to devise methods for summarizing the information in the whole complex of axes, sometimes going up to more than 20. I spoke, yesterday, of the technique of applying cluster analysis to a matrix of squared generalized distance; Charles Oxnard has spoken today of other techniques. I am certain that it is upon these, as an essential supplement to simple bivariate canonical plots, that we shall have to concentrate in the future.

* Ashton, E. H., Healy, M. J. R. & Lipton, S. (1957). The descriptive use of discriminant functions in physical anthropology. *Proc. R. Soc.* (B) **146**: 552–572.

Symp. zool. Soc. Lond. (1973) No. 33, 301–337.

COMPARATIVE ANATOMY AND PRIMATE SYSTEMATICS

R. D. MARTIN

Department of Anthropology, University College, London, England

SYNOPSIS

Comparative anatomical studies still provide the major source of information relating to primate evolution. The tradition to which Sir Grafton Elliot Smith made such a notable contribution is still very much alive, and developments in this field continue to add significantly to our data. It is true that the only *concrete* evidence for primate evolution is provided by the fossil record; but fossil evidence can only be interpreted in terms of a sound body of comparative anatomical literature, based on comprehensive studies of living forms. Biochemical techniques for assessing phylogenetic affinity are now developing rapidly, though for the time being a framework of comparative morphological studies of living and fossil forms is still necessary to test and calibrate the evolutionary models proposed. Hopefully, with theoretical and practical developments in both fields we shall eventually achieve broad agreement between comparative anatomical and biochemical studies. At this stage, however, there is sometimes conflict between the results from morphological and biochemical studies of primate evolution. It is argued here that this is partly due to the fact that comparative anatomical studies are open to certain phylogenetic misinterpretations which can be analysed at a purely theoretical level.

There is a traditional tendency in comparative anatomy to take "primitive" living species as models for discussing evolution, rather than to extrapolate back into the past in an attempt to reconstruct actual evolutionary stages. With the Primates, this leads to a second source of error stemming from oversimplified treatment of living primates as a "Scala naturae". Various living primate species are regarded as models for rungs on an evolutionary ladder leading to man, with the bottom rung occupied by the living Insectivora. Taking modern developments of Elliot Smith's classical studies of the primate brain, one can demonstrate how this oversimplification has in some ways tended to block progress from the foundation provided by his work. Although recent work on fossil primate endocasts has provided a significant new source of data, interpretation of this evidence is dependent upon a sound theoretical approach.

THE THEORETICAL APPROACH

Analysis of anatomical similarities with a view to reconstructing evolutionary history dates back to Darwin's time; the broad evolutionary significance of anatomical similarities between species was generally appreciated for the first time with the publication of Darwin (1859). Indeed, Huxley's comparative study (1864) of certain anatomical features of man and other primates (particularly the apes) was a decisive factor in ensuring wide acceptance of Darwin's evolutionary hypotheses. Discussion of primates, their anatomy and their evolution has continued to occupy a central place in the debate over phylogeny which has taken

301

place throughout the last century. It is vital to remember, however, that the Linnaean "natural classification" of plants and animals (1758) was also based on the assessment of anatomical similarities, though without the evolutionary hypothesis. The very fact that Linnaean classifications were generally very similar to those still used today demonstrates the degree to which anatomical similarity as such reflects evolutionary relationship without any further analysis. The "natural classification" of animals was undoubtedly a prerequisite for the formulation of evolutionary hypotheses, since the arrangement of animals according to "fundamental" anatomical similarities greatly facilitated recognition of evolutionary groupings. In the period 1758–1859, animals were classified in a manner which closely reflected evolutionary history, yet without any explicit statement of the broad evolutionary principle. One concomitant of this was the inclusion, by Linnaeus, of various fossils in the *geological* section of his classification. Fossils gradually came to be recognized as the remains of extinct organisms, but it was only after Darwin that they were recognized as samples from interconnected lineages in the continuous phylogenetic tree of organisms. This underlines the fact that fossils can only be incorporated into an evolutionary tree on the basis of hypotheses about relationships between the living forms, though knowledge about the first fossils considered may influence the allocation of fossils discovered later on. The secondary place occupied by fossils in the development of any phylogenetic tree is dictated by a number of factors. In the first place, there is the historical factor that natural classification of living species according to anatomical characters preceded by a number of decades the corresponding allocation of fossils. Secondly, there is the undeniable fact that living forms yield vastly more anatomical information than fossils. Thirdly, the interpretation of (typically fragmentary) fossils usually involves explicit or implicit consideration of the anatomy of living forms. Finally, it must be remembered that any phylogenetic tree is an *hypothesis*. There are many references nowadays to "the fact of evolution"; but strictly speaking the theory of evolution is an hypothesis which can be supported only by circumstantial evidence, though it has undeniably proved to be the most viable hypothesis. The construction of evolutionary trees, as an extension of this hypothesis, is even more speculative, and there is virtually no way of "testing" a phylogenetic tree with reference to its two main features: (1) the pattern of branching; (2) the time levels of the branching nodes. Biochemical comparisons between species are now beginning to provide an independent test for patterns of branching (see Cook & Hewett-Emmett, in press; Barnabas, Goodman & Moore, 1972); but the time factor can at present only be introduced by reference to the fossil record. Accordingly, the following 3-step procedure would seem to be preferable in constructing and checking any phylogenetic tree:

1. Comparison of living forms, followed by construction of an hypothetical phylogenetic tree with specified branching nodes. The probability of accuracy of the phylogenetic tree should increase with: (a) the number of species considered, (b) the uniformity of the comparisons, (c) the number of characters considered (both anatomical and physiological, with due reference to behaviour).

2. Allocation of fossil forms to the phylogenetic tree, with due reference to hypotheses about the characters present at each of the branching nodes. If a particular fossil does not accord with the hypotheses already formed, this may mean that: (a) the fossil species concerned is not related to the other forms included in the tree, or (b) the tree is faulty, and certain hypotheses must be revised.

By incorporation of fossil species, and modification of manifestly faulty hypotheses, the phylogenetic tree is rendered more plausible, and the all-important *time-factor* can be introduced on the basis of reliable fossil dating.

3. Testing of the phylogenetic tree through biochemical comparisons conducted between the various living species which are included. Although present techniques largely relate to the occurrence of branching-points, reliable relative checks on the times of branching may soon be possible (see Goodman's paper in this volume).

It could, of course, be argued that one could construct a phylogenetic tree by relating known fossils to one another and by subsequently adding living forms to the fossil tree thus produced. However, there does not seem to be a single case where any group of mammals with living representatives has been approached in this way, for the reasons outlined above. In any case, such an approach would be heavily biased by inevitable concentration on a restricted number of (primarily dental) characters. At the present time, the incorporation of new fossil forms into phylogenetic trees (at least within the Class Mammalia) mainly involves a process of accretion, rather than re-organization, and there must inevitably come a time when sufficient new fossil evidence has emerged to permit a serious re-examination of any given phylogenetic tree. This examination can have two effects: (1) modification of certain hypotheses on which the original tree was based, (2) exclusion of certain fossils which were erroneously included at some time because of inadequate information. It would seem that the time has now come for a re-examination of the hypothetical phylogenetic tree of the Primates, with a critical reconsideration of the fossil forms which have traditionally

been included. As an integral part of such reconsideration, it is essential
to pay attention to methodology (e.g. see Martin, 1968).

The combined efforts of Darwin and Huxley provided the foundation
for a long tradition of comparative anatomical studies of the Primates
with a view to understanding their evolution. Following Mivart's initial
definition of the Primates (1873), there has been a long series of publications greatly adding to our knowledge of primate anatomy and providing an extremely broad basis for considering primate evolution.
Particular landmarks have been provided by the publications of Gregory
(1916), Elliot Smith (1927), Le Gros Clark (1934, 1962), Wood Jones
(1929a, b), Simpson (1940, 1945, 1962), Hill (1953), Napier (1971) and
Simons (1963, 1972). The studies of these authors, and many others,
have produced an enormous quantity of detailed information about
both living and fossil primates. Indeed, the rapid growth in our knowledge has itself led to the present situation where it is necessary to stand
back and take stock of the information, to re-consider the methods
used for analysing primate evolution, and to re-organize the material to
establish yet another hypothetical phylogenetic tree in the continuing
search for greater accuracy. There is therefore no implied criticism of
previous authors attached to proposals for modifying methodology and
the phylogenetic trees which are suggested as new hypotheses. Indeed,
Huxley—in a typically perceptive manner—foresaw this growth process
(1864: p. 58):

> "In a well-worn metaphor, a parallel is drawn between the life of
> man and the metamorphosis of the caterpillar into the butterfly; but the
> comparison may be more just as well as more novel, if for its former term
> we take the mental progress of the race. History shows that the human
> mind, fed by constant accessions of knowledge, periodically grows too
> large for its theoretical coverings and bursts them asunder to appear in
> new habiliments, as the feeding and growing grub, at intervals, casts its
> too narrow skin and assumes another, itself but temporary."

It is unavoidable that development in knowledge must render some
hypotheses obsolete, and the present time is particularly favourable for
a review of Elliot Smith's contribution to primate comparative anatomy,
especially since some of the developments which have taken place since
his *Essays on the Evolution of Man* (1927) have tended to obscure, rather
than to clarify, the relationships which he was attempting to determine.

Revision of current primate phylogenetic trees is partly necessary
because of new information; but it is also necessary because some of this
new information has indicated flaws in the methodology which is
generally accepted for constructing phylogenetic trees. There have been

a number of notable discussions of animal systematics (e.g. Mayr, 1942: 175–298; Hennig, 1950; Cain, 1954; Simpson, 1961), but these have largely been concerned with the manner in which a phylogenetic tree relates to a formal classification, rather than with the way in which the phylogenetic tree is developed from the basic zoological data. This is associated with the fact that there has been a traditional confusion (since Mivart, 1873) between classifications and phylogenetic reconstructions. Classification is an arbitrary (though controlled) division of the continuum of living things into categories; but classifications are also used as a basis for discussing phylogenetic trees, since it is virtually impossible to discuss the evolution of a group of animals until they have been arranged into appropriate sub-groups (cf. Martin, 1972). This is one of the major reasons for the generally accepted requirement that classifications should reflect phylogenetic relationships. However, the arbitrary nature of classifications is such that several different classifications may be compatible with a given phylogenetic tree. When one is discussing evolutionary relationships, therefore, it is essential to talk in terms of the phylogenetic tree and not in terms of ambiguous classificatory categories. Simpson's classification of the Primates (1945) has withstood the test of time because it has generally remained compatible with the consensus of opinion about primate evolution, even though several important changes have taken place with respect to hypotheses about specific branching-points. However, it must not be assumed that Simpson's classification faithfully expresses all of the evolutionary relationships between the various primate species; the classification was not (and could not be) constructed to do this.

Confusion between classifications and phylogenetic trees has tended to obscure the fact that there has been very little detailed discussion of the methods used to establish broad evolutionary relationships between living forms and associated fossils. As has been pointed out previously, students of systematics learn largely by imitation, rather than according to an established methodology. Now might be an excellent time, therefore, to take stock of the enormous mass of new data and re-organize it according to an explicit, generally acceptable methodology.

METHODOLOGICAL CONSIDERATIONS

The present author has already attempted to clarify some of the methodological considerations involved in constructing primate evolutionary trees (Martin, 1968), though this attempt was necessarily restricted by its limited scope and the heavy concentration on theoretical issues. It will suffice to summarize the main points here before proceeding to analyse a particular case (the primate brain) which has occupied an important place in discussions of primate evolution:

1. As the quantity of information about living and fossil mammals has increased, there has been an increasing tendency to treat the adaptive radiation of mammals as a set of sub-units. Discussions of primate evolution are now frequently restricted to a comparison between the two Orders Primates and Insectivora. This restriction is based on the assumption that the Order Insectivora includes living forms which closely resemble ancestral placental mammals. Cain (1954: 17), for example, effectively summarized a prevalent view when he stated:

> "For example, in the great group of primates (tree-shrews, lemurs, monkeys, apes and man) it is almost impossible to find a single diagnostic character (running through the whole group) which is not so basic that it says only that all are mammals, and therefore fails to divide them from hedgehogs and other insectivores."

This statement also brings out a common point of confusion: although ancestral placental mammals (if and when they are found) would probably be included in the Order Insectivora, this does not mean that living insectivores, such as hedgehogs, can be equated with ancestral placental mammals (see later). It is not generally recognized that a wide knowledge of all living placental mammals—and of living marsupials—is necessary in order to define reliably the characters of the "ancestral placental stock". Similarly, a wide knowledge of living primates is necessary in order to define the characters of the "ancestral primate stock" (Fig. 1). One of the main exercises involved in constructing an evolutionary tree of the Primates is adequate definition of specific characters which were developed between the "ancestral placental stock" and the "ancestral primate stock", and which probably influenced the subsequent adaptive radiation of primates.

As has already been stated above, the greater the scope of anatomical (and other) information available, the greater is the probable accuracy of any suggested phylogenetic tree. Elliot Smith made a specific point of the necessity for broad comparisons among mammals when considering homologies of certain structural features of mammalian brain, such as the sulci formed on the dorsal surface of the cerebral hemispheres (Elliot Smith, 1902a: 397). Apart from a few notable exceptions (e.g. Campbell, 1966b), however, many subsequent authors have abandoned this broad approach.

2. As Cain (1954) has pointed out, classifications of animals are produced by evaluation of similarities and differences between the animal species considered, and allocating by similarity is a more pro-

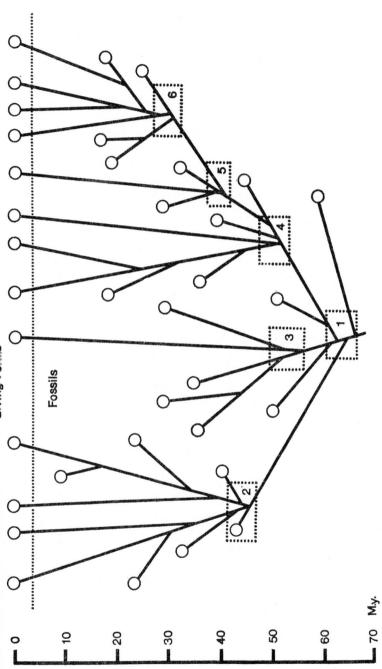

FIG. 1. Schematic "phylogenetic tree", illustrating the evolutionary relationships between a number of living and fossil species (circles). The time-scale is illustrated on the left in millions of years. The boxes enclosed by the dotted lines (1–6) represent arbitrary "ancestral stocks", selected because of their relationship to major points of evolutionary radiation. For this particular tree, box 1 represents the "basal ancestral stock", and boxes 2–6 represent "later ancestral stocks". Each stock covers a time-bracket of several million years, and in each case, the box must include the ancestral species population from which all subsequent forms are derived. *Note*: The living species can generally be defined as non-interbreeding populations (biospecies), but the fossil species (palaeospecies) can only be defined arbitrarily, usually on the basis of convenient gaps in the fossil record.

ductive exercise than grouping by difference. It is also common practice
to distinguish between *homologous* similarities and *analogous* similarities
(Simpson, 1961: 78). Homologous characters found in any two animal
species are defined as those which have been inherited from a common
ancestor which already possessed the character concerned, whilst
analogous characters are defined as those which, though similar in
appearance, have been independently developed since the last common
branching-point shared by the two species (Fig. 2). Whenever the
attempt is made to construct a classification of any group of animals
reflecting hypotheses about their phylogenetic relationships, it is
necessary to exclude analogous similarities from assessments of affinity.
This distinction has been summarized by Simpson (1945: 9):

> "Animals may resemble one another because they have inherited like
> characters, homology, or because they have independently acquired
> like characters, convergence. On the average, two animals with more
> homologous characters in common are more nearly related, their
> ancestral continuity is relatively more recent, than two animals with
> fewer."

This distinction is extremely important whenever one is constructing
either a classification or a phylogenetic tree. However, the distinction
does not go far enough. In the first place, it is rarely emphasized that
the homologous/analogous division is circular, in that it is necessary to
distinguish homology from analogy when constructing an evolutionary
tree, yet the distinction can only be reliably made when the outlines of
the tree (i.e. the basic pattern of ancestral branching-points) are already
known. Inheritance from a common ancestor can only be defined when
the existence of that common ancestor has been recognized and its
probable characters have been established. This is one of the major
reasons why the construction of evolutionary classifications is still more
of an art than a science. Secondly, one must go on to distinguish the
levels at which particular characters appear in any evolutionary tree
(Fig. 2). In any tree purporting to illustrate the evolutionary relationships between a number of related animal species, there is a basic
ancestral stock from which all of those species are derived. This is an
axiomatic feature dependent upon the assumption that all animal
species can ultimately be traced to a common ancestor. Since the
existence of this basal ancestral stock is axiomatic, characters retained
from that stock (although they are homologous) provide no information
about later subdivisions involving later common stocks. Such later
stocks can only be defined by the occurrence of specific characters
developed *after* their independent derivation from the basal ancestral

RADIATION GIVING RISE TO VARIOUS LIVING FORMS

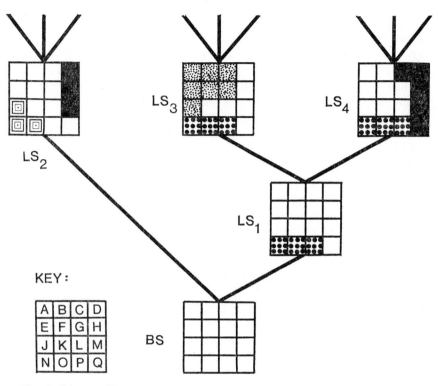

FIG. 2. Diagram illustrating the way in which the characters of descendent forms may be retained from the "basal ancestral stock" (BS) or modified in various later stocks (LS_1, LS_2, LS_3, LS_4) in the evolution of any group of related animals (adapted from Martin, 1968—Fig. 2). Characters developed in each stock influence the subsequent evolution of descendants from that stock. The following types of character similarities between stocks can be defined for any given evolutionary tree: (1) Basal stock characters (white squares). (2) Later stock characters (e.g. large black dots in squares N, O and P of LS_1). (3) Convergent characters (black squares D, H and M of LS_2 and LS_4). LS_2 shares 10 similarities with LS_4, but only 3 with LS_3. LS_3 shares 5 similarities with LS_4; but 3 of these similarities are particularly significant in that they are based on homologous characters retained from the "later ancestral stock" LS_1. Although LS_2 is superficially very similar to LS_4, 7 of the similarities are based on basal stock homology, and the other 3 are due to convergent evolution (presumably resulting from functional analogy).

stock. Thus, one must distinguish *later stock homology* from *basal stock homology* in any given comparison. For example, in discussing the phylogenetic relationships of the tree-shrews (Tupaiidae), there is no value in stressing similarities between tupaiids and primates when such similarities are based on the retention of ancestral placental mammal characters. It has often been maintained that the occurrence of the sublingua in tree-shrews and in some primates provides evidence for their phylogenetic affinity, since the sublingua is lacking in Insectivora. Since there is good reason to suppose that the sublingua was present in ancestral placental mammals (Sonntag, 1925), the presence of this organ in tree-shrews and prosimians is not, in itself, an indicator of affinity, as has been pointed out by Le Gros Clark (1962: 287). Yet the sublingua is almost certainly an homologous feature in tree-shrews and those primate species which have retained it.

It is therefore extremely important in discussing primate evolution to have a wide knowledge of primates and of other mammals, and to establish clear-cut hypotheses about the occurrence of particular branching points ("ancestral stocks") and about the points at which specific characters are thought to have appeared in the evolutionary tree. In many cases, the evolution of the Primates has been discussed without this broad coverage and without precise reference to a hypothetical tree. It might be said that vagueness in stating evolutionary hypotheses is to be commended, since it honestly reflects uncertainty; but this only applies to discussions of classification and not to construction of evolutionary trees. If an evolutionary tree is proposed, one must assume that the author has some idea of the characters which are present in each of the species represented and in each of the major branching-points of the tree. Scientific discussion of primate evolution is only possible if the premises and conclusions are clearly stated.

PRIMITIVE SURVIVORS AND THE SCALA NATURAE

In the discussion of any complex pattern of inter-relationships, it is useful to have a simple model which can be used as a first step in organization of the data. In general approaches to animal evolution, this simple model finds expression in the concept of the "primitive survivor"; that is, an animal species which has remained relatively close to the ancestral condition. Within any group of animal species, the choice of such a "primitive survivor" does of course involve a circular argument, in that one must have some idea of the overall evolution of the group in order to decide which species has remained the most primitive. Once such a choice has been made, it is common practice to refer to the chosen species as "primitive" in a general sense. However, it is wiser in discussing evolution to refer to primitive *characters*, rather than to primitive *species*, since blanket use of the term "primitive" implies that the species concerned ceased at some stage (i.e. quite close

to the "basal ancestral stock" of the group concerned) to be affected
by evolutionary processes which continued to bring about modifications
in other species of the group. It is, of course, generally accepted that
animals rarely, if ever, cease to evolve in this abrupt and absolute
manner. However, common reference to "primitive animals" (e.g. to
"primitive mammals" such as living Insectivora, when discussing
primate evolution) does lead to misunderstanding, and often to implicit
acceptance of concepts which prove to be suspect on closer examination.
The best example for this is provided by a quotation from Simpson
(1951: 167), since the view is all the more unusual in that it comes from
a scientist who is extremely rigorous in his approach and has made a
major contribution to evolutionary theory and to mammalian classification:

> "Even primate re-radiation would be entirely possible if present
> primates were wiped out and their empty ecological niches continued to
> exist, because living tree shrews have no specializations that would
> clearly exclude approximate repetition of such a radiation."

Quite apart from the fact that this statement is based on the questionable assumption that Primates evolved from a tree-shrew-like ancestor,
there is also the apparent assumption that tree-shrews have undergone
no significant evolutionary modifications since the time at which the
Order Primates began to develop (late Cretaceous?). This is the comparative anatomical concept of the "primitive survivor" in its purest
form, since Simpson must have based this judgement entirely upon a
small number of fossilizable morphological characters, with the added
drawback that there is virtually no fossil record for the evolution of the
tree-shrews (see van Valen, 1964, 1965 for summary). But the expression
of this concept by Simpson is not unusual when viewed in the light of
the general literature on evolution. The idea of the "primitive survivor"
has been a mainstay of biological theory for some time, and it has
undeniably been of great value in producing preliminary evolutionary
models. What we must now ask is whether this idea is adequate for the
process of further refinement of phylogenetic trees.

The concept of the "primitive survivor" has been used in many
different contexts in the analysis of animal evolution. In fact, it forms
part of a general philosophical approach to the evolution of life on earth.
The various forms of animal life are often interpreted as graded replicas
of the stages through which evolution passed prior to the origin of man.
This is the notion underlying references* to the "Great chain of being"

---

* It should be emphasized that the concept of the "Great chain of being" and that of the
"Scala naturae" were developed long before the theory of evolution by natural
selection.

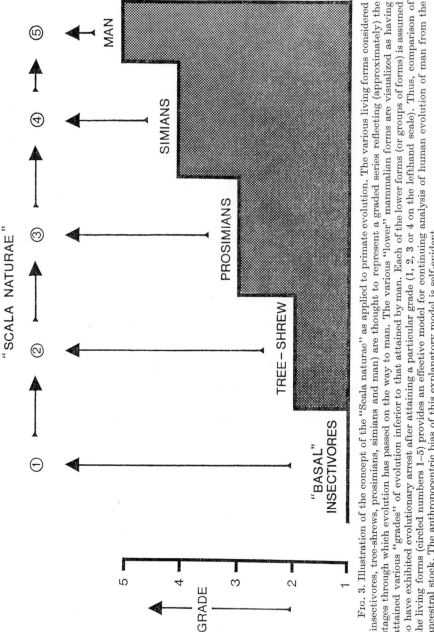

Fig. 3. Illustration of the concept of the "Scala naturae" as applied to primate evolution. The various living forms considered (insectivores, tree-shrews, prosimians, simians and man) are thought to represent a graded series reflecting (approximately) the stages through which evolution has passed on the way to man. The various "lower" mammalian forms are visualized as having attained various "grades" of evolution inferior to that attained by man. Each of the lower forms (or groups of forms) is assumed to have exhibited evolutionary arrest after attaining a particular grade (1, 2, 3 or 4 on the lefthand scale). Thus, comparison of the living forms (circled numbers 1–5) provides an effective model for continuing analysis of human evolution of man from the ancestral stock. The anthropocentric bias of this explanatory model is self-evident.

(Lovejoy, 1936), also referred to as the "Scala naturae" or the "phylogenetic scale" by various authors (Fig. 3). Recently use of the concept of the "Scala naturae" in comparative psychology has been effectively criticized by Hodos & Campbell (1969: 337), who state:

> "The concept that all living animals can be arranged along a continuous 'phylogenetic scale' with man at the top is inconsistent with contemporary views of animal evolution."

In fact, Hodos & Campbell were largely concerned with the extension of the phylogenetic scale model to comparative psychology, and they did indicate some acceptance of this approach with respect to comparative anatomy (Hodos & Campbell, 1969):

> "The earliest primates, the prosimians, appear to have developed as a specialization of the line of the insectivores. The living prosimians (tarsiers, lorises, and lemurs) retain some insectivore characteristics (Le Gros Clark, 1959). A comparison could therefore be made between living insectivores, prosimians, cercopithecoid (Old World) monkeys, pongids (great apes) and hominids (men) which would give some clue to patterns of evolution in the human lineage."

Yet the key word here is "clue", since the authors were well aware of the dangers of rigid interpretations of comparisons made between living forms. Simpson (1958) has also emphasized these dangers. Bearing this in mind, the following statements made by Hodos & Campbell can be taken as a warning to comparative anatomists, as well as to comparative psychologists (Hodos & Campbell, 1969: 339):

> "An important feature of the *Scala naturae* is the concept of a smooth continuity between living animal forms rather than the discontinuities implicit in the theory of evolution as a result of the divergence of evolutionary lines and the extinction of many intermediate forms."
>
> "Another characteristic of the *Scala naturae*, which seems to be implicit in discussions of a phylogenetic scale, is the notion that man is the inevitable goal of the evolutionary process and that once he has evolved, the phylogenetic process ends."

The idea of the "phylogenetic scale" has an obvious appeal, and it has served a useful purpose in enabling us to prepare an approximate outline of evolution within the Animal Kingdom. One invaluable service was performed in the early development of evolutionary theory, when Huxley (1864) made use of this simple model as a device to convince a largely sceptical public that human evolution had actually occurred. However, Huxley would have been one of the first to admit that this was a very simple model, and that it should not be taken literally. One

must not forget the value of the "Scala naturae" concept as an explanatory device and as a source of hypotheses about evolution; but, equally,
one must not slip into the circular (nay, *spiral*) argument of testing the
hypothesis in terms of the explanatory model. At this stage of our
knowledge, in order to guarantee a fruitful confrontation with the
results of independent biochemical assessments of primate evolution,
we must attempt to go beyond our simple explanatory model and to
construct hypotheses about actual ancestral stages. One can only agree
wholeheartedly with the final statement made by Hodos & Campbell
(1969: 349) with respect to comparative psychology:

> "These goals can best be attained by ridding ourselves of concepts
> like the phylogenetic scale, higher and lower animals, unrepresentative
> behavioural typologies, and other notions which have had the effect of
> oversimplifying an extremely complex field of research."

THE PRIMATE BRAIN

One of the central themes of discussions about primate evolution
has concerned the structure of the brain, following Huxley's particular
interest in this organ and its evolutionary significance. Indeed, one of
Huxley's statements (1864: 96) has long provided one of the main
justifications for use of the "phylogenetic scale" concept in analysing
primate evolution:

> "As if to demonstrate, by a striking example, the impossibility of
> erecting any cerebral barrier between man and the apes, Nature has
> provided us, in the latter animals, with an almost complete series of
> gradations from brains little higher than that of a rodent to brains
> little lower than that of Man".

Since Huxley's time, comparative anatomical investigations of the
primate brain have been concerned with two main aspects: (1) the size
of the brain relative to that of the body and (2) specific features of the
organization of the brain. Elliot Smith (1901, 1902a, 1903, 1908, 1927)
provided a foundation of comparative anatomical studies which brought
analysis of these features to an extremely high level and exerted an
undeniable influence on subsequent studies, particularly in the field of
cerebral organization. Indeed, one of Elliot Smith's statements (1902a:
425) is still generally appropriate as a summary of the cerebral developments which characterized primate evolution:

> "The brain of the Primates was derived from some Insectivore-like
> type, the cerebral hemispheres of which attained a precocious develop-
> ment and, as one of the expressions of their greatness, bulged backwards

over the cerebellum. In consequence of this great extension of the 'physical organ of the associative memory of visual, auditory, and tactile sensations', the sense of smell lost the predominance which it exercised in the primitive mammal (and in all the Orders of recent mammals), and the olfactory parts of the brain rapidly dwindled. This early Primate developed its distinctive type of calcarine sulcus and 'Sylvian fissure', the lateral, coronal and orbital sulci, and the characteristic central sulcus.''

One of the main extensions of Elliot Smith's work in this field was made through Le Gros Clark's detailed studies of the brain in various insectivores and primates (summarized by Le Gros Clark, 1962), and Le Gros Clark (1962: 227) has provided one of the most succinct summaries of the importance of the brain in primate evolution:

"Undoubtedly the most distinctive trait of the Primates, wherein this order contrasts with all other mammalian orders in its evolutionary history, is the tendency towards the development of a brain which is large in proportion to the total body weight, and which is particularly characterized by a relatively extensive and often richly convoluted cerebral cortex. It is true that during the first half of the Tertiary epoch of geological time the brain also underwent a progressive expansion in many other evolving groups of mammals, but in the Primates this expansion began earlier, proceeded more rapidly, and ultimately advanced much further."

Throughout the investigations conducted by Huxley, Elliot Smith and Le Gros Clark, the concept of the "phylogenetic scale" has been explicitly utilized. The result has been that many fundamental features of evolutionary change in the development of the primate brain have been broadly identified and subjected to detailed examination. However, it is not at all clear how far the actual course of evolution within the Primates has been reconstructed, since there has been relatively little attempt to move away from the comparison of living forms and the simple explanatory model of the "Scala naturae". In order to move beyond the foundation provided by Elliot Smith's work, one must attempt to reconstruct actual evolutionary stages and to examine the resulting hypotheses through reference to endocranial casts available from the fossil record.

Size relationships

One simple approach to development of the brain is to compare its *size* relative to that of the body. Brief reference to this was made by Huxley (1864), Elliot Smith (1902a) and Le Gros Clark (1962). In fact, Huxley (1864: 77)—in addition to indicating the possible value of brain size as such—also provided the earliest warning of the crudeness of this measure, with reference to the human brain:

> "The heaviest brain (1872 gm) was, however, that of a woman; next to it comes the brain of Cuvier (1861 gm), then Byron (1807 gm), and then an insane person (1783 gm)."

However, if fairly large samples are taken from several different species, general inter-specific comparisons can be made to provide a broad indication of differences in relative development of the brain. The basis for such studies was provided by Snell (1892) and Dubois (1897), who developed the empirical formula:

$$C = b. B^\alpha$$

(where C = brain weight; B = body weight; b = constant for a natural group of vertebrates; α = approximately 0·63).

This initial work has been greatly expanded by the meticulous studies carried out by R. Bauchot, H. Stephan and their collaborators (see Stephan, 1959; Bauchot & Stephan, 1964; Stephan & Bauchot, 1965; Stephan, 1972). These two authors have established tentative characteristic brain and body weights for many insectivore and primate species, and have subjected the figures obtained to detailed analysis. In particular, they have expressed the relationships between brain and body weight for four different groups ("basal insectivores"; "other insectivores"; prosimians; simians) on a double logarithmic scale (Fig. 4; see also Stephan, 1972), which indicates approximately linear arrangements following the formula:

$$\log C = \log b + \alpha \log B$$

This presentation is extremely useful, in that it shows:

1. That there would appear to be a regular inter-specific relationship between brain and body weight for animals of different sizes within each of the four "natural" groups selected, and

2. That there is a considerable difference between the four groups in the size of the brain for any given body size. (On average, "other insectivores" have a brain size approximately 1·5× that of "basal insectivores" of similar size, whilst prosimians and simians have brains 4·0× and 6·5×, respectively, larger than those of "basal insectivores" of comparable size.)

Such statistical treatments are also useful in that they permit direct comparison with other mammal groups (e.g. see Stephan, 1972). These comparisons show that relative brain sizes similar to those found in prosimians occur in many non-primate mammals, and that similar values to those found in simians can be found in a number of mammal

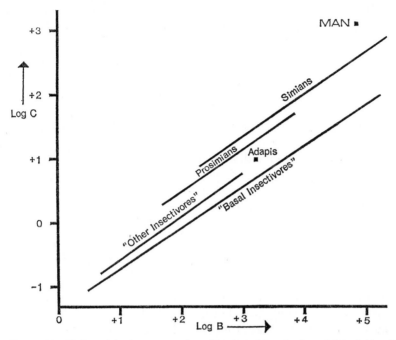

Fig. 4. Double logarithmic presentation of brain weight : body weight relationships, derived from Stephan (1972: 157). Note that all four lines have approximately the same slope (α approximately equal to 0·63). The value for *Adapis*, based on an estimated body weight of 1½ kg and a brain weight of 10 g (cf. Table I), lies close to the line for the "other insectivores".

groups (e.g. Pinnipedia; Cetacea). The only primate which is outstanding among mammals generally in terms of relative brain size is *Homo sapiens*, and Stephan (1972) points out that the difference between Man and simians is as great as that between simians and "basal insectivores". Thus, large brain size as such is not a unique feature of the Order Primates, and the prosimians in fact have relatively small brain sizes in comparison to those of a number of other mammal groups.

Nevertheless, one can discuss evolution of the brain *within* the Order Primates, provided that the species under consideration have been firmly established as primates on other grounds. The tree-shrews, for example, cannot be regarded as primates merely on the grounds of their relatively large brains, and they should not be considered unless their primate status has been established through independent evidence.

In discussing primate brain evolution, Stephan utilizes the "basal insectivores" as a base-line of mammalian development, with the implication that these insectivores (some Tenrecidae, Erinaceidae and

Soricidae) have brains identical in relative size and structure to those of Cretaceous ancestral placental mammals. Although it is explicitly stated by Stephan (1972) that the "Ascending Primate Scale" does not represent a direct sequence of evolutionary stages, his discussion of brain evolution is based on two assumptions:

1. Lower extant forms correspond well in their brain patterns with phylogenetic stages.

2. Reverse evolution (reduction of neocortex) is unlikely under natural conditions.

Since discussion of mammalian brain evolution has been based, almost without exception, on extant forms arranged on a "phylogenetic scale", the circularity of these assumptions is unavoidable in any discussion which does not take account of the fossil record. Thus, one cannot unhesitatingly accept Stephan's claim (1972: 163) that:

> "The graph based on the indices of neo-cortical progression (Fig. 6–6) shows that, within the Primates many different evolutionary stages are preserved in extant species."

Such statements can only be verified by demonstration that certain extant species do really have brains identical in relative size and form to those of ancestral forms. It must be shown that living "basal insectivores" have brains identical to those of Cretaceous ancestral placentals (see Fig. 5), and that living prosimians have brains identical to those of the ancestral primates which gave rise both to extant prosimians and to extant simians.

Surprisingly, there is relatively little reference in the literature to fossil endocasts, which provide the only reliable information about brain size and gross structure in extinct forms. Edinger's classical paper (1929) on the brains of fossil forms has rarely been cited in general discussions of mammalian brain evolution, with the notable exception of Piveteau (1958), and there has been little work to extend Edinger's preliminary conclusions. Recently, however, Jerison (1961, 1970) and Radinsky (1967, 1970) have conducted detailed investigations of mammalian endocasts and have extended considerations of *relative* brain size to the fossil record. (In order to conduct a meaningful discussion of brain evolution, it is necessary to include considerations of body size so that one can compare extant and fossil forms in terms of standardized brain : body relationships.) Jerison (1970), for example, has extended the double logarithmic presentation of brain weight : body weight relationships to include values for certain small-brained early Tertiary (Palaeocene/Eocene) mammals (condylarths, amblypods and creodonts). This presentation indicates that these "archaic mammals"

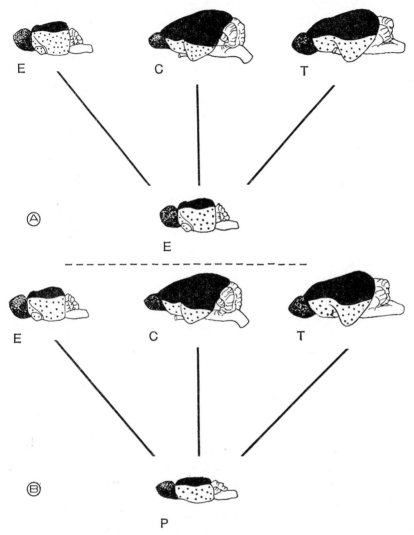

FIG. 5. Two alternative hypotheses for the evolution of brain size and structure in a living "basal insectivore" (E = *Erinaceus europaeus*), a lemur (C = *Cheirogaleus medius*) and a tree-shrew (T = *Tupaia belangeri*). The lemur and the tree-shrew are similar in body-weight (brains drawn to the same scale), whilst the brain of *Erinaceus* has been reduced in proportion to its body-weight, following Stephan's data (1959). *Hypothesis A:* The brain of *Erinaceus* has remained unchanged from the ancestral mammalian condition. In both the tree-shrew and the lemur, the olfactory bulbs (fine stippling) have been reduced in size, and the neopallium (black) has expanded to cover the archipallium (heavy stippling). *Hypothesis B:* The ancestral mammalian brain (P) was smaller and structurally less developed than in any of the three living forms. Expansion of the olfactory bulbs has occurred in both *Erinaceus* and *Tupaia*, whilst the olfactory areas of the *Cheirogaleus* brain have remained relatively stable in extent. Expansion of the neopallium has occurred to a limited extent in *Erinaceus* and to a far greater extent in *Tupaia* and *Cheirogaleus*.

had similar brain : body weight relationships to those found in modern
"basal insectivores". The values quoted by Jerison for the creodont
Cynohyaenodon cayluxi, for example, fall almost exactly on the "basal
insectivore" line shown in Fig. 4. Jerison implies that this indicates the
validity of Stephan's assumption that living "basal insectivores"
possess brains which are not advanced beyond the ancestral placental
condition. But this is to forget that the placental mammals are derived
from a common ancestor which existed in the mid-Cretaceous, and not
from an early Tertiary ancestor. The oldest early Tertiary forms
examined by Jerison (1970) date back to 60 million years ago, whereas
the ancestral placental stock may have existed as far back as 100 million
years ago (or more). It is logical to assume that the relative brain size
of the ancestral placental mammals was even smaller than that in
extant "basal insectivores" and in "archaic" early Tertiary mammals,
and that the latter underwent a definite (though restricted) expansion
in brain-size in the 40 million years (or so) which elapsed after the initial
emergence of the placental mammals.

Some evidence for this latter view is provided by the fact that the
Jurassic mammal *Triconodon mordax* (Simpson, 1927) had an extremely
small, poorly elaborated brain (see also Martin, 1968). It is reasonable
to assume that the relative brain size and development of ancestral
placental mammals in the Cretaceous was intermediate between that
of Jurassic forms and that found in Jerison's "archaic" early Tertiary
mammals. Support for this assumption can be derived from information
on the endocasts of fossil Leptictidae (Edinger, 1929). The Leptictidae
are generally regarded (e.g. Romer, 1966) as an ancient family of the
Order Insectivora in which there was little modification from the
ancestral placental mammal condition. Edinger (1929: 147) indicates
that, for comparable skull lengths, the brain-case of the Oligocene
leptictid *Ictops* was probably somewhat shorter (3 cm versus 3·5 cm)
than in the extant insectivore *Erinaceus europaeus*. Until detailed
volumetric comparisons have been carried out, this difference cannot
be regarded as conclusively significant, but it does indicate that even
in Oligocene leptictids the brain may have been of noticeably smaller
volume (by 20–40%) than the brain of an extant "basal insectivore"
(*Erinaceus*) of comparable skull-size (see Fig. 5).

Within the Order Primates, Radinsky (1970) has given an admirable
review of prosimian endocasts, showing that it is even possible to carry
out a limited comparison of sulcal patterns. Radinsky also makes
reference to a technique (see also Radinsky, 1967) for plotting brain
volume against body size when only the skull is available for examination. It is claimed that the cross-sectional area of the foramen magnum

can be taken as a measure of body size, and Radinsky (1970: 222) provides a graph showing that the logarithmic ratio of endocranial volume to foramen magnum area in *Adapis parisiensis* is similar to that found in living lemurs. Radinsky concludes that the relative brain size of *Adapis* falls within the range of modern prosimians. This is surprising in view of the fact that *Adapis parisiensis* has an extremely small brain-case relative to overall skull size, as was recognized by Le Gros Clark (1945) in his original paper on the endocranial cast of this species. It is also obvious from Gregory's monograph (1920) on the closely-related Notharctinae that the most striking difference between the skull of *Notharctus osborni* and a skull of comparable size from the extant genus *Lemur* lies in the extremely small brain-case of the former.

An alternative interpretation of Radinsky's graph is that the area of the foramen magnum is more strongly correlated with brain-size than with body-size in the forms compared, and that *Adapis* shows relation-ships similar to those in extant prosimians primarily for this reason. One way of checking this is to compare the cranial capacity of the *Adapis parisiensis* skull directly with that of skulls of similar size taken from extant *Lemur* species (Table I; Figs 6 & 7). This comparison shows that the cranial capacity of the Eocene *Adapis* skull is approximately half that of an extant *Lemur* skull which is comparable in its essential dimensions (Table I). Unfortunately, the post-cranial skeleton of *Adapis parisiensis* is insufficiently known to permit direct comparison of the body-size with that of *Lemur mongoz*, though it seems reasonable to assume that the body-sizes were quite comparable. For the time being one can at least conclude that *Adapis* had a considerably smaller brain weight : body weight ratio than do extant prosimians. Assuming that *Adapis parisiensis* did have a body-size comparable to that of *Lemur mongoz*, it can be calculated (Fig. 4) that *Adapis* falls within the range of Stephan's "other insectivores" in terms of its relative brain-size. Since the skull of *Adapis parisiensis* dates back only to about 40 million years ago (cf. Radinsky, 1970), it is open to question whether the ancestral primates of 65–80 million years ago had relative brain sizes much larger than those of extant "basal insectivores". Until studies can be made of endocranial casts of Palaeocene or Late Cretaceous primates, this question must remain largely unanswered. But it has at least been established that one cannot simply equate the relative brain size of extant "basal insectivores" with that of Cretaceous ancestral placental mammals and that even in those living prosimian species which have (relatively) the smallest brains there has been a marked expansion in brain-size (by a size factor of at least 2) since their divergence from the ancestral primate stock.

P

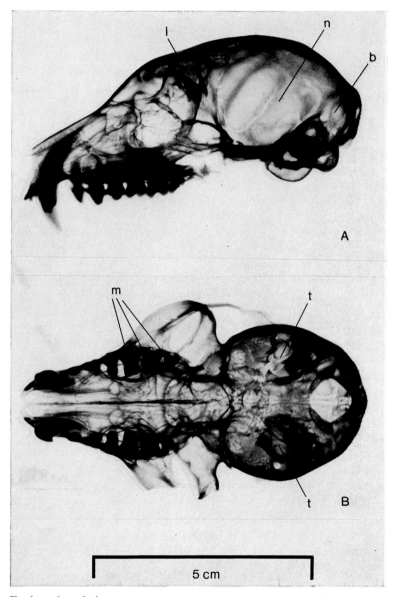

A

B

5 cm

For legend see facing page.

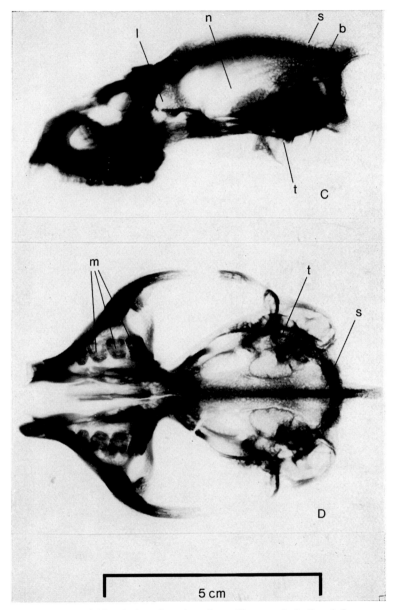

Fɪɢ. 6. Prints of identical scale taken from X-rays of skulls of *Lemur mongoz*
(BM.35.1.8.54: A = lateral view; B = view from above) and of *Adapis parisiensis*
(BM.M.1345: C = lateral view; D = view from above). l, olfactory bulb cavity;
n, cerebral cavity; b, cerebellar cavity; m, upper molars 1, 2 and 3; t, free ectotympanic
ring (characteristic of adapids and Malagasy lemurs); s, marked sagittal crest of *Adapis*,
indicating powerful temporal musculature. Note the far smaller cerebrum in *Adapis*.

TABLE I

Comparison of major cranial dimensions* and cranial capacities† in Lemur and Adapis

		Lemur fulvus rufus BM 39.3944 (B.M.N.H.)	Lemur mongoz coronatus BM 35.1.8.54 (B.M.N.H.)	Adapis parisiensis M.1345 (B.M.N.H.)
Skull length	Anterior extremity of premaxilla to lower lip of foramen magnum	75·0 mm	66·0 mm	72·0 mm (approx.)
	Anterior extremity of premaxilla to upper lip of foramen magnum	82·5 mm	74·0 mm	74·0 mm (approx.)
	Maximum cranial length	86·0 mm	76·0 mm	83·0 mm (approx.)
Skull width (Maximum width between outer faces of zygomatic arches)		52·0 mm	45·0 mm	54·5 mm
Foramen magnum dimensions	Maximum height	8·5 mm	10·5 mm	7·0 mm
	Maximum width	11·0 mm	10·0 mm	8·5 mm
Left cheek tooth row (premolars and molars)	Maximum length	29·0 mm‡	25·0 mm‡	25·5 mm§
	Maximum width	6·5 mm	6·0 mm	5·0 mm
Orbit size‖	Maximum height	9·0 mm	8·5 mm	5·0 mm
	Maximum width	10·0 mm	9·0 mm	4·0 mm
Cranial capacity†		22·2 cc	17·6 cc	8·8 cc

* All dimensions to nearest 0·5 mm.
† Cranial capacities to nearest 0·2 cc, measured with mustard seed packed into the brain case (see also Fig. 7).
‡ 3 premolars; 3 molars.
§ 4 premolars; 3 molars.

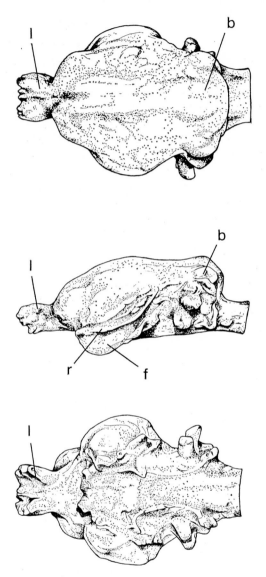

Fig. 7. Illustrations of the endocranial cast of *Adapis parisiensis* BM.M.1345 (prepared from photographs taken by the author). Note the well-developed temporal lobe (f), the relatively large olfactory bulbs (l), the relatively low position of the rhinal fissure (r; indicated by the vascular canal on the temporal lobe), and the dorsal exposure of the cerebellum (b).

Gross brain morphology

Elliot Smith (1902a, 1927) and Le Gros Clark (1962) directed most of their attention, in discussing primate brain evolution, to structural features of the brain. Bauchot and Stephan have also paid attention to those features, in so far as they have compared the sizes of various component parts of the brain to body-size. However, these latter studies are subject to the same limitation as that applying to consideration of overall brain-size: one cannot discuss evolution of the primate brain in terms of its derivation from the extant "basal insectivore" condition without relying almost exclusively on circular arguments.

It is generally regarded as an established fact of primate evolution that there has been increasing elaboration of those areas of the brain concerned with vision, hearing and touch, accompanied by a decrease in the importance of the sense of smell (Elliot Smith, 1927; Le Gros Clark, 1962). However, it is not immediately clear whether decrease in the importance of the sense of smell has merely led to *arrest* of the evolution of the olfactory areas within the Order Primates, or has actually brought about *physical reduction* of these areas (Martin, 1968). Both Elliot Smith (1902a, 1927) and Le Gros Clark (1962) have given the impression that the latter is the case; but their assumptions are based on comparisons with living insectivores (cf. Fig. 5). If the "basal insectivores" have undergone *expansion* of olfactory areas of the brain since their derivation from the ancestral placental mammal stock, such comparisons must inevitably lead to confusion. In view of the habits of extant "basal insectivores", such further development of the olfactory structures would not be surprising.

In fact, even assuming that the "basal insectivores" do exhibit olfactory bulbs similar in absolute size (for a given body-size) to those of ancestral placentals, there is only slight evidence for "reduction" of the olfactory bulbs in the evolution of the lemurs and lorises, as was recognized by Elliot Smith (1902a). (Though it should be noted, in passing, that Elliot Smith proposed that the lemurs and lorises had first undergone reduction of the olfactory bulbs in evolution and had subsequently undergone retrogressive expansion to regain the original condition.) Stephan (1972: 164) shows that the palaeocortex and olfactory bulb are only slightly smaller (relatively speaking) in extant prosimians than in extant "basal insectivores", whilst these regions of the brain are considerably smaller in simians (and *Tarsius*). Thus, if one accepts the possibility that the extant "basal insectivores" have undergone enlargement of the olfactory regions of the brain, it could be proposed that the absolute size of these areas has remained generally stable

in lemurs and lorises (Strepsirhini), whilst there has been actual reduction of the olfactory regions in the evolution of the tarsiers and simians (Haplorhini). This proposition is supported by the fact that the olfactory bulbs are approximately of the same absolute size in *Adapis parisiensis* and the extant *Lemur mongoz*, despite the fact that the living form has a much larger brain overall. On the other hand, Radinsky (1970) indicates that the relative size of the olfactory bulbs was considerably larger in the fossil *Tetonius* and *Necrolemur* than in the extant *Tarsius*, though he also suggests that the size of the olfactory bulbs was "reduced" in the Eocene tarsioid *Tetonius*. However, until a precise comparison has been made between *Tetonius* and other mammals of similar body-size, it is not clear whether this early tarsioid merely had small olfactory bulbs relative to its overall brain-size or whether actual reduction had occurred. It does seem that the fossil tarsioids may have had relatively large brain weight : body weight ratios and relatively small olfactory areas compared to Eocene lemuroids such as *Adapis* and *Notharctus*; but no figures are available in the literature to substantiate this subjective impression.

Apart from the question of the size of the olfactory areas of the brain, there has been little attempt since Elliot Smith's monograph (1902a) to define specific features of the primate brain. Le Gros Clark (1962, 227–264) does state that the primate brain "shows certain features of its intrinsic organization which also distinguish it from that of other mammals"; but he does not actually list particular features in which the primate brain differs from that of all other mammals. Most of the characteristics of the primate brain to which Le Gros Clark refers are general developments associated with increasing emphasis on vision, hearing and touch and with a relative decrease in the importance of the sense of smell. With respect to these developments, Le Gros Clark did, in fact, provide some definition of the primate brain (by implication) in listing the specific aspects in which the "tree-shrew brain" (viz. that of *Tupaia minor*) differs from that of the living Insectivora and resembles that of the Primates (Le Gros Clark, 1962: 241);

1. The relative size of the brain as a whole.

2. Expansion of the neopallium, with accompanying downward displacement of the rhinal sulcus.

3. Formation of a distinct temporal pole of the neopallium.

4. Backward projection of the occipital pole.

5. The presence of a calcarine sulcus.

6. Well-marked lamination and cellular richness of the neopallial cortex, and the degree of differentiation of the several cortical areas.

7. Pronounced elaboration of the visual apparatus of the brain, especially in the cortex.

8. Advanced degree of differentiation of nuclear elements of the thalamus.

9. Well-defined cellular lamination of the lateral geniculate nucleus. This list has already been effectively criticized by Campbell (1966a,b), and his comments need only be summarized here. The major criticism is that all of the characteristics listed occur in various other mammal groups, particularly in association with arboreal habits. Indeed, many of the above characteristics are found in the arboreal marsupial *Trichosurus vulpecula*, so developments of this kind are not even restricted to placental mammals. For example, Campbell cites evidence that *T. vulpecula* has a far more prominent and extensive calcarine sulcus than does *Tupaia*. Thus, one cannot exclude the possibility that the brain of *Tupaia minor* has developed the above "primate-like" features independently, in adaptation to arboreal habits. In addition, one must question Le Gros Clark's reference to *Tupaia minor* as representative of the "tree-shrew brain", since—by his own admission—many of the stated "primate-like features" are lacking in the brain of *Ptilocercus lowii*. Since the common ancestor of *Ptilocercus lowii* and *Tupaia minor* undoubtedly possessed an even more primitive central nervous system than the least specialized of the living forms (*Ptilocercus*), there can be no doubt that the "primate-like" features of the *T. minor* brain represent independent developments.

The calcarine sulcus is of particular interest, since the presence of this feature in the primate brain was listed by Mivart (1873) as a defining characteristic of the Order Primates. Le Gros Clark's interpretation of the presence of a poorly-developed calcarine sulcus in *Tupaia minor* as a character indicating affinity with Primates can be related to this definition. However, as Campbell has pointed out (1966b), a calcarine sulcus occurs in many mammals—marsupials as well as placentals—and this feature is therefore not exclusively diagnostic of Primates. In addition, Le Gros Clark noted the absence of a calcarine sulcus in *Ptilocercus*, and it can be assumed that ancestral tree-shrews lacked this feature. Most important of all is Elliot Smith's observation (1902a) that the formation of a calcarine sulcus is correlated with overall brain-size. In small-bodied marsupials and bats, for example, the calcarine sulcus is absent, whereas it is present in larger representatives of these two groups. Elliot Smith in fact regarded the calcarine sulci of various mammals as *homologous*, though it is unlikely that their small-brained (probably small-bodied) common ancestor actually exhibited this sulcus. Such expanded use of the homology concept

occurs repeatedly in Elliot Smith's monograph (1902a), whereas modern writers would generally prefer to use the word *parallelism* (e.g. Simpson, 1961) for the separate appearance of similar structures in several different descendent lines. Nevertheless, one can accept Elliot Smith's basic implication that the structure of the ancestral marsupial/placental brain was such that size-increase would automatically lead to formation of a calcarine sulcus. In addition, it has been overlooked that Elliot Smith regarded the formation of a *triradiate* calcarine sulcus complex as characteristic of Primates. This triradiate complex (consisting of calcarine, retrocalcarine and paracalcarine sulci) was apparently present in all of the primate brains examined by Elliot Smith (1902a) and lacking in all other mammalian brains which he studied. Such a triradiate complex has never been reported in tree-shrews.

Thus, there are apparently few specific structures which characterize the primate brain. If one were to attempt a distinction between the ancestral placental mammal brain and the ancestral primate brain, there is very little that can be said. One could perhaps suggest that there was a triradiate calcarine complex (following Elliot Smith, 1902a) and a well-developed pyramidal tract system, with the tracts located in the lateral funiculi of the spinal cord and descending to a lumbar or sacral level (following Campbell, 1966b); but it is impossible to confirm the existence of these features in fossil primates. The other specific features suggested by Elliot Smith as typical of the primate brain must be ruled out as unlikely *ancestral* features, either because these features are known not to have occurred in early fossil primates such as *Adapis* (e.g. there is no overlap of the cerebellum by the occipital cortex), or because certain characters (such as the lateral, coronal, orbital and central sulci) can be reliably interpreted as independent, parallel developments in various later, larger-bodied Primates. Elliot Smith did regard the "sylvian sulcus" as characteristic of primate brains; but it should be emphasized that he regarded this sulcus, in Primates, as the product of "a peculiar union" (Elliot Smith, 1902a: 404) between the suprasylvian and pseudosylvian sulci found in other mammals. This definition would appear to be applicable to the "sylvian sulcus" of *Adapis* and of *Necrolemur* (for example), and it could be postulated that the presence of this distinctive sulcus was characteristic of ancestral primates. Although there have been occasional reports that tree-shrews have a shallow "sylvian fossa" (e.g. Le Gros Clark, 1962), there is no evidence that this feature is homologous with the "peculiar" sylvian sulcus identified in Primates by Elliot Smith. In sum, of the few features which may be distinctively typical of the Primate brain, not one is found in tree-shrews.

CONCLUSIONS

When the evolution of the primate brain (in terms of size and gross structure) is considered in relation to actual ancestral stages—rather than purely in the light of comparisons made between living mammals— a somewhat novel picture emerges. Although primates are generally characterized by relatively large brains, most primates differ from other mammals only in that the process of brain-enlargement may have occurred earlier in evolutionary history. (For example, there is some evidence for precocious development of the temporal lobe in primate evolution—cf. Le Gros Clark, 1962; & Fig. 7.) There are a small number of discrete structural features which may provisionally be regarded as distinctive of the brains of living primate species; but it is a moot point whether these were actually present in the ancestral primate stock. Certainly, there is so little that is unique in the gross evolution of the brain within the Order Primates as a whole that cerebral features are of little systematic value in their own right. For example, resemblances in brain structure and size between tree-shrews and primates may be traced to retention of ancestral mammalian features and to convergent elaboration of certain areas (primarily visual) in relation to arboreal habits.

Within the Order Primates, however, differential developments of the central nervous system may be of crucial significance to an understanding of primate adaptive radiation. Elliot Smith (Fig. 8) regarded the tarsiers as more closely related to the simians than to the prosimians, and his stated views (Elliot Smith, 1902a, 1927) support Hill's division (1953) of the Primates into two basic groups—the Strepsirhini (lemurs and lorises) and the Haplorhini (tarsiers, monkeys, apes and man). This basic division is primarily linked to the retention of the rhinarium in Strepsirhini and its loss in Haplorhini. New evidence on *Microcebus murinus* (Schilling, 1970: 214–215) indicates that the rhinarium may be linked in many Strepsirhini with the operation of the Jacobson's organ. In correlation with this, living Strepsirhini typically exhibit a gap between the upper incisors, marking the backward passage of the ventral rhinarium, which bears a central groove leading to the apertures of the Jacobson's organ. This gap between the upper incisors is also found in Notharctinae and Adapinae. In living Haplorhini, however, the upper incisors are not separated by a gap, and the same seems to apply to fossil haplorhines such as *Necrolemur*, *Tetonius* and various early simian forms. Since Jacobson's organ serves an accessory olfactory role, one could postulate (following Elliot Smith) that the olfactory function was reduced at a very early stage in haplorhine evolution. In

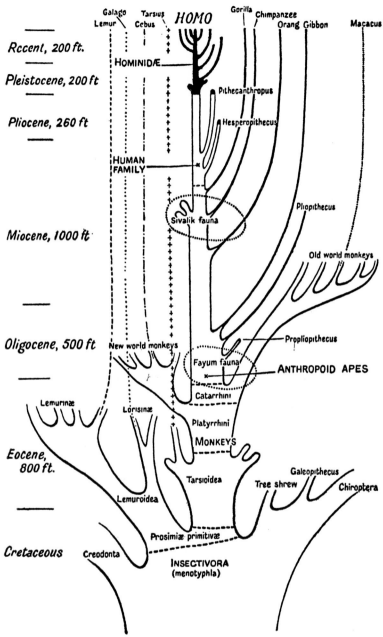

FIG. 8. Elliot Smith's (1927; Fig. 2) phylogenetic tree of the Primates (from *Evolution of Man* by G. Elliot Smith, republished with the kind permission of the Oxford University Press). Note the suggestion of an ancestral relationship between tarsiers and simians, and the departure from the usual anthropocentric representation of Man as the "most advanced form" on the far right of the tree.

strepsirhines, on the other hand, the olfactory function was generally maintained at its primitive level, with reduction occurring sporadically in some recent (mainly diurnal) lemurs. This distinction between Strepsirhini and Haplorhini in relative emphasis on olfaction is inversely correlated with visual developments, which are more marked in tarsiers and simians than in lemurs and lorises. This, in turn, can be correlated with the fact that lemurs and lorises can reasonably be traced to a nocturnal ancestor in which a tapetum had been developed for more effective vision under conditions of restricted light intensity (Martin, 1972). The Haplorhini—with the exception of *Tarsius* and *Aotus*—are diurnal in habits, and there is reason to believe (Le Gros Clark, 1962; Wolin & Massopust, 1970) that the two exceptional nocturnal haplorhine genera are secondarily derived from a diurnal ancestor possessing a mixed cone and rod retina incorporating a fovea, as an adaptation to normal daylight vision. One can therefore suggest that the ancestral primates gave rise to a predominantly nocturnal branch (Strepsirhini) in which the olfactory sense was generally maintained at a primitive level, and to a predominantly diurnal branch (Haplorhini) in which the olfactory sense was reduced in importance at an early stage and primarily visual developments gradually exerted an overwhelming influence on cerebral organization. This interpretation is broadly in accord with the evidence and opinions published by Elliot Smith, and it would also explain why—in a very loose sense—living prosimians exhibit a lower "grade" of cerebral organization than do living simians.

ACKNOWLEDGEMENTS

I am extremely grateful to Prof. N. A. Barnicot, whose advice and encouragement have been invaluable in developing and refining the principles expressed in this paper. Thanks also go to the staff of the British Museum of Natural History (Sub-Department of Anthropology) for access to key specimens and for assistance, especially from Miss Theya Molleson, who suggested and carried out X-ray photographs of the *Adapis* and *Lemur* skulls. I would also like to thank the staff of the audiovisual aids section of the Anatomy Department, University College London, for their assistance in preparing prints of the X-rays. Thanks are also due to my wife, Anne-Elise, for help in preparing the text figures, and to the Curator of the Royal College of Surgeons (Miss Dobson) for advice relating to Elliot Smith's work on the vertebrate brain.

The concepts set out in this paper have been gradually formulated in the course of behavioural and morphological studies of tree-shrews and lemurs. The former were studied at the Max-Planck-Institut für Verhaltensphysiologie, Seewiesen (thanks to Prof. K. Z. Lorenz and Dr I. Eibl-Eibesfeldt) and at the Department of Zoology, Oxford (thanks to Prof. N. Tinbergen). The latter were studied at the Muséum National d'Histoire Naturelle, Ecologie Générale, Brunoy, France (thanks to Prof. C. Delamare-Deboutteville and Dr J.-J. Petter). Financial support for these studies was provided by grants from the German Academic Exchange Service and by the Science Research Council (London). A laboratory study of the Mouse-lemur (*Microcebus murinus*) and the Fat-tailed dwarf lemur (*Cheirogaleus medius*), which has yielded relevant information, is at present being conducted with the aid of a grant from the Medical Research Council (London).

REFERENCES

Barnabas, J., Goodman, M. & Moore, G. W. (1972). Descent of mammalian Alpha Globin chain sequences investigated by the maximum parsimony method. *J. molec. Biol.* **69**: 249–278.

Bauchot, R. & Stephan, H. (1964). Le poids encéphalique chez les Insectivores malgaches. *Acta zool., Stockh.* **45**: 63–75.

Cain, A. J. (1954). *Animal species and their evolution.* London: Hutchinson University Library.

Campbell, C. B. G. (1966a). Taxonomic status of tree shrews. *Science, N.Y.* **153**: 436.

Campbell, C. B. G. (1966b). The relationships of the tree shrews: the evidence of the nervous system. *Evolution* **20**: 276–281.

Cook, C. & Hewett-Emmett, D. (in press). The uses of protein sequence data in systematics. In *Prosimian biology.* Martin, R. D., Doyle, G. A. & Walker, A. C. (eds). London: Duckworth.

Darwin, C. (1859). *On the origin of species.* London: John Murray.

Dubois, E. (1897). Über die Abhängikeit des Hirngewichts von der Körpergrösse bei den Säugetieren. *Arch. Anthrop.* **25**: 1–28.

Edinger, T. (1929). Die fossilen Gehirne. *Ergebn. Anat. EntwGesch.* **28**: 1–249.

Elliot Smith, G. (1901). The natural subdivision of the cerebral hemisphere. *J. Anat. Physiol.* **35**: 431–454.

Elliot Smith, G. (1902a). On the morphology of the brain in the Mammalia, with special reference to that of the lemurs, recent and extinct. *Trans. Linn. Soc. Lond.* (Zool.) **8**: 319–432.

Elliot Smith, G. (1902b). *Catalogue of the physiological series of comparative anatomy.* London: Royal College of Surgeons.

Elliot Smith, G. (1903). Further notes on the lemurs, with special reference to the brain. *J. Linn. Soc. Lond.* (Zool.) **29**: 80–89.

Elliot Smith, G. (1908). On the form of the brain in the extinct lemurs of Madagascar, with some remarks on the affinities of the Indrisinae. *Trans. zool. Soc. Lond.* **18**: 163–177.

Elliot Smith, G. (1927). *The evolution of man: essays* (2nd Ed.). London: Oxford University Press.

Gregory, W. K. (1916). Studies on the evolution of Primates. *Bull. Am. Mus.. nat. Hist.* **35**: 239–356.

Gregory, W. K. (1920). On the structure and relations of *Notharctus*, an American Eocene Primate. *Mem. Am. Mus. nat. Hist.* (n.s.) **3**: 59–243.

Hennig, W. (1950). *Grundzüge einer Theorie der phylogenetischen Systematik.* Berlin: Deutscher Zentralverlag.

Hill, W. C. O. (1953). *Primates.* **1**: *Strepsirhini.* Edinburgh University Press.

Hodos, W. & Campbell, C. B. G. (1969). *Scala naturae*: Why there is no theory in comparative psychology. *Psychol. Rev.* **76**: 337–350.

Huxley, T. H. (1864). *Evidence as to man's place in nature.* London: Williams and Norgate.

Jerison, H. J. (1961). Quantitative analysis of evolution of the brain in mammals. *Science, N.Y.* **133**: 1012–1014.

Jerison, H. J. (1970). Gross brain indices and the analysis of fossil endocasts. In *The Primate brain*: 225–244. Noback, C. R. & Montagna, W. (eds). New York: Appleton Century Crofts.

Le Gros Clark, W. E. (1934). *Early forerunners of man.* London: Bailliere, Tindall & Cox.

Le Gros Clark, W. E. (1945). Note on the palaeontology of the lemuroid brain. *J. Anat.* **79**: 123–126.

Le Gros Clark, W. E. (1959). *The antecedents of man.* Edinburgh: University Press.

Le Gros Clark, W. E. (1962). *The antecedents of man.* (Revised edition). Edinburgh: University Press.

Linnaeus, C. (1758). *Systema naturae per regna tria naturae, secundam classes, ordines, genera, species cum characteribus, synonymis, locis.* Stockholm: Laurentii Salvii.

Lovejoy, A. O. (1936). *The great chain of being.* Cambridge Mass.: Harvard University Press.

Martin, R. D. (1968). Towards a new definition of Primates. *Man* **3**: 377–401.

Martin, R. D. (1972). Adaptive radiation and behaviour of the Malagasy Lemurs. *Phil. Trans. R. Soc.* (B) **26**: 295–352.

Mayr, E. (1942). *Systematics and the origin of species.* New York: Columbia University Press.

Mivart, St.-G. J. (1873). On *Lepilemur* and *Cheirogaleus* and on the zoological rank of the *Lemuroidea. Proc. zool. Soc. Lond.* **1873**: 484–510.

Napier, J. R. (1971). *The roots of mankind.* London: George Allen & Unwin.

Piveteau, J. (ed.) (1958). *Traité de Paléontologie:* **7**(2): Mammifères, Évolution. Paris: Masson et Cie.

Radinsky, L. B. (1967). Relative brain size: a new measure. *Science, N.Y.* **155**: 838.

Radinsky, L. B. (1970). The fossil evidence of prosimian brain evolution. In *The Primate brain*: 209–224. Noback, C. R. & Montagna, W. (eds). New York: Appleton Century Crofts.

Romer, A. S. (1966). *Vertebrate palaeontology.* (3rd Ed.) Chicago: University of Chicago Press.

Schilling, A. (1970). L'organe de Jacobson du lémurien malgache *Microcebus murinus* (Miller, 1777). *Mém. Mus. natn. Hist. nat. Paris* (n.s.), Ser. A. (Zool.) **61**: 203–280.

imons, E. L. (1963). A critical reappraisal of Tertiary primates. In *Evolutionary and genetic biology of Primates:* 165–129. Buettner-Janusch, J. (ed). New York and London: Academic Press.

Simons, E. L. (1972). *Primate evolution: an introduction to man's place in nature.* New York: Macmillan & Co.

Simpson, G. G. (1927). Mesozoic mammals. 9: The brain of Jurassic mammals. *Am. J. Sci.* **14**: 259–268.

Simpson, G. G. (1940). Studies on the earliest primates. *Bull. Am. Mus. nat. Hist.* **77**: 185–212.

Simpson, G. G. (1945). The principles of classification and a classification of mammals. *Bull. Am. Mus. nat. Hist.* **85**: 1–350.

Simpson, G. G. (1951). *The meaning of evolution.* New York: Mentor Books.

Simpson, G. G. (1958). Behaviour and evolution. In *Behaviour and Evolution.* Roe, A. & Simpson, G. G. (eds). New Haven: Yale University Press.

Simpson, G. G. (1961). *Principles of animal taxonomy.* London: Oxford University Press.

Simpson, G. G. (1962). Primate taxonomy and recent studies of non-human primates. *Ann. N.Y. Acad. Sci.* **102**: 497–514.

Snell, O. (1892). Die Abhängigkeit des Hirngewichts von dem Körpergewicht und den geistigen Fähigkeiten. *Arch. Psychiatr. Nervenkrankh.* **23**: 436–446.

Sonntag, C. F. (1925). Comparative anatomy of the tongues of the Mammalia: Summary, classification and phylogeny. *Proc. zool. Soc. Lond.* **1925**: 701–762.

Stephan, H. (1959). Vergleichend-anatomische Untersuchungen an Insektivoren-gehirnen. 3: Hirn-Körpergewichtsbeziehungen. *Morph. Jb.* **99**: 853–880.

Stephan, H. (1972). Evolution of primate brains: a comparative anatomical investigation. In *The functional and evolutionary biology of primates:* 155–174. Tuttle, R. (ed.). Chicago: Aldine-Atherton.

Stephan, H. & Bauchot, R. (1965). Hirn-Körpergewichtsbeziehungen bei den Halbaffen (Prosimii). *Acta. zool., Stockh.* **46**: 209–231.

van Valen, L. (1964). A possible origin for rabbits. *Evolution* **18**: 484–491.

van Valen, L. (1965). Treeshrews, Primates and fossils. *Evolution* **19**: 137–151.

Wolin, L. R. & Massopust, L. C. (1970). Morphology of the primate retina. In *The Primate brain:* 1–27. Noback, C. R. & Montagna, W. (eds). New York: Appleton Century Crofts.

Wood Jones, F. (1929a). Some landmarks in the phylogeny of the Primates. *Human Biol.* **1**: 214–228.

Wood Jones, F. (1929b). *Man's place among the mammals.* London: Edward Arnold & Co.

DISCUSSION

DIAMOND: I am very grateful to Dr Martin for his lucid discussion of the relation between the Scala naturae and phylogenetic history.

My question is, does he see some use for the Scala naturae under certain circumstances? For example, I recall the remarkable passage in Darwin on the advancement in organization. It seemed to me that one could argue, for example, that the three-chambered heart is intermediate between the two-chambered and the four-chambered heart, and for the purpose of studying that "advancement in organization" the species that you select need not represent ancestral types. This would be an example, perhaps, of the use of the scale of nature without any effort at dealing with the question of the reconstruction of phylogenetic history.

MARTIN: I think that is an extremely good point. I did try to say that I thought the Scala naturae is a valuable concept in certain respects, provided that one states the purpose for which one is using it. I noticed in your paper that you were extremely explicit about how you were using this comparison, and I think that this is the way it should be done. I am sure that this kind of approach can give us a lot of information, but not about evolution, only about functional similarities and comparisons between living forms.

The example you quote is, in fact, the best example of the way Scala naturae, in evolutionary terms, has misled evolutionary theory, in that for many years it was believed that the three-chambered heart must be intermediate between the two-chambered heart of the fish and the four-chambered heart of the mammals. Because fish are on a lower "scale" than amphibians, and amphibians, which have a three-chambered heart, are on a lower "scale" than the mammals, this was thought to be the position. It was not until comparatively recently that it was realized that the three-chambered heart of the amphibians is a highly specialized structure derived from something like a primitive four-chambered heart. So, provided we are talking about function and comparing living forms for particular purposes, I think this is fine. If we are going to talk about evolution, loose use of the Scala naturae can only lead to misconceptions about the actual process of evolution.

ZUCKERMAN: We have had from Dr Martin an extraordinarily salutary analysis of some of the fallacies which have misled people, depending, of course, upon which books they have read. We all know that one starts with the basic natural unit, which is a species population, and that the rest is classification; that has been said time and time again. The inferences we make about phylogeny are our own—and heaven alone knows how many of us have said, time and time again, that taxonomy is not phylogeny. But people go on believing that you can reconstruct phylogenetic pictures which correspond to a taxonomic picture. But what you have, in effect, said is that on the basis of pure logic, it would be impossible to deny the proposition that man was derived from some basal stock from which other lines—e.g. pongids or australopithecines—sprang as well.

May I just put to you one question? You have talked about brain size and body size, a subject in which I was once interested. Do you know anything about the genetical factors, as well as the selective factors, which determine human brain size?

MARTIN: The last question I simply cannot answer. I have never read anything which treats that subject. We are working on an assumption here that the brain has evolved, but I think that at the basic level of genetic change we simply have very little information about it.

ZUCKERMAN: There is a condition known as microcephaly and the genetics underlying microcephaly are known. They have nothing whatever to do with human evolution, but there are some facts which we ought to bear in mind.

MARTIN: I should like to come back to the point you made originally. I think that man does share a lot of primitive characters with the ancestral mammalian stock, and therefore with living mammals that have retained those same characters. Yet I think that one can quite well define characters which man retains from later stocks yielding certain living forms and, therefore, one can logically exclude the possibility that man has come up straight from the ancestral stock without any relationship, say, to the apes. This is a field where there is some conflict between palaeontologists and anatomists at the moment over the biochemical evidence, in that the palaeontologists are tending to place man's split from the apes at about 25 million years ago, while some biochemists would place this split at about 5 million years ago, and others say possibly 10 million years ago. This is a case in point where I think the methodology of the comparative anatomist has to be analysed to make sure that the difference is due to a fault in the biochemical analysis rather than in the comparative anatomical one.

ZUCKERMAN: I agree.

Symp. zool. Soc. Lond. (1973) No. 33, 339–375

THE CHRONICLE OF PRIMATE PHYLOGENY
CONTAINED IN PROTEINS

MORRIS GOODMAN

Department of Anatomy, Wayne State University, Medical Research Building,
Detroit, Michigan, U.S.A.

SYNOPSIS

The amino acid sequences of proteins from man and other animals are evolutionary documents. These documents can be deciphered using the codon catalog of the genetic code to translate the amino acid sequences into nucleotide sequences from which gene phylogenetic trees are constructed. This is done by the parsimony principle, i.e. by seeking those codon ancestors and branching topologies which require the fewest nucleotide replacements in the descending branches. In confirmation of views advocated by Elliot Smith, the gene phylogenetic trees position the gibbons as an ancient branch of Hominoidea, then separate the line to orang-utan from the common ancestor of gorilla, man, and chimpanzee.

Additional molecular-genetic evidence from immunological data on cross reacting protein antigens and DNA hybridization data on polynucleotide homologies also demonstrate the close phylogenetic relationship of African apes to man and the successively more distant relationship of orang-utan, gibbons, and old world monkeys. In further agreement with the views of Elliot Smith, the protein and polynucleotide data separate the tree shrew from the line to lemuroids, lorisoids, *Tarsius* and Anthropoidea. Then in the early Primates the ancestors of lemuroids, lorisoids, *Tarsius*, and Anthropoidea branch apart, with lorisoids and lemuroids showing a slightly more recent common ancestor and, similarly, *Tarsius* and Anthropoidea. Ceboidea emerges as closer to catarrhines than to any living prosimian group.

Thus the molecular record of primate phylogeny depicts cladistic relationships which complement the fossil record. However, if we assume that molecular changes accumulated at a uniform rate, the molecular evidence either depicts much more recent branching times for catarrhine primates or alternatively much more ancient branching times for basal eutherian mammals than the times depicted by fossil evidence. If we accept the fossil evidence, i.e. do not attempt to use proteins and polynucleotides as evolutionary clocks, then it becomes apparent that rates of molecular evolution decelerated in the stem line to man as the Tertiary progressed. Such a pattern of evolutionary change can be predicted from the views of Elliot Smith.

"Any one who is familiar with the anatomy of Man and the Apes must admit that no hypothesis other than that of close kinship affords a reasonable or credible explanation of the extraordinarily exact identity of structure that obtains in most parts of the bodies of Man and Gorilla. To deny the validity of this evidence of near kinship is tantamount to a confession of the utter uselessness of the facts of comparative anatomy as indications of genetic relationships, and a reversion to the obscurantism of the Dark Ages of biology. But if any one still harbours an honest doubt in the face of this overwhelming testimony from mere structure, the reactions of the blood will confirm the teaching of anatomy; and the susceptibility of the Anthropoid Apes to the infection of human diseases, from which other Apes and mammals in general are immune, should complete and clinch the proof for all who are willing to be convinced." G. Elliot Smith.

INTRODUCTION

In recent years investigations directed at prying loose the record of evolution in proteins and in DNA sequences have provided fresh evidence on man's place in the phylogeny of the Primates. The purpose of my paper is to describe this evidence. It will be seen that the picture of primate phylogeny from the perspective of proteins and poly-nucleotides is remarkably similar to that which had already been drawn by Elliot Smith from the older evidence of comparative anatomy. In the tentative scheme proposed by Elliot Smith (1924), the basal Primates, originating at the end of the Cretaceous from tree shrew like parents, split into Lemuroidea and Tarsioidea, with lemurs and lorisoids evolving from the former and *Tarsius* and Anthropoidea from the latter. Next the Anthropoidea separated into Platyrrhini and Catarrhini with the latter splitting into old world monkeys and anthropoid apes. Then, from the anthropoid ape line, represented as the main trunk of the ascending primate tree, the gibbons were the first to branch off, next the orang-utan, and finally chimpanzee followed by gorilla separated from the stem to *Homo*. The present day molecular evidence unequivocally supports this Darwinian view of an especially close kinship between the African apes and man. However, with regard to finer details, the molecular evidence tends to portray man as closer to chimpanzee than to gorilla.

Even in Elliot Smith's day there was a body of data from the perspective of proteins demonstrating the close kinship between ape and man. This data was due to the classic study of Nuttall (1904) who immunized rabbits with serum proteins from different donor species and then used the precipitin reaction to determine a crude index of "chemical blood relationship" among Primates and other mammals. Elliot Smith was ahead of his time in appreciating that these "reactions of the blood" measured genetic affinities, even though he could not have known that the genes were segments of DNA and that the nucleotide sequence of the DNA determined the detailed structures of proteins. We now know, of course, that the sequence of amino acids in the polypeptide chains of proteins is specified by the sequence of codons in the corresponding genes, each codon being a triple of nucleotide bases. Immunological cross reactions gauge relative degrees of structural divergence among related proteins in that the protein sites which the antibodies react with, i.e. the antigenic sites, are shaped by amino acid groups at the protein surfaces. Exact numbers of codon differences among related genes are found by chemically determining the actual amino acid sequences of the proteins specified by the genes. From Nuttall's time to the present, voluminous data on the immunological cross reactions of serum proteins among primate species have been gathered (e.g. Wolfe, 1939; Boyden, 1942, 1958; Kramp, 1956; Goodman, 1962, 1963, 1965, 1967a, b; Goodman & Moore, 1971; Picard, Heremans & Vandesbroek, 1963; Sarich, 1970). Moreover during the past decade amino acid sequence data have begun to accumulate (e.g. Matsuda et al., 1968; Margoliash & Fitch, 1968; Romero Herrera & Lehman, 1971; Wooding & Doolittle, 1972; Tashian et al., 1972). In addition the phylogenetic relationships of different primates are now being estimated in terms of the genetic material itself by DNA hybrid reassociation experiments (Hoyer, McCarthy & Bolton, 1964; Hoyer, Bolton, McCarthy & Roberts, 1965; Hoyer & Roberts, 1967; Hoyer, van de Velde, Goodman & Roberts, 1972; Martin & Hoyer, 1967; Kohne, Chiscon & Hoyer, 1972).

METHODS OF DEPICTING PHYLOGENY FROM MOLECULAR DATA

The divergence approach

Once a body of protein or DNA data has been gathered it needs to be interpreted by suitable concepts or hypotheses on evolution to reveal its phylogenetic information. The most frequently used hypothesis states that the more ancient is the common ancestor for a pair of species the greater is the genetic distance for that pair of species. This divergence concept can be illustrated by the data on amino acid differences among hominoids given in Tables I and II.

Table I lists the protein substances so far compared in at least two or more of the five hominoids (chimpanzee, man, gorilla, orang-utan, and gibbon), and Table II, presented as a matrix, gives in the upper half the total number of differing amino acids per number of shared amino acid positions for each pair of hominoids and in the lower half the percent of differing amino acids. Despite the fact that the same protein substances were not compared uniformly in all five hominoids, the dissimilarity values in the matrix for comparisons involving gibbon or orang-utan are so much larger than for comparisons not involving these Asiatic apes, that we would have to conclude that the African apes and man have diverged much less from one another than from the Asiatic apes. Furthermore, we would take these dissimilarity values as evidence that after the ancestral separation of gibbon and orang-utan lineages from the remaining hominoids a common ancestor still existed for chimpanzee, man, and gorilla. The immunological data on proteins and the DNA data also demonstrate

TABLE I

Protein substances compared for amino acid differences in hominoids

Chimpanzee	αHb*	βHb*	δHb*	γHb*	CaI*	FibA–B	Cyt C	
Man	αHb	βHb	δHb*	γHb	CaI	FibA–B	Cyt C	Myo
Gorilla	αHb*	βHb*	δHb*			FibA–B		
Orang-utan	αHb†	βHb†			CaI*	FibA–B		
Gibbon	pαHb‡	βHb*	δHb*			FibA–B		Myo*

* Inferred sequence from amino acid composition of peptide fragments and homology with known sequences.

† Amino acid composition (Buettner-Janusch, Buettner-Janusch & Mason, 1969).

‡ Sequence on the first 31 positions from the N-terminal end of the alpha haemoglobin chain (Boyer, Noyes, Timmons & Young, 1972).

Sequence for chimpanzee gamma haemoglobin chain is from DeJong (1971). See the legend of Fig. 4 for references to other haemoglobin chain sequences, the legend of Fig. 6 for references to the myoglobin sequences. The carbonic anhydrase I sequences are from Tashian *et al.* (1972) and the cytochrome c sequences are from Margoliash & Fitch (1968). The fibrinopeptide A and B sequences are from Wooding & Doolittle (1972).

TABLE II

The amino acid difference matrix for hominoids

Upper half of matrix: Number of differing amino acids per number of shared amino acid positions

Lower half of matrix: Percent of differing amino acids

	Chi	*Man*	*Gor*	*Ora*	*Gib*
Chimpanzee		2/828	2/463	11/432	10/352
Man	0·24		3/463	12/432	12/505
Gorilla	0·43	0·65		7/322	10/352
Orang-utan	2·54	2·78	2·17		5/175
Gibbon	2·84	2·38	2·84	2·85	

a much greater divergence of the Asiatic apes from chimpanzee, gorilla, and man than of chimpanzee and gorilla from man. Thus this divergence data, even in its raw state, would make us conclude that the African apes showed a closer kinship to man than to the Asiatic apes.

When dealing with a dissimilarity matrix for a large series of species, the order of ancestral branching for all the species in the series is not always obvious. It then becomes necessary to subject the dissimilarity values to certain calculating procedures to construct a plausible phylogeny for these species. Moore (1971), on developing a mathematical model for the divergence hypothesis, demonstrated that if rates of molecular evolution were relatively uniform in all lines of descent, several computer algorithms would be suitable for constructing phylogenetic trees, the most suitable being the unweighted pair group method of Sokal & Michener (1958) in that the phylogeny it depicted was less affected by deviations from the uniform rate condition than in the case of the other algorithms. The unweighted pair group method builds a tree from the smallest to the largest branches in a series of pairwise clustering cycles; in each cycle grouping together the two members of the set (either singleton species or joined species from a previous cycle) with the smallest average dissimilarity value between them. However such dissimilarity values can lead to fallacious phylogenetic groupings if there are marked nonuniformities in evolutionary rates in the descent of the different species. Thus the unweighted pair group method cannot be guaranteed to produce a correct phylogenetic tree from a dissimilarity matrix, but it does produce a useful first approximation.

Taxonomic antigenic distance tables

Dissimilarity values among primates, in the form of taxonomic antigenic distance tables, have been gathered by quantitative immunological techniques (e.g. Hafleigh & Williams, 1966; Sarich & Wilson, 1966, 1967; Wang, Shuster, Epstein & Fudenberg, 1968) and in my laboratory by an immunodiffusion technique (Goodman & Moore, 1971). In quantitative immunological work, antiserum is produced to a purified protein from a particular animal species (the homologous species); then the degrees of antigenic divergence of the homologs of this protein in a series of other animal species (the heterologous species) are measured by the cross reactions of the antiserum and represented directly as a taxonomic antigenic distance table. The work in my laboratory is carried out with antisera to purified proteins and to mixtures of proteins such as antisera produced against whole serum of the homologous species. With both types of antisera we compare heterologous to homologous species by our immunodiffusion technique, utilizing trefoil Ouchterlony plates. Such a plate consists of three wells which circumscribe a centre field of agar. Antiserum is put into the bottom well and antigen preparations from two different species are placed in the two top wells, respectively. The reactants diffuse towards each other in the agar producing precipitin lines, generally a separate precipitin line for each reactive antigenic protein. If antigen from the homologous species is in one of the top wells and from a heterologous species in the other top well, the precipitin lines of the homologous reactions will extend beyond the precipitin lines of the corresponding heterologous reactions provided the two species do not share all the antigenic sites to which there are antibodies; these extensions are called spurs. The fewer the antigenic sites shared by the two species, the longer the spurs. A comparison network consists not only of homologous to heterologous but also of heterologous to heterologous species comparisons. The spurs are scored according to length and intensity, and then the scores from a network of species comparisons are transformed on the basis of set theoretical logic (Moore & Goodman, 1968; Goodman & Moore, 1971) by a computer program into a taxonomic antigenic distance table. Tables III through VI in the present paper were produced this way.

Minimum mutation distances

As pointed out earlier, the antigenic sites of a protein which yield *in vitro* reactions with antibody are restricted to the configurations of amino acid groups at the surface of the protein. Thus several or more amino acid substitutions might have to take place in the protein before even one of its antigenic sites changes in specificity. On the other hand, comparisons of the amino acid sequences of diverging proteins can detect many of the mutations which separate these

TABLE III

Taxonomic antigenic distance table
Rabbit anti-squirrel monkey serum

Saimiri	0·00	Hylobates	3·03
Callimico	1·62	Macaca	3·19
Alouatta	1·67	Homo	3·35
Cebus	1·72	Pongo	3·35
Oedipomidas	1·79	Presbytis	3·55
Aotes	1·86	Tarsius	5·61
Chiropotes	1·89	Perodicticus	6·14
Lagothrix	1·91	Galago	6·14
Ateles	2·12	Loris	6·25
Cacajao	2·17	Nycticebus	6·46
Callicebus	2·18	Tupaia	6·72

proteins. Here it may be pointed out that data on amino acid differences between proteins from amino acid composition studies are less revealing than data on actual amino acid sequence differences. For example, inclusion in Table I of orang-utan alpha and beta haemoglobin chains, for which only amino acid compositions were known, forced us to restrict the calculations in Table II purely to numbers of amino acid differences. Such numbers are likely to grossly underestimate the genetic distance between long separated species. However, when all the data for a set of species consist of amino acid sequences we can estimate genetic distances somewhat more fully by converting the amino acid differences into minimum mutation distances (Fitch & Margoliash, 1967).

To calculate these minimum mutation distances the sequences in a set of homologous polypeptide chains are aligned. This is done by matching residue positions so as to maximize the amino acid homologies throughout the set. Then the minimum mutation distances are calculated over all pairs of the aligned sequences. These calculations determine at each alignment position for each amino acid pair the minimum number of nucleotides that would need to be changed in order to convert a codon for one amino acid into a codon for the other. Since there are three nucleotide positions in each codon up to three nucleotide substitutions can be detected over each pair of amino acids; however, multiple substitutions at the same nucleotide positions and substitutions producing synonymous codons for the same amino acid can not be detected.

The additive approach

A dissimilarity matrix of minimum mutation distances like any dissimilarity matrix can be interpreted from the standpoint of the divergence hypothesis. Moreover, it is especially well suited, in comparison to matrices of antigenic distances or even of numbers of amino acid differences, to interpretation by the *additive* hypothesis of molecular evolution. The principal condition of the additive hypothesis (Cavalli-Sforza & Edwards, 1967; Moore, Goodman & Barnabas, 1973) is that the dissimilarity value observed between any two species in a comparison matrix is exactly proportional to the number of mutations fixed in descent since the time those two species last shared a common ancestor. While the pairs of recently separated sequences in a comparison matrix are apt to have the same dissimilarity values whether represented as minimum mutation distances or numbers of amino acid differences, the pairs of the more anciently separated sequences will have dissimilarity values more closely approaching the additive condition when estimated by the minimum mutations distance procedure rather than by just counting differing amino acids.

Assuming that a dissimilarity matrix of minimum mutation distances conforms to the additive condition, the dissimilarity values can be apportioned into lengths on the putative phylogenetic tree so that when the distances over all pairs of species in the matrix are calculated from the lengths derived from the apportioning procedure they exactly equal the minimum mutation distances of the original matrix, provided the putative tree is the correct tree. However, if the putative tree is not the correct tree, the two sets of values will deviate from each other. This deviation is measured by a coefficient such as the one of Fitch & Margoliash (1967) or the one of Moore, Goodman et al. (1973), the latter having properties suited perfectly to the assumptions of the additive hypothesis. On the basis of this type of coefficient, a search is conducted for that tree topology in which the value of the coefficient is at a minimum, and this tree topology is considered the correct or near correct one. With ideal data which conform to the additive hypothesis the coefficient for the correct tree will be zero, but for real amino acid sequence data, no tree topology is likely to have a zero coefficient. This is because the conditions of the additive hypothesis are often violated by the dissimilarity matrices for these sequence data. Firstly, since multiple substitutions at a nucleotide site increase with evolutionary time, yet cannot be detected from the pairwise comparisons of the contemporary species, the minimum mutation values will underestimate the true distance between anciently separated species proportionately more than between recently separated species, and secondly, since parallel and back mutations may not be evenly distributed in all lines of descent the minimum mutation values may be falser for certain pairs of species than for others. Nevertheless, if mutations have accumulated in different lineages at markedly dissimilar rates and if this is reflected in the dissimilarity matrix, the tree with the lowest coefficient value, i.e. the "additive" tree, may approximate phylogeny at least as well if not somewhat better than the "divergence" tree produced by the unweighted pair group method.

The maximum parsimony approach

We can also search for the correct phylogenetic tree on the basis of the *parsimony* hypothesis of molecular evolution (Fitch, 1971; Moore, Barnabas & Goodman, 1973). Initial experience indicates that this hypothesis, which takes full advantage of our knowledge of the genetic code, is the most powerful of the three; it permits a detailed decoding of the story of evolution contained in proteins. The parsimony hypothesis assumes that evolutionary changes take place in the fewest steps needed to account for them. With the set of sequenced protein chains, the search procedure by the parsimony principle for the correct evolutionary tree starts directly with the aligned sequences and with a given branching arrangement representing a plausible phylogeny for the protein chains. Then *maximum parsimony* ancestral codons are constructed at the forks of the tree, i.e. codons which yield fewer nucleotide replacements or mutations over the entire tree than any other codons would yield. Maximum parsimony ancestral codons are constructed for many alternative tree topologies and the number of nucleotide replacements for each topology is counted to give the tree of that topology its length. The search is directed by those changes in topology which reduce the mutation length; it stops when the tree of minimum number of nucleotide replacements appears to have been discovered. The parsimony assumption which underlies this search is likely to be close to true when many evolutionary forks separated consecutively by short spans of evolutionary time describe the descent of the contemporary sequences, because then the mutation rate itself would have minimized the possible number of nucleotide substitutions between adjacent ancestors and descendants. Moreover, selection against new mutations would further limit the number of codon changes in descent. Even so with the present amino acid sequence data in which a good proportion of the sequences are connected through few evolutionary forks over long stretches of evolutionary time, the parsimony assumption in the strict sense may be false; i.e. more nucleotide replacements are likely to have occurred than the minimum number needed to account for the evolutionary changes observed. Nevertheless, the parsimony approach yields a fuller and more correct estimate of mutation lengths in descending lineages than can be calculated from minimum mutation distance matrices. Furthermore, the empirical evidence on the present sets of amino acid sequence data suggest that maximum parsimony trees depict phylogeny more accurately than either divergence or additive trees.

Figures 1 through 7 show phylogenetic trees of mammalian fibrino-peptides A and B, alpha and beta-like haemoglobin chains and myo-globins constructed by the maximum parsimony method (Moore, Barnabas *et al.*, 1973) using computer programs PSLNG, PSITR, and PANCS designed by Dr G. W. Moore to execute the method. First, however, for each of these sets of sequence data, divergence and additive trees were constructed using computer program MMUTD (Barnabas,

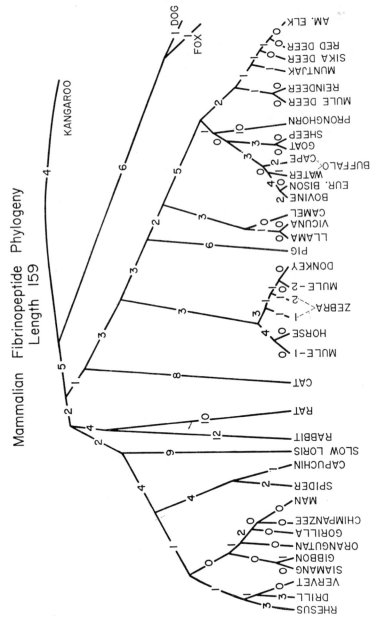

Fig. 1. Maximum parsimony tree of mammalian fibrinopeptide A and B sequences. In the alignment employed, that of Wooding & Doolittle (1972), 27 amino acid residue positions are shared by most species, and the numbers of nucleotide replacements recorded on the links (the link lengths) are for these 27 positions except on the terminal links to kangaroo, horse and mule-1, cape buffalo, and rhesus monkey in which the numbers are for 25, 25, 23, and 25 residue positions respectively. The amino acid sequences of the 12 primate species are given in Wooding & Doolittle (1972). References to the sequences of the other species are cited in Goodman et al. (1971).

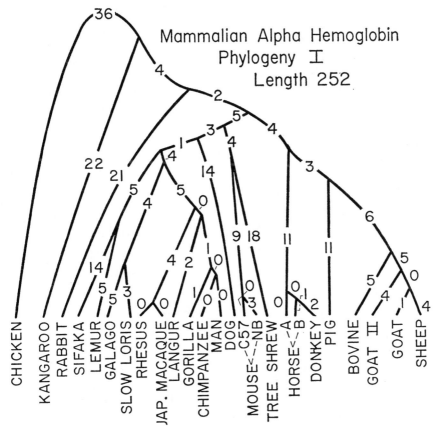

FIG. 2. Maximum parsimony tree of chicken and mammalian alpha haemoglobin chain sequences. See legend of Fig. 4 for references to the amino acid sequences of the different haemoglobin chains constituting the contemporary species in Figs 2–5.

Goodman & Moore, 1971; Goodman, Barnabas, Matsuda & Moore, 1971) to create from the aligned sequences the dissimilarity matrix of minimum mutation distances, computer program UWPGM to generate the divergence tree by the unweighted pair group method of Sokal & Michener (1958), and computer programs DENDR and ITERA to find the additive tree (Moore, Goodman *et al.*, 1973). These computer programs were also designed by Dr G. W. Moore. The divergence and additive trees served as initial starting points in the search for the maximum parsimony trees.

The computer runs using programs PSLNG, PSITR, and PANCS require as input data a file of aligned amino acid sequences and a

given tree topology. In the execution of each of these programs the
tree is first transformed into a network, because this facilitates the
length calculations (Moore, Barnabas et al., 1973). A network is a
tree whose most ancient common ancestor (the *roct*) has been removed.
There are two kinds of points in a network: *exterior points* (corresponding to contemporary species) and *interior points* (corresponding to
hypothetical ancestors or evolutionary forks in the original tree). Each
exterior point links to just one adjacent point; i.e. it has just one
nearest-neighbour (always an interior point). Each interior point links
to exactly three adjacent points; i.e. it has three nearest-neighbours
(at least one of which must be another interior point in any network

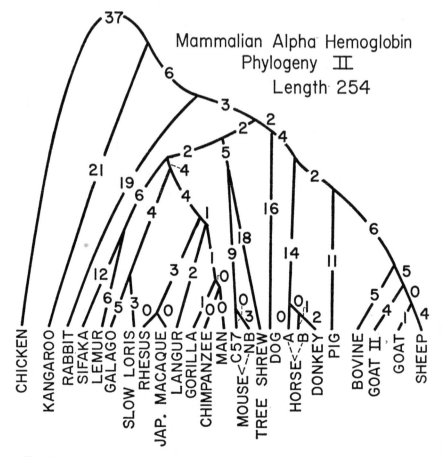

Fig. 3. A near-maximum parsimony tree of the chicken and mammalian alpha
haemoglobin chain sequences.

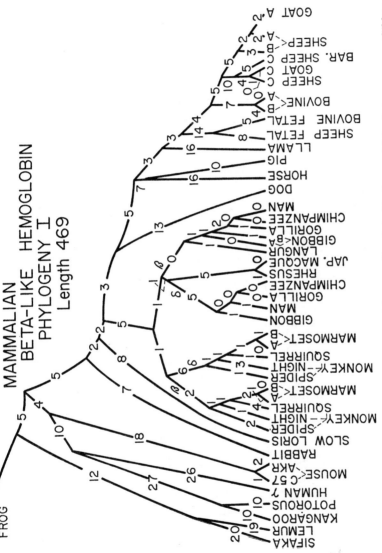

FIG. 4. Maximum parsimony tree of frog and mammalian beta-like haemoglobin chain sequences. Gibbon beta and delta, chimpanzee and gorilla delta, and the various ceboid beta and delta sequences are taken from Boyer et al. (1971). Dog alpha and beta sequences from Jones, Brimhall & Duerst (1971). Kangaroo alpha sequence from Beard & Thompson (1971). Potorous beta. sequence from Thompson & Air (1971). Slow loris (Nycticebus coucang) and langur (Presbytis entellus) alpha and beta sequences from G. Matsuda (pers. comm. of recently completed sequences). References to all other sequences are cited in Goodman et al. (1971).

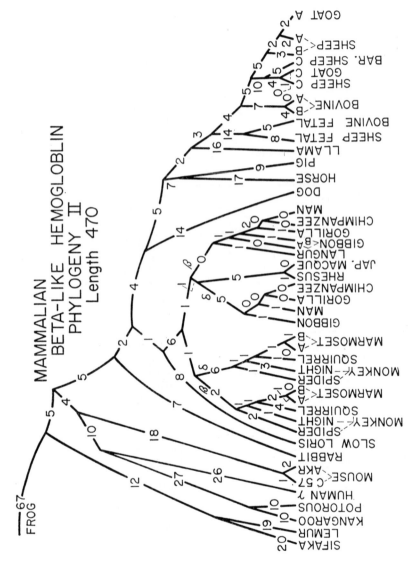

Fig. 5. A near-maximum parsimony tree of the frog and mammalian beta-like haemoglobin chain sequences.

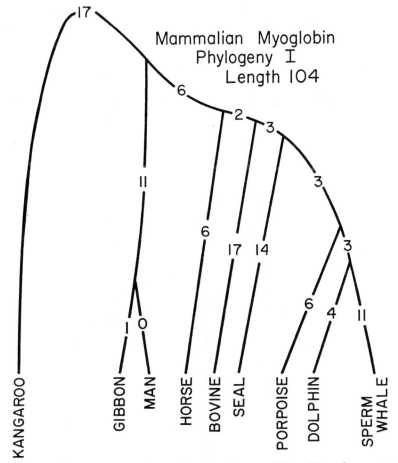

FIG. 6. Maximum parsimony tree of mammalian myoglobins. The references to the amino acid sequences constituting the contemporary species of this figure are as follows: Kangaroo—Air, Thompson, Richardson & Sharman (1971); Gibbon—Romero Herrera & Lehmann (1971a); Man—Romero Herrera & Lehmann (1971b); Horse—Dautrevaux, Boulanger, Han & Biserte (1969); Bovine—Han, Dautrevaux, Chaila & Biserte (1970); Harbour seal—Bradshaw & Gurd (1969); Porpoise—Bradshaw & Gurd (1969); Sperm whale—Edmundson (1965); Dolphin—Karadjova, Nedkov, Bakardjieva & Genov (1970); Kluh & Bakardjieva (1971).

containing more than three exterior points). A network with N exterior points has exactly N-2 interior points.

Program PSLNG calculates the mutation length (minimum number of nucleotide replacements over the entire tree) at each aligned amino acid position and sums these lengths for the total maximum parsimony length of the given tree.

Program PSITR also calculates the mutation length of the initial
tree; then examines all one-step nearest-neighbour changes in tree
topology. In a tree of N exterior points, exactly $2(N\text{-}2)$ tree topologies
(the number of different one-step nearest-neighbour changes in topology which can be made) are examined. The topology which lowered
the length of the tree the most is used as the start for the next cycle
of one-step nearest-neighbour changes. This iterative search procedure
stops when the mutation length of a tree cannot be lowered by further
one-step changes in topology.

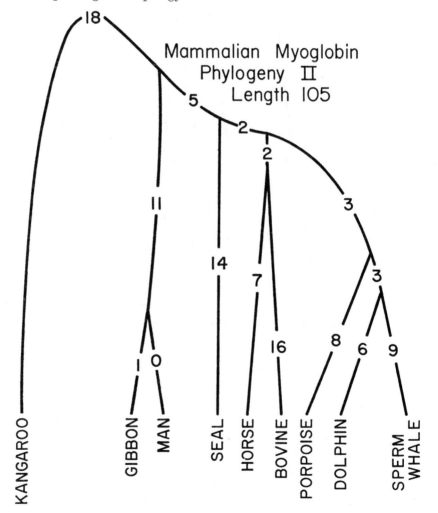

Fig. 7. A near-maximum parsimony tree of mammalian myoglobins.

Program PANCS, like program PSLNG, calculates the mutation length at each aligned amino acid position. In addition it constructs the maximum parsimony ancestral codons at the interior points of the tree and prints out alternative maximum parsimony solutions when they exist at an aligned amino acid position. For each solution at each position the program tallies the number of mutations between the root (the most ancestral point assigned to the tree) and each contemporary species and sums this tally over all contemporary species. The solution of minimum value is called the "A-solution", the solution of maximal value, the "B-solution". A mutation which may appear in a lineage leading either to few or many contemporary species will choose the few-lineage in the A-solution, since this decreases the number of times the mutation is counted among lineages between the root and each contemporary species. Conversely the mutation will choose the many-lineage in the B-solution, since this increases the number of times the mutation is counted. Considered side by side, these two solutions summarize the range of possibilities encompassed by maximum parsimony solutions. The idea for these two solutions was developed in a previous study (Barnabas, Goodman & Moore, 1972).

Although the maximum parsimony method finds mutations which occur *at* evolutionary forks, it misses multiple mutations occurring *between* evolutionary forks. Moreover, not all evolutionary forks are represented when contemporary sequence data are incomplete, i.e. obtained from a scattering of species. This contributes a systematic bias to estimates of evolutionary change by the maximum parsimony method. Regions in a phylogenetic tree which are sparsely represented by contemporary sequence data surely underestimate evolutionary change more substantially than regions well represented by contemporary sequence data. The least biased maximum parsimony solution is the A-solution because it reduces the tendency to underestimate evolutionary change in the sparse regions of the tree. In turn the most biased maximum parsimony solution is the B-solution because it increases this tendency.

The true phylogenetic tree may be a *near* parsimonious tree but not necessarily the *most* parsimonious tree. Figures 1, 2, 4, and 6 represent the most parsimonious trees discovered respectively for fibrinopeptides A and B, alpha haemoglobin chains, beta-like haemoglobin chains, and myoglobins after hundreds of tree topologies were examined. Figures 3, 5, and 7 are near parsimonious trees found for the three sets of globin data, in each case just one or two mutations longer than the most parsimonious trees. These near parsimonious trees are even more similar than are the most parsimonious trees to

phylogenies derived from comparative anatomical and palaeontological evidence on the contemporary species represented by the
sequences. However other near parsimonious trees could be shown
which are not as similar to phylogenies from the traditional evidence.
Only the link lengths from the less biased maximum parsimony
solutions, the A-solutions, are shown in Figs 1 through 7. Comparable
results gained previously in a much more limited search of tree topologies by Dr John Barnabas before computer programs PSLNG,
PSITR, and PANCS were operational are reported elsewhere (Goodman,
Barnabas & Moore, 1972).

EVIDENCE ON CLADISTIC RELATIONSHIPS OF THE PRIMATES

From amino acid sequence data

Fibrinopeptides A and B sequences

The maximum parsimony tree shown in Fig. 1 for mammalian
fibrinopeptides depicts the same cladistic relationships among primate
lineages as that deduced by Wooding & Doolittle (1972). The chimpanzee and gorilla cannot be distinguished from man. This African ape-man
complex diverges by two mutations from orang-utan, three to four
mutations from the gibbons (*Symphalangus* and *Hylobates*), five to
eight mutations from old world monkeys, nine to ten mutations from
new world monkeys, and 17 mutations from slow loris, the only prosimian in the series. When the root of the tree is arbitrarily placed
on the link to kangaroo, as was the case in calculating the link lengths
by the A-solution in the tree shown in Fig. 1, a branch to rabbit and
rat is cladistically closer to Primates than to other mammals. However,
the divergence tree produced by the UWPGM algorithm places the
root on the link to rabbit and then separates a rat-primate branch from
the other mammals. Thus both trees agree with the view of McKenna
(1969) and Wood (1962) from palaeontological evidence which has the
rodents evolve from early primates or proto-primates.

The order of cladistic branching for gorilla, chimpanzee, and man
cannot be distinguished by the maximum parsimony tree, because
the maximum parsimony African ape-man ancestor has the same
sequence as the chimpanzee-man ancestor. Similarly the hominoid
ancestor cannot be distinguished from the catarrhine ancestor; thus
the maximum parsimony tree of fibrinopeptides does not tell us if
the gibbons are closer cladistically to other hominoids or to old world
monkeys. The answer to this comes from other molecular data.

Alpha and beta haemoglobin chain sequences

Among the alpha chains, chicken, kangaroo, rabbit, mouse C-57, slow loris, rhesus, Japanese macaque, langur, man, dog, horse slow component (designated horse A in Figs 2 and 3), pig, and bovine have been either completely or over two thirds sequenced; among the beta-like chains, kangaroo, potorous human gamma, rabbit, slow loris, langur, rhesus, Japanese macaque, man, dog, horse, bovine B, bovine foetal sheep C, and sheep B have also been either completely or about two thirds or more sequenced; all remaining sequences have been in good measure inferred from the amino acid compositions of peptide fragments by homology with known sequences. The alignments used in the present analysis were as reported by the original authors and by Dayhoff (1969).

The maximum parsimony and the near-maximum parsimony trees for alpha globins (Figs 2 and 3, respectively) join tree shrew with mouse and place this branch closer to primates than to most other mammals, thus further supporting the belief (McKenna, 1969; Wood, 1962) that rodents originated from proto-primates. The lorisoids (slow loris and galago) are depicted as closer by one evolutionary fork to catarrhine primates than to lemuroids (lemur and sifaka). However in another near-maximum parsimony tree (length 255 compared to lengths 254 and 252) in which tree shrew joins the union of mouse-primate and dog-ungulate branches, the union of lorisoids with lemuroids before these prosimians joined the line to catarrhines is as parsimonious as when lorisoids first join with catarrhines. This particular tree as well as the ones shown in Figs 2 and 3 are in iterative valleys in that in each case a further cycle of one-step nearest-neighbour changes in topology fails to lower the mutation length.

In the maximum parsimony alpha tree (Fig. 2) the langur-hominoid ancestor has the same sequence as the macaque-langur-hominoid ancestor. However in the near-maximum parsimony alpha trees (Fig. 3 and the other one described above) and also in the divergence (UWPGM) tree the langur is cladistically closer to hominoids than to macaques. It is quite possible that the gene phylogeny depicted in these several trees is in error. On the other hand assuming that the gene phylogeny is not in error, and that colobines (langur) and cercopithecines (macaques) are closer to each other cladistically than to hominoids as demonstrated by immunodiffusion data on serum proteins (Goodman, 1967a & b; Goodman & Moore, 1971), it would then seem that the evolutionary fork in the tree from which langur and hominoid alphas descended coincided with the most recent common animal ancestor of cercopithecoids and hominoids, whereas the gene ancestors of macaque and langur alphas had already separated in the basal catarrhine population. The latter could have resulted from the presence of allelic alpha genes at polymorphic frequency, or alternatively, of duplicated loci for alpha genes in this basal catarrhine population. In this connection it is worth noting that duplicated alpha loci exist in macaques (Barnicot, Wade & Cohen, 1970; Wade, Barnicot & Huehns, 1970), hominoids (Boyer et al., 1971), and other mammals, e.g. mouse (Hilse & Popp, 1968), goat (Huisman, Wilson & Adams, 1967), and horse (Kilmartin & Clegg, 1967).

Human and chimpanzee alpha sequences are identical and diverge from gorilla alpha sequence by one mutation. Similarly human and chimpanzee betas are identical and diverge from gorilla beta by one mutation. In turn chimpanzee and gorilla deltas are identical and diverge from human delta by one mutation. With respect to alpha, beta, and delta chains, no cladistic difference of chimpanzee and gorilla from each other and from man can be proven by the maximum parsimony solutions (Figs 2–5). Furthermore, as with the maximum parsimony fibrinopeptide tree (Fig. 1), the maximum and near-maximum parsimony alpha and beta-like trees (Figs 2–5) depict chimpanzee, gorilla, and man to be cladistically closer to one another than to other animals. However, in contrast to the fibrinopeptide tree, the beta-like trees (Figs 4 and 5) provide evidence that the gibbon is closer cladistically to hominoids than cercopithecoids; less parsimonious trees for the beta-like sequences are obtained when the gibbons are joined to cercopithecoids.

Both Figs 4 and 5 depict two independent beta gene duplications in the Anthropoidea, one after the platyrrhine–catarrhine splitting apart producing in the basal catarrhines the ancestral hominoid delta gene line and the other producing in parallel in the basal platyrrhines the ancestral ceboid delta-like gene line. The delta chain replaces the beta chain in the minor adult haemoglobin, called A_2 haemoglobin, of hominoids. A comparable haemoglobin is also present in ceboids. The observation by Barnicot & Hewett-Emmett (in press) of an A_2-like haemoglobin in tarsier but not in lorisoids or lemuroids suggests that an original beta proto-delta duplication had occurred in a primitive tarsioid line ancestral to Tarsius and Anthropoidea. Such a duplication could have set the stage for further duplications by the mechanism of unequal, homologous crossing over. Secondary origins of the delta genes in ceboids on the one hand, and in hominoids, on the other, by this mechanism would account for the fact that glycine is present at β^5 and δ^5 in ceboids against proline at β^5 and δ^5 in hominoids and still account for the presence of arginine and aspargine in positions 116 and 117 respectively in all the delta chains as against histidine and histidine respectively at these positions in the ceboid and catarrhine beta chains.

In both the maximum and near-maximum parsimony beta-like trees the lemuroid sequences link to the most ancient mammalian evolutionary fork and the mouse sequences link to a nearby evolutionary fork closer to the ancestor of marsupial beta and human gamma than to the ancestor of typical eutherian betas. This unexpected branching arrangement might simply be due to "noise" in the inferred mouse and lemuroid sequences. Nevertheless, there seems little question that the mouse and lemuroid sequences diverge markedly from typical eutherian beta chains. It is as if they descended from beta-like gene loci distinct from the ancestral locus for typical eutherian betas. This is further emphasized by marsupial beta joining human gamma and then being joined by mouse. It is known that human gamma, the beta-like chain in human foetal haemoglobin, is coded by a different gene locus than the locus coding for the beta chain in adult human haemoglobin. Thus the results suggest that a burst of gene duplications in a beta-like locus occurred in the basal proto-therian mammals and gave rise to these different beta-like lineages, only one of which descended to typical eutherian betas. The evolutionary fork to which the rabbit sequence links can be taken as the eutherian beta ancestor. Similarly in the alpha phylogeny (Figs 2 and 3) the evolutionary fork to which rabbit links can be taken as the eutherian alpha ancestor. These results agree with the proposal of McKenna (1969) that the most ancient splitting in the Eutheria separated Lagomorpha from all other extant eutherian orders.

Myoglobin and other protein chain sequences

The inferred sequence of gibbon myoglobin differs from the amino acid sequence of human myoglobin by only one mutation. Yet as can be calculated from either the maximum or near-maximum parsimony trees (Figs 6 and 7, respectively) each of these primate sequences differs on the average from any other eutherian myoglobin sequence by about 31 to 32 mutations. The smallness of the divergence between human and gibbon myoglobins can be contrasted with the much more substantial divergence between bovine and sheep myoglobins (Han et al., 1972). The sheep myoglobin sequence, which we became aware of after constructing the myoglobin phylogenetic tree (Figs 6 & 7), differs from its closest sequenced relative, bovine myoglobin, by eight mutations. While the myoglobin data do not tell us much about cladistic relationships within the Order Primates since only two primate species are represented, they emphasize the exceptionally high genetic correspondence which exists between hominoid branches.

If we assume that myoglobin had evolved in different lineages at a uniform rate and then attempt to use myoglobin as an evolutionary clock, we obtain results which contradict the fossil record. For example, if we consider the branching time between Primates and other eutherian orders to be in the range of 90 million years ago (about as ancient a time as the fossil record permits (McKenna, 1969)), then the branching time for gibbon and man would calculate at about 3 million years ago (90/31 × 1). However the existence of gibbon-like fossils in the Miocene (Le Gros Clark, 1960; Simons, 1967) rules such a recent separation out. It would seem that myoglobin evolution has not been at a uniform rate in different eutherian lineages but has markedly decelerated in the Hominoidea.

Amino acid sequence data from three other proteins bear on the cladistic and genetic relationships of higher Primates. These are the sequence data for cytochrome C (Margoliash & Fitch, 1968; Fitch & Markowitz, 1970) and for carbonic anhydrases I and II (Tashian et al., 1972). Chimpanzee and human cytochrome C are identical (findings of Dr S. B. Needleman reported in Margoliash & Fitch, 1968) and diverge from rhesus monkey by only one mutation, but from any other eutherian mammal by an average of 13 mutations (Fitch & Markowitz, 1970).

Comparisons of the partially sequenced carbonic anhydrases I and II have been carried out on 115 aligned amino acid positions (Tashian et al., 1972). With respect to carbonic anhydrase I at these aligned positions, chimpanzee and man are separated from orang-utan by three to four mutations and from old world monkeys (vervet, baboon, and rhesus) by five to eight mutations, but vary from each other by only one mutation. Sequence data on carbonic anhydrase I from other animals are not yet available; thus broader cladistic comparisons cannot be made. With respect to carbonic anhydrase II at the 115 aligned positions, man is the same as vervet, and varies by only two mutations from rhesus monkey, but by 47 mutations from sheep. The very small divergencies of carbonic anhydrase II among species of catarrhines, which prevents meaningful conclusions on the cladistic relationships of these catarrhines to one another, contrasts with the very marked divergence from sheep carbonic anhydrase II. As will be argued in the concluding section of this paper, such results speak not so much for a recent separation of the major hominoid and cercopithecoid branches but rather for a deceleration of protein evolution in the higher Primates.

From immunodiffusion data

A large mass of species comparisons in trefoil Ouchterlony plates developed with antisera to the serum proteins of various catarrhine species were used to draw conclusions on the cladistic relationships of the catarrhine primates (Goodman & Moore, 1971; Darga, Goodman & Weiss, in press). These conclusions, drawn from the standpoint of the divergence hypothesis, will be briefly recapitulated here. The catarrhine primates behave as a monophyletic group and diverge from other eutherian groups in the following cladistic order: ceboids, tarsier, strepsirhines (lorisoids and lemuroids), tree shrew, and non-primate eutherians. The Catarrhini itself subdivides into Cercopithecoidea and Hominoidea, the colobines clearly grouping with cercopithecines in the Cercopithecoidea and the hylobatines clearly grouping with other hominoids in the Hominoidea. Within the Hominoidea, the hylobatines separate first from other hominoids, then the ancestral line to orang-utan separates from the line to the common ancestor of gorilla, man, and chimpanzee. These conclusions are in perfect agreement with those of Elliot Smith (1924).

The closeness of the immunological picture of primate phylogeny to Elliot Smith's views is further demonstrated by the species comparisons developed with antisera to the serum proteins of ceboids and prosimians. Examples of the taxonomic antigenic distance tables obtained from these comparisons and the cladistic branching arrangement which can be generated from the antigenic distance data are given in Tables III through VI and in Fig. 8. As can be noted, after the Tupaioidea separates from the Primates, the Primates separate into Haplorhini and Strepsirhini, with the haplorhines subdividing into Tarsioidea and Anthropoidea and the strepsirhines subdividing into Lemuroidea and Lorisoidea. Finally the Anthropoidea subdivides into Platyrrhini and Catarrhini.

Another aspect of this data is that the catarrhine primates appear not to have changed as much as the ceboids from the ancestral (presumably more tarsioid) state of the Anthropoidea. This is suggested by tarsier showing less divergence from catarrhine primates than from ceboids (Table IV). An indication of more conservative evolution in catarrhine primate lineages than in ceboid lineages is also given by the phylogenetic trees for fibrinopeptide sequences (Fig. 1) and beta-like haemoglobin sequences (Figs 4 and 5). It can be noted in these trees that from the common ancestor of catarrhines and ceboids, there were on the average two or three more mutations in descent to the contemporary ceboids than to the contemporary catarrhines.

CLADISTIC RELATIONSHIPS AND AVERAGE ANTIGENIC
DISTANCE (IN PARENTHESES) BETWEEN MAJOR PRIMATE
BRANCHES

(Calculated from comparisons developed by rabbit antisera to *Galago,
Nycticebus, Loris, Tarsius, Saimiri,* and *Aotes*)

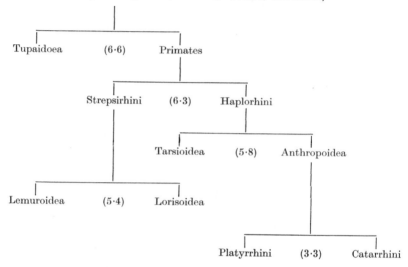

Fig. 8. Each taxonomic antigenic distance table was normalized to have an antigenic
distance of 6·6 for tree shrew from the homologous species. Then from the normalized
tables the branching topology and the average antigenic distance between branches was
determined by the divergence procedure.

From DNA data

Earlier DNA experiments by Dr Bill H. Hoyer and his co-workers
focused on the repeated polynucleotide sequences in primate genomes
and measured the degree of complementarity between DNA of heterologous and homologous species (Hoyer *et al.*, 1964; Hoyer *et al.*, 1965;
Hoyer and Roberts, 1967). Results were obtained from the standpoints
of man and rhesus monkey and agreed with protein data in dividing
the catarrhines into hominoids and cercopithecoids. Gibbon DNA was
not quite as similar to human DNA as chimpanzee but more so than
rhesus and baboon DNAs. In turn rhesus resembled baboon more
than man, chimpanzee, or gibbon. Catarrhine groups were similar
to each other more than to ceboids (capuchin and night monkey
DNAs) and similar to ceboid DNA more than to prosimian DNAs.
The order of divergence after ceboids was tarsier, lorisoids (galago,
slow loris, and potto), lemur, tree shrew, mouse, hedgehog and chicken.

More recently the DNA reassociation experiments have focused on the nonrepeating polynucleotide sequences in primate genomes (Kohne *et al.*, 1972). From the human standpoint the order of divergence was chimpanzee, gibbon, old world monkeys (vervet and rhesus monkey), capuchin, and galago; from the vervet standpoint the order of divergence was rhesus monkey, hominoids (man, chimpanzee and gibbon), capuchin, and galago. These results agreed with those obtained on the repeating DNA sequences as well as with the protein data.

TABLE IV

Taxonomic antigenic distance table

Rabbit anti-tarsius plasma

Tarsius	0·00	Cebus	5·04
Presbytis	4·13	Ateles	5·16
Theropithecus	4·19	Galago	5·32
Erythrocebus	4·25	Tupaia	5·36
Pan	4·25	Cacajao	5·40
Gorilla	4·25	Lemur	5·49
Pongo	4·25	Perodicticus	5·51
Cercocebus	4·25	Loris	5·62
Hylobates	4·25	Nycticebus	5·67
Cercopithecus	4·25	Citellus	6·17
Papio	4·25	Dasypus	6·17
Macaca	4·53	Bos	6·41
Homo	4·65	Bradypus	6·78
Saimiri	4·90	Petrodromus	6·78
Aotes	4·99	Atelerix	6·96

TABLE V

Taxonomic antigenic distance table

Rabbit anti-galago plasma

Galago c.	0·00	Loris	2·93
Galago s.	0·12	Lemur	6·24
Galagoides	0·28	Homo	6·80
Nycticebus	2·24	Saimiri	6·95
Perodicticus	2·40	Tupaia	7·52
Arctocebus	2·63		

TABLE VI

Taxonomic antigenic distance table
Rabbit anti-slow loris plasma

Nycticebus	0·00	*Perodicticus*	3·42
Loris	1·34	*Lemur*	7·69
Galago c.	2·73	*Homo*	8·02
Galago s.	2·86	*Saimiri*	8·10
Arctocebus	3·25	*Tupaia*	8·83

Dr Bill H. Hoyer and his co-workers are now carrying out a detailed investigation of the evolution of nonrepeating DNA in the Hominoidea. So far results have been obtained from the human and orang-utan standpoints (Hoyer *et al.*, 1972). The divergence between homologous and heterologous species was measured in terms of the thermostability properties of the reassociated DNA complexes. From the human standpoint the divergence values are man 0, chimpanzee 0·7, gorilla 1·4, gibbon 2·7, orang-utan 2·9, and vervet 5·7. From the orang-utan standpoint the divergence values are orang-utan 0, chimpanzee 1·8, man 1·9, gorilla 2·3, gibbon 2·4, and vervet 4·3. These results provide further evidence that chimpanzee and gorilla are cladistically closer to man than to orang-utan and other primates. The results also suggest that gorilla DNA may have diverged slightly more from the ancestral state than chimpanzee or human DNA, since orang-utan does not diverge quite as much from chimpanzee or man as from gorilla. Although the present molecular evidence leaves little room to doubt that the African apes and man constitute a monophyletic branch within the Hominoidea, the question still remains as to whether gorilla is cladistically closest to chimpanzee or closest to man or whether, indeed, as seems more likely from the present evidence, chimpanzee is cladistically closest to man. A decisive answer to this question should be forthcoming when DNA comparisons from chimpanzee and gorilla standpoints are completed.

RATES OF PROTEIN EVOLUTION IN THE MAIN STREAM OF DEVELOPMENT

In the introduction to his book of essays on the evolution of man, Elliot Smith (1924) presented a figure diagramming the phylogeny of the Primates. This figure, reproduced here as Fig. 9, agrees in almost all respects with the evidence from proteins and polynucleotides on

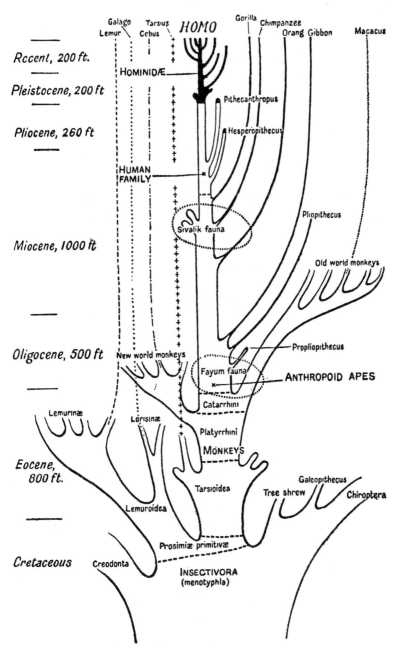

FIG. 9. Diagram of Elliot Smith for man's place in the phylogeny of the Primates. Republished from Elliot Smith (1924) with the kind permission of the Oxford University Press.

cladistic relationships among contemporary Primates, except perhaps in placing gorilla rather than chimpanzee closer to man. An aspect of Elliot Smith's diagram and philosophy which can be indirectly supported by molecular evidence is the portrayal of a process in which man's ancestors are always at the forefront of the main stream of evolutionary development. Let me quote Elliot Smith on this:

> "The object of this diagram, like the previous one, is to indicate the fact that all of these Lemurs, Monkeys, and Apes, which have become specialized in one way or another, should be regarded as having departed from the main stream of development that leads straight up to Man, and by doing so lost something of the primitive structure and plasticity that were necessary for the attainment of the high powers of adaptation which represent one of the most distinctive characteristics of the Human Family."

This view of Elliot Smith contrasts sharply with an opposing idea in which the lineage to man on occupying a new adaptive zone after it separates at an evolutionary fork from other lineages diverges more markedly from the ancestral state than these other lineages. At the molecular-genetic level this opposing idea is not supported. The protein evidence suggests that there has been a pronounced retention of ancient genetic structure at least in genes coding for proteins, in the lineage to man. This evidence, based on analysis of amino acid sequences by the maximum parsimony approach, presently comes from rather meagre data, but surely in the future with the automation of amino acid sequencing procedures the accumulated sequence data will permit definitive analyses to be carried out.

It may be recalled that the maximum parsimony method applied to sets of amino acid sequences from contemporary species reconstructs plausible ancestors at evolutionary forks and measures the amounts of change from these ancestors through the intervening forks to the contemporary species. Such measurements, however, are biased towards underestimates of the real amount of evolutionary change especially in regions of a phylogenetic tree sparsely represented by contemporary species. Thus only those regions of a tree represented by similar numbers of species and of ancestors encompassing the same spans of evolutionary time should be compared in determining if lineages differ in the rate at which they change. Moreover to calculate rates, maximum or near-maximum parsimony trees should be employed which are similar to trees derived from fossil evidence, because then palaeontological dating times can be assigned to at least some of the evolutionary forks on the trees.

Primates and ungulates are about equally represented in the phylogenetic trees for fibrinopeptides A and B, alpha haemoglobin, and beta-like haemoglobin sequences (Figs 1, 3, and 5). Among contemporary primate species, slow loris, rhesus monkey, gorilla, chimpanzee,

TABLE VII

Comparison of mutational change in primates and ungulates

	FibA–B	*αHb*	*βHb*	*Total*
Eutherian ancester to primate (slow loris-man) ancestor	9	11	3	23
Eutherian ancestor to ungulate (horse-goat) ancestor	9	9	11	29
Primate (slow loris-man) ancestor to slow loris	9	7	8	24
Primate (slow loris-man) ancestor to rhesus monkey	9	7	14	30
Primate (slow loris-man) ancestor to gorilla	8	7	12	27
Primate (slow loris-man) ancestor to chimpanzee	8	6	11	25
Primate (slow loris-man) ancestor to man	8	6	11	25
Ungulate (horse-goat) ancestor to horse	7	14	24	45
Ungulate (horse-goat) ancestor to pig	9	13	16	38
Ungulate (horse-goat) ancestor to bovine	20	13	18	51
Ungulate (horse-goat) ancestor to sheep	14	17	23	54
Ungulate (horse-goat) ancestor to goat	14	14	23	51

Mutational change found by summing the lengths of the appropriate links in the phylogenetic trees shown in Figs 1, 3, and 5 for fibrinopeptides A and B, alpha haemoglobin, and beta haemoglobin sequences respectively. In Fig. 1 the interior point to which the dog-fox branch links is taken as the eutherian ancestor, and in Figs 3 and 5 the interior point to which rabbit links is taken as the eutherian ancestor.

and man are present in these three phylogenetic trees; similarly among contemporary ungulates, horse, pig, bovine, sheep, and goat are present in the three trees. The evolutionary fork for the most recent common ancestor of slow loris and man can be considered to have existed at about the same time level as the evolutionary fork for the most recent common ancestor of horse and goat; the early primitive Primates and the early primitive ungulates represented by these ancestors are usually placed by palaeontologists at the Cretaceous–Paleocene boundary (Romer, 1966; McKenna, 1969). The scattering of evolutionary forks in these trees is such that among the Primates, the largest bias to underestimating evolutionary change is in the lineage to slow loris and the least bias is in the lineage to man and chimpanzee; among the ungulates the largest bias is in the lineage to horse and the least bias in the lineage to sheep and goat. On putting together the results on mutation lengths from the three phylogenetic trees (Table VII) we found that 23 mutations occurred on going from the eutherian ancestor to the early primate ancestor and 29 mutations from the eutherian ancestor to the early ungulate ancestor. From the

TABLE VIII

Rates of molecular evolution in descent

Approximate span of time	Number of detected nucleotide replacements	Replacement rate per 10^7 years
(a) 90×10^6 to 65×10^6 years ago or 25×10^6 years	23	9·2
(b) 65×10^6 to 40×10^6 years ago or 25×10^6 years	17	6·8
(c) 40×10^6 to 15×10^6 years ago or 25×10^6 years	8	3·2
(d) 15×10^6 years ago to present or 15×10^6 years	0	0

(a) From eutherian ancestor to primate (slow loris-man) ancestor.
(b) From primate ancestor to catarrhine ancestor.
(c) From catarrhine ancestor to African ape-man ancestor.
(d) From African ape-man ancestor to man.
(For 314 codon positions of alpha and beta globins and fibrinopeptides a and b).

early primate ancestor to the contemporary primates the amounts of mutational change were slow loris 24, rhesus monkey 30, gorilla 27, chimpanzee 25, and man 25; from the early ungulate ancestor to the contemporary ungulates the amounts of mutational change were horse 45, pig 38, bovine 51, sheep 54, and goat 51. Thus the primate lineages evolved only about half as fast as the ungulate lineages. Furthermore in the line to man it appears (Table VIII) that this slower change was due to a decelerating rate as the Tertiary progressed. If we place the eutherian ancestor at about 90 million years ago (McKenna, 1969), the early primate ancestor at about 65 million years ago (McKenna, 1969; Romer, 1966), the catarrhine ancestor at about 40 million years ago (Le Gros Clark, 1960; Simons, 1967), and the chimpanzee-man ancestor at about 15 million years ago (Pilbeam, 1968) the rate of mutational change decelerated in the successive time periods from 9·2 to 6·8 to 3·2 to 0 nucleotide replacements per every 10^7 years (Table VIII). This pattern of conservative change agrees nicely with Elliot Smith's diagram of a main stream of development leading straight to man.

Acknowledgements

It is a pleasure to acknowledge the close collaboration of Drs G. William Moore and John Barnabas in developing with me the approach used for analysing the evolutionary content of molecular data. I also thank Dr Genji Matsuda for sending me the amino acid sequence data on alpha and beta haemoglobin chains of slow loris and langur. The expert assistance of Mr Walter Farris and the cooperation of the Wayne State University Computing Center is gratefully acknowledged. This research was supported by NSF grants GB-7426 and GB-15060.

References

Air, G. M., Thompson, E. O. P., Richardson, B. J. & Sharman, G. B. (1971). Amino-acid sequences of kangaroo myoglobin and haemoglobin and the date of marsupial-eutherian divergence. *Nature, Lond.* **229**: 391–394.

Barnabas, J., Goodman, M. & Moore, G. W. (1971). Evolution of haemoglobin in primates and other therian mammals. *Comp. Biochem. Physiol.* **39**: 455–482.

Barnabas, J., Goodman, M. & Moore, G. W. (1972). Descent of mammalian alpha globin chain sequences investigated by the maximum parsimony method. *J. molec. Biol.* **69**: 249–272.

Barnicot, N. A. & Hewett-Emmett, D.(in press). Electrophoretic studies on red cell and serum proteins of prosimians. In *Prosimian biology*. Martin, R. D., Doyle, G. A., Walker, A. C. (eds). London: Duckworth.

R

Barnicot, N. A., Wade, P. T. & Cohen, P. (1970). Evidence for a second haemo-globin α-locus duplication in *Macaca irus*. *Nature, Lond.* **228**: 379–381.

Beard, J. M. & Thompson, E. O. P. (1971). Studies on marsupial proteins V. Amino acid sequence of the α-chain of haemoglobin from the grey kangaroo, *Macropus giganteus*. *Austr. J. biol. Sci.* **24**: 765–786.

Boyden, A. (1942). Systematic serology: a critical appreciation. *Physiol. Zool.* **15**: 109–145.

Boyden, A. (1958). Comparative serology: aims, methods and results. In *Serological and biochemical comparisons of proteins*: 3–24. Cole, W. (ed.) New Brunswick: Rutgers University Press.

Boyer, S. H., Crosby, E. F., Noyes, A. N., Ruller, G. F., Leslie, S. E., Donaldson, L. J., Vrablik, G. R. Schaefer, E. W. Jr. & Thurmon, T. F. (1971). Primate hemoglobins: some sequences and some proposals concerning the character of evolution and mutation. *Biochem. Genet.* **5**: 405–448.

Boyer, S. H., Noyes, A. N., Timmons, C. F. & Young, R. A. (1972). Primate hemoglobins: polymorphisms and evolutionary patterns. *J. Human Evol.* **1**: 515–543.

Bradshaw, R. A. & Gurd, F. R. N. (1969). Comparison of myoglobin from harbor seal, porpoise, and sperm whale. V. The complete amino acid sequences of harbor seal and porpoise myoglobins, *J. biol. Chem.* **244**: 2167–2181.

Buettner-Janusch, J., Buettner-Janusch, V. & Mason, G. A. (1969). Amino acid compositions and amino-terminal end groups of α and β chains from poly-morphic hemoglobins of *Pongo pygmaeus*. *Archs Biochem. Biophys.* **133**: 164–170.

Cavalli-Sforza, L. L. & Edwards, A. W. F. (1967). Phylogenetic analysis: models and estimation procedures. *Evolution, Lancaster, Pa.* **21**: 550–570.

Darga, L. L., Goodman, M. & Weiss, M. L. (in press). Molecular evidence on the cladistic relationships of the Hylobatidae. In *Gibbon and siamang* 2. Rumbaugh, D. M. (ed) Basel: Karger.

Dautrevaux, M., Boulanger, Y., Han, K. & Biserte, G. (1969). Structure covalente de la myoglobine de cheval. *Eur. J. Biochem.* **11**: 267–277.

Dayhoff, M. O. (1969). *Atlas of protein sequence and structure* 4. Silver Springs: National Biomedical Research Foundation.

DeJong, W. W. W. (1971). Chimpanzee foetal haemoglobin: structure and heterogeneity of the γ chain. *Biochim. biophys. Acta* **251**: 217–226.

Edmundson, A. B. (1965). Amino-acid sequence of sperm whale myoglobin. *Nature, Lond.* **205**: 883–887.

Elliot Smith, G. (1924). *Evolution of Man*. pp. 13–15; 23–24. London: Oxford University Press.

Fitch, W. M. (1971). Toward defining the course of evolution: minimum change for a specific tree topology. *Syst. Zool.* **20**: 406–416.

Fitch, W. M. & Margoliash, E. (1967). Construction of phylogenetic trees. *Science, N.Y.* **155**: 279–284.

Fitch, W. M. & Markowitz, E. (1970). An improved method for determining codon variability in a gene and its application to the rate of fixation of mutations in evolution. *Biochem. Genet.* **4**: 579–593.

Goodman, M. (1962). Evolution of the immunologic species specificity of human serum proteins. *Human Biol.* **34**: 104–150.

Goodman, M. (1963). Serological analysis of the systematics of recent hominoids. *Human Biol.* **35**: 377–436.

Goodman, M. (1965). The specificity of proteins and the process of primate evolution. In *Protides of the biological fluids* **12**: 70–86. Peeters, H. (ed.) Amsterdam: Elsevier.

Goodman, M. (1967a). Deciphering primate phylogeny from macromolecular specificities. *Am. J. phys. Anthrop.* **26**: 255–275.

Goodman, M. (1967b). Effects of evolution in Primate macromolecules. *Primates* **8**: 1–22.

Goodman, M., Barnabas, J., Matsuda, G. & Moore, G. W. (1971). Molecular evolution in the descent of man. *Nature, Lond.* **233**: 604–613.

Goodman, M., Barnabas, J. & Moore, G. W. (1972). Man, the conservative and revolutionary mammal. Molecular findings on this paradox. *J. Human Evol.* **1**: 663–686.

Goodman, M. & Moore, G. W. (1971). Immunodiffusion systematics of the Primates I. The Catarrhini. *Syst. Zool.* **20**: 19–62.

Hafleigh, A. S. & Williams, Jr., C. A. (1966). Antigenic correspondence of serum albumins among the primates. *Science, N.Y.* **151**: 1530–1535.

Han, K., Dautrevaux, M., Chaila, X. & Biserte, G. (1970). The covalent structure of beef heart myoglobin. *Eur. J. Biochem.* **16**: 465–471.

Han, K., Tetaert, D., Moschetto, Y., Dautrevaux, M. & Kopeyan, C. (1972). The covalent structure of sheep heart myoglobin. *Eur. J. Biochem.* **27**: 585–592.

Hilse, K. & Popp, R. A. (1968). Gene duplication as the basis for amino acid ambiguity in the alpha-chain polypeptides of mouse hemoglobin. *Proc. natn. Acad. Sci. U.S.A.* **61**: 930–936.

Hoyer, B. H., Bolton, E. T., McCarthy, B. J. & Roberts, R. B. (1965). Evolution of polynucleotides. In *Evolving genes and proteins*. Bryson, V. & Vogel, H. J. (eds) New York and London: Academic Press.

Hoyer, B. H., McCarthy, B. J. & Bolton, E. T. (1964). A molecular approach in the systematics of higher organisms. *Science, N.Y.* **144**: 959–967.

Hoyer, B. H. & Roberts, R. B. (1967). Studies on DNA homology by the DNA agar technique. In *Molecular genetics*, Part 2: 425–479. Taylor, H. (ed.) New York and London: Academic Press.

Hoyer, B. H., van de Velde, N. W., Goodman, M. & Roberts, R. B. (1972). Examination of hominoid evolution by DNA sequence homology. *J. Human Evol.* **1**: 645–649.

Huisman, T. H. J., Wilson, J. B. & Adams, H. R. (1967). The heterogeneity of goat hemoglobin: Evidence for the existence of two non-allelic and one allelic alpha chain structural genes. *Archs Biochem. Biophys.* **121**: 528–530.

Jones, R. T., Brimhall, B. & Duerst, M. (1971). Amino acid sequence of the α and β chains of dog hemoglobin. *Fedn Proc. Fedn Am. Socs exp. Biol.* **30**. Pt. 2 of two volumes, abstract 1207.

Karadjova, M., Nedkov, P., Bakardjieva, A. & Genov, N. (1970). Differences in amino acid sequences between dolphin and sperm whale myoglobins. *Biochim. Biophys. Acta* **221**: 136–139.

Kilmartin, J. V. & Clegg, J. B. (1967). Amino acid replacements in horse haemoglobins. *Nature, Lond.* **213**: 269–271.

Kluh, I. & Bakardjieva, A. (1971). Primary structure of N-terminal part of molecule of dolphin myoglobin. *FEBS Letters* **17**: 31–34.

Kohne, D. E., Chiscon, J. A. & Hoyer, B. H. (1972). Evolution of primate DNA sequences. *J. Human Evol.* **1**: 627–644.

Kramp, V. P. (1956). Serologische Stammbaumforschung. In *Primatologia*. I. *Systematik Phylogenie Ontcgenie*. Hofer, H., Schultz, A. H. & Stark, D. (eds) Basel: S. Karger.

Le Gros Clark, W. E. (1960). *The antecedents of Man*. Chicago: Quadrangle Books.

Margoliash, E. & Fitch, W. M. (1968). Evolutionary variability of cytochrome C primary structures. *Ann. N. Y. Acad. Sci.* **151**: 359–381.

Martin, M. A. & Hoyer, B. H. (1967). Adenine plus thymine and guanine plus cytosine enriched fractions of animal DNAs as indicators of polynucleotide homologies. *J. molec. Biol.* **27**: 113–129.

Matsuda, G., Maita, T., Takei, H., Ota, H., Yamaguchi, M., Miyanchi, T. & Migita, M. (1968). The primary structure of adult hemoglobin from *Macaca mulatta* monkey. *J. Biochem., Tokyo* **64**: 279–282.

McKenna, M. C. (1969). The origin and early differentiation of therian mammals. *Ann. N. Y. Acad. Sci.* **167**: 217–240.

Moore, G. W. (1971). *A mathematical model for the construction of cladograms*. Ph.D. Thesis, North Carolina State University.

Moore, G. W., Barnabas, J. & Goodman, M. (1973). A method for constructing maximum parsimony ancestral amino acid sequences on a given network. *J. theor. Biol.* **38**: 459–485.

Moore, G. W. & Goodman, M. (1968). A set theoretical approach to immunotaxonomy: analysis of species comparisons in modified Ouchterlony plates. *Bull. math. Biophys.* **30**: 279–289.

Moore, G. W., Goodman, M. & Barnabas, J. (1973). An iterative solution from the standpoint of the additive hypothesis to the dendrogram problem posed by molecular data sets. *J. theor. Biol.* **38**: 423–457.

Nuttall, G. H. F. (1904). *Blood immunity and blood relationship*. Cambridge, England: Cambridge University Press.

Picard, T., Heremans, J. & Vandesbroek, G. (1963). Serum proteins found in primates. Comparative analyses of the antigenic structure of several proteins. *Mammalia* **27**: 285–299.

Pilbeam, D. (1968). The earliest hominids. *Nature, Lond.* **219**: 1335–1338.

Romer, A. S. (1966). *Vertebrate paleontology*. Chicago: The University of Chicago Press.

Romero Herrera, A. E. & Lehmann, H. (1971a). The myoglobin of primates. I. *Hylobates agilis* (gibbon). *Biochim. biophys. Acta* **251**: 482–487.

Romero Herrera, A. E. & Lehmann, H. (1971b). Primary structure of human myoglobin. *Nature, Lond. (New Biol.)* **232**: 149–152.

Sarich, V. M. (1970). Primate systematics with special reference to old world monkeys—a protein perspective. In *Old World monkeys, evolution, systematics, and behavior*. Napier, J. R. & Napier, P. H. (eds). London and New York: Academic Press.

Sarich, V. M. & Wilson, A. C. (1966). Quantitative immunochemistry and the evolution of primate albumins: micro-complement fixation. *Science, N.Y.* **154**: 1563–1566.

Sarich, V. M. & Wilson, A. C. (1967). Immunological time scale for hominid evolution. *Science, Wash.* **158**: 1200–1203.

Simons, E. L. (1967). The earliest apes. *Scient. Am.* **217**: 28–35.

Sokal, R. R. & Michener, C. D. (1958). A statistical method for evaluating systematic relationships. *Kans. Univ. Sci. Bull.* **38**: 1049–1438.

Tashian, R. E., Tanis, R. J., Ferrell, R. E., Stroup, S. K. & Goodman, M. (1972).

Differential rates of evolution in the isozymes of primate carbonic anhydrases. *J. Human Evol.* **1**: 545–552.

Thompson, E. O. P. & Air, G. M. (1971). Studies on marsupial proteins VI. Evolutionary changes in β-globins of the Macropodidae and the amino acid sequence on β-globin from *Potorous tridactylus*. *Aust. J. biol. Sci.* **24**: 1199–1217.

Wade, P. T., Barnicot, N. A. & Huehns, E. R. (1970). Structural studies on the major and minor haemoglobin of the monkey *Macaca irus*. *Biochim. Biophys. Acta* **221**: 450–466.

Wang, A. C., Shuster, J., Epstein, A. & Fudenberg, H. H. (1968). Evolution of antigenic determinants of transferrin and other serum proteins in primates. *Biochem. Genet.* **1**: 347–358.

Wolfe, H. R. (1939). Standardization of the precipitin technique and its application to studies of relationships in mammals, birds and reptiles. *Biol. Bull. mar. biol. Lab. Woods Hole* **76**: 108–120.

Wood, A. E. (1962). The early Tertiary rodents of the family Paramyidae. *Trans. Am. phil. Soc.* (n.s.) **52**: 1–261.

Wooding, G. L. & Doolittle, R. F. (1972). Primate fibrinopeptides: evolutionary significance. *J. Human Evol.* **1**: 553–563.

DISCUSSION

BARNICOT (CHAIRMAN): Thank you, Professor Goodman. I am very sorry that you have had so little time to develop this interesting theme. Perhaps you might like to comment on the point that the construction of these protein trees depends on the assumption that evolution has taken the shortest path. Much computer time is expended in seeking the "best" tree on this criterion, and yet, as far as I know, there is no clear biological justification for the minimal path assumption. As it happens protein trees have not, on the whole, proved to be outrageously discrepant with phylogenies based on morphology. No doubt this is comforting but it is not entirely logical that it should be. Presumably we must go on sequencing a variety of proteins from a variety of species hoping that this will lead to consistent patterns of relationship.

GOODMAN: I think you have raised an excellent point. Certainly the parsimony assumption can be fallacious, but the nature of the fallacy is such that it gets less and less the more voluminous our data become. The reason is that if there is a very dense network with many evolutionary forks, where the adjacent forks are separated by short spans of evolutionary time rather than by long spans, then it is less likely that the parsimony assumption will be violated. This is because the mutation rate itself becomes a limiting factor if the amount of evolutionary time is not all that great. More important perhaps is the role of selection in weeding out most mutations that occur and only allowing a few to get through the sieve, so to speak, and be fixed in the line of descent.

So the main error in these kinds of trees, a bias towards under-estimating the true amount of evolutionary change, is in those branches where long spans of evolutionary time go by with few evolutionary forks. The parts of the tree that are densely represented by evolutionary forks will suffer from this problem less. We could also have situations due to parallel mutations and the like which would not be properly detected by the maximum parsimony method. However, when we combine the data from different protein sequences, the errors due to such situations should cancel out. Thus we need data on many more protein sequence chains, on the one hand, and, on the other hand, we need many more species represented.

In the meantime we are trying to see all that we can with the available data, and these certainly have shown the close relationship of the African apes to man claimed by Darwin and Elliot Smith.

Note added after discussion: The point raised by Professor Barnicot that there is no clear biological justification for the minimum path assumption has troubled my colleague, Dr G. William Moore, and myself for quite some time. Recently we have formulated (or better, stumbled upon, since we claim no priority) some ideas which do offer a more complete, but still heuristic explanation for why maximum parsimony trees of protein sequences capture the evolutionary history of these sequences. In brief, here is our thinking:

The procedure which constructs both those sequence ancestors and that tree topology yielding together fewer mutations than any other ancestors and branching topologies ensures that the maximum number of amino acid identities at aligned positions in the descendant sequences are due to inheri-tance from common ancestors and, conversely, the fewest number to parallel and back mutations. Accepting this proposition (it has not yet been phrased in mathematical terms and proven formally, but we feel it can be), the following probabilistic argument can be given for why the maximum parsi-mony method captures the phylogeny of protein sequences. Since there are 20 different amino acids, one would expect that many mutations would have to occur at an aligned position in the descent of two sequences before a parallel or back mutation in one or the other descendant line produced the same amino acid in the two sequences if they had previously diverged. In other words, it seems more probable to us that among descendant sequences the amino acid identities which are due to inheritance from common ancestors are much more frequent than those due to parallel or back mutations. Thus in reconstructing the phylogeny of the sequences, one should use a method which maximizes the number of identities due to inheritance from common ancestors. We feel the maximum parsimony method does just this.

Our particular maximum parsimony method (Moore, Barnabas & Goodman, 1973)* first translates the contemporary protein amino acid

---

* See list of references.

sequences into codon sequences, including all necessary alternative synonymous codons among the 61 codons which specify the 20 amino acids, and then seeks those ancestral-descendant configurations of codons which minimize mutations over the tree. Often different codons for the same amino acid can be distinguished; a leucine, e.g., in one region of the tree at an aligned position may have to be specified, owing to the configuration of amino acids in surrounding species, by a different codon from that used for a leucine in another region of the tree to yield the minimum number of mutations. This discriminating power of the maximum parsimony method supports the above argument, for it is even more probable when dealing with codon rather than just amino acid identities in descendant species that these are due to common inheritance rather than to parallel and back mutations.

Symp. zool. Soc. Lond. (1973) No. 33, 377–404.

DISTRIBUTION OF MALARIA PARASITES IN PRIMATES, INSECTIVORES AND BATS

P. C. C. GARNHAM

Department of Zoology, Imperial College of Science and Technology, London, England

SYNOPSIS

A study of the evolution of a parasite entails equal consideration of the host; in malaria parasites two hosts, vertebrate and invertebrate, are involved, and genetic traits are important in both. Equally, the three components of malaria have each to be studied in relation to the zoogeography. Three stages occur in the life cycle of malaria parasites: in the insect, the liver and the blood. The blood stages were the last to have evolved and are represented by two types, asexual and sexual. The latter are the earlier and some parasites are limited to this stage (e.g. *Hepatocystis*). *Hepatocystis* has never become established in the Hominoidea, nor is it present in the lower primates in the New World.

There are three subgenera of *Plasmodium*. The most recently evolved is *Laverania*, confined to man and African apes; the preceding is *Plasmodium*, widespread throughout the Anthropoidea; the earliest is *Vinckeia* of lemurs and other arboreal mammals.

Man and chimpanzee share three species of *Plasmodium*: *P. (L.) falciparum, P. (P.) vivax* and *P. (P.) malariae*. Both hosts can be cross infected with each other's parasites, but full susceptibility is only exhibited to the (primitive) liver stages; splenectomy however renders the heterologous host susceptible also to the blood stages. Four species of *Plasmodium* occur in gibbons and two in orang-utans ;the parasites of the latter are more like the former than those of man or chimpanzee.

The cradle of primate malaria was probably the primaeval forest of Southern Asia, where great speciation has occurred; from here it spread to the African apes and thence in recent centuries to American monkeys via man. *Plasmodium (Vinckeia)* is absent in tarsiers, slow loris, *Tupaia* and galagos; present in lemurs; rare in bats and squirrels, but locally common in African tree-rats. *Hepatocystis* is common in monkeys, fruit-bats and squirrels.

Speculations are made on the course of evolution of these parasites and the phylogenetic implications regarding their respective hosts.

INTRODUCTION

Most studies on evolution start at the bottom of the tree and ascend to the top. The absence of fossils in the parasitic protozoa means that there is a total absence of direct evidence of the course of their evolution and we have to depend on circumstantial evidence. Even this is difficult to obtain and our conclusions are inevitably speculative. It is therefore desirable to abandon the usual order and begin at the top where the data are immediately available.

A second fundamental problem is the existence of three separate entities in malaria: the parasite itself, the vertebrate host and the insect host. Each of these has its own scale of evolution; yet the parasite is utterly dependent on its two hosts.

Malaria parasites belong to the Suborder Haemosporidiidea (Table I), and they are characterized by two cycles in the vertebrate host—one in the organs and the other in the blood cells, while a single cycle takes

TABLE I

Genera of Haemosporidia in primates

(Family)	Plasmodiidae	Haemoproteidae
(Genus)	*Plasmodium*	*Hepatocystis*
(Subgenus)	*Plasmodium Laverania Vinckeia*	

place in the invertebrate. In the adjoining Suborder, Eimeriidea, the parasites are largely confined to the organs and only in a few of the more highly evolved examples do they invade the blood and acquire an invertebrate (blood sucking) host.

In man and other primates the malaria parasites belong to the genus *Plasmodium*, which is divided into three subgenera. The most highly evolved is the subgenus *Laverania*, which has only two species. The next subgenus is *Plasmodium*, which is typical of the group and is widespread in man and most supralemuroid primates. The third subgenus is *Vinckeia* and includes two species in lemurs and 14 species in a rather narrow range of lower mammals.

The genus *Hepatocystis* includes a number of species of so-called malaria parasites (originally placed in the genus *Plasmodium*) which is of a less evolved type in that (like the few eimeriid parasites which have managed to break through as gametocytes into the blood) no multiplication (schizogony) occurs in the blood. There are other major differences which indicate that this genus arose at an earlier date than *Plasmodium*; the large (macroscopic) size of the liver forms (resembling the avian genera in the Leucocytozoidae), the rapidity of microgamete formation, the nature of the invertebrate host (*Culicoides*, instead of mosquitoes which evolved probably at a later date than the former). *Hepatocystis* is absent in the Hominoidea (except for one report in *Hylobates*) and is absent in the primates and other mammals of the New World. On the other hand, *Hepatocystis* is extremely common in the Old World monkeys and other tropical mammals e.g. fruit bats, squirrels and other arboreal creatures.

MALARIA PARASITES OF MAN

The geographical distribution of human malaria parasites is probably of little significance as the migratory habits of man (as of birds) have led to the dissemination of the parasite into whichever regions possess suitable climatic features e.g. a high enough temperature to permit the development of the parasite in the mosquito, and a suitably humid environment for the breeding of the insect and its survival as an adult. The situation in the New World is interesting because there is some evidence that malaria was imported into the New World as recently as the date of arrival of the Conquistadores.

It is simplest to consider the distribution of human malaria at that date in history which just preceded the introduction of DDT and the modern synthetic antimalarial drug, i.e. about 1945. Since that time malaria has disappeared from Europe, the U.S.S.R., the United States of America and from a few places in the Tropics. Up to 1945 the disease had spread and even infiltrated into highland areas which hitherto had been free from malaria. The natural governing factors, *viz.* temperature and humidity, were bypassed as a result of the opening up of new lands, and local movements of the population introduced the parasites into the virgin territory. The spread in modern times has been well documented (*vide* Garnham, 1971); it can occur under many different situations (e.g. soldiers returning from tropical theatres of war; tourist travel; mass expulsion of unwanted races, etc.). The evidence however for the introduction of malaria into fresh lands during the more ancient migrations is flimsy. It has been suggested (Bruce-Chwatt, 1965) that malaria possibly reached the Americas by tribes crossing the Behring Straits 50 000 or more years ago, and that when they reached a latitude suitably far South, transmission of the infection by the local anopheline mosquitoes took place. But the distances involved in such travel are too great to make this theory acceptable.

Coatney, Collins, Warren & Contacos (1971) have discussed in some detail the possibility that the distribution of primate malaria in the Old World is the result of the following occurrences. The cradle of primate malaria was the forest covering much of South and South-east Asia, where many species of *Plasmodium* evolved in the simian inhabitants. Into this environment, intruded "peripatetic" man from Africa, perhaps in the early Pleistocene Period. He became infected with the parasites, and his descendants in due course returned to Africa, where they spread the infection among the human inhabitants and the great apes. This theory seems far fetched because such migrations a million or more years ago seem very unlikely to have taken place; but it is mentioned here in order to give a possible explanation of the curious distribution of malaria parasites in African primates (see pp. 385–388; 398–399).

Four species of *Plasmodium* occur in man: *Plasmodium (Laverania) falciparum*, *P. (Plasmodium) vivax*, *P. (P.) malariae* and *P. (P.) ovale*. The incidence of *P. falciparum* is primarily governed by a minimum temperature of 20°C and thus this species is essentially a parasite of tropical and subtropical regions. *P. falciparum* does not multiply readily in the blood of man when he possesses the "sickling trait"; the infections are therefore less severe and rarely fatal. Thus, this trait confers an advantage on its carriers who are able to withstand the infection; natural selection operates, and a high incidence of the sickle cell trait in a population indicates the existence of past or present holoendemic falciparum malaria (Allison, 1961). Allison & Clyde (1961) suggested also that another genetic trait—deficiency of glucose–6–phosphate dehydrogenase—had a similar effect, but the evidence for this is much less convincing.

Plasmodium vivax is essentially a parasite of temperate regions, though its distribution is high in the Indian sub-continent and in some other parts of the tropics. Its northern limit is determined by the summer isotherm of 16°C. Like *P. falciparum*, *P. vivax* also is affected by an adverse genetic factor; the factor itself is unknown but it is present in the true Negro of West Africa (i.e. between Senegal and the Congo River) and in his descendants in the New World. This race is almost totally resistant to *P. vivax* and to vivax-like parasites of apes and monkeys. They include *P. schwetzi* of the chimpanzee, and *P. cynomolgi cynomolgi* and *P. cynomolgi bastianellii* of monkeys. On the contrary, the Negro is fully susceptible to *P. ovale*. This response of the Negro to the former group of parasites is possibly indicative of their mutual phylogenetic relationships and of the lack of a relationship between them and *P. ovale*.

Plasmodium malariae has a curiously patchy distribution, though it is probably commonest in tropical Africa. Sporogony occupies a minimum duration of two weeks; thus the mosquito must be capable of living at least three weeks if transmission is to occur.

Plasmodium ovale is the species most rarely found in humans, but it is relatively common on the west coast of Africa; it occurs also in most tropical parts of the continent, and as isolated instances elsewhere in the warmer parts of the world. The explanation of such sporadic infections has never been found, but neither in the case of *P. ovale* nor in that of *P. malariae* does a special vector mosquito provide the answer, because experimentally both these parasites can be transmitted by many species of *Anopheles*.

Fig. 1. Distribution of malaria parasites in non human primates in Africa.

//// = *Hepatocystis* sp. in monkeys.
\\\\ = *Plasmodium* spp. in monkeys.
|||| = *Plasmodium* spp. in apes.

MALARIA PARASITES OF APES OF AFRICA

The distribution of the malaria parasites of the great apes (Fig. 1) nearly coincides with that of their vertebrate hosts, but the details are not very precisely known, at least in the interior of tropical Africa (Bray, 1963, 1964). From the coastal forests of Liberia and Sierra Leone and at the mouth of the Congo River, the infection in the chimpanzee passes inland through the Congo forest and is still quite common as far East as Stanleyville where Schwetz (1933) was the first to redescribe the three species in chimpanzees after Reichenow's original observations in

Cameroun. But at the eastern limit of the distribution of these apes, beyond Lake Kivu, the infection is apparently rare or absent.

There are three species of malaria parasites in chimpanzees, which closely resemble *Plasmodium falciparum* (*viz. P. reichenowi*), *P. vivax* (*viz. P. schwetzi*) and *P. malariae* (*viz. P. rodhaini*). On morphological characters and reciprocal susceptibility, the differences between the parasites of man and ape are greatest in the *P. falciparum: reichenowi* and least in the *P. malariae: rodhaini* pairs; in fact the latter pair are usually considered to be identical. Yet *P. falciparum* is thought to have evolved last and *P. malariae* is considered to be the most primitive; if these suppositions were true it would seem more likely that *P. rodhaini* which has parasitized man longest would have become more instead of less differentiated, while the supposedly more recent *P. falciparum* should bear the closest resemblance to its counterpart in the chimpanzee. The chimpanzee parasite fails to develop in the common vectors, i.e. *Anopheles gambiae* and *A. funestus* of the human parasites, and the former parasite seems to be strictly host specific.

Only two species of *Plasmodium* (*P. schwetzi* and *P. reichenowi*) have been detected in Lowland gorillas; the few observations on the Mountain gorilla have been negative. For what it is worth (for surveys were few), the apparent *absence* of *P. rodhaini* in the gorilla may have some significance, though the incidence of this species in the chimpanzee and indeed of *P. malariae* in man is also lower than that of the two other parasites.

The percentage infection rates in the chimpanzee are as follows (Garnham, 1966, from figures supplied by R. S. Bray): *P. reichenowi* 30%, *P. schwetzi* 13%, *P. rodhaini* 8%.

The absence of a *P. ovale* type of parasite in chimpanzees is strange, as West Africa is the region, *par excellence*, where this human species occurs. Coatney (1968) concluded (in the writer's opinion, wrongly) that *P. ovale* represents a zoonosis derived from *P. schwetzi*.

MALARIA PARASITES OF APES OF ASIA

(Figure 2.) The orang-utan is found today only in Borneo and the northern half of Sumatra, and its numbers are declining so fast that the species is in danger of extinction. *Plasmodium pitheci* has been known to occur in the orangutan since 1907 (Halberstaedter & von Prowazek) and several records of its existence have been made subsequently. A second species was discovered early last year in the Sepilok Forest of Sabah, East Borneo by Garnham, Rajapaksa, Peters & Killick-Kendrick (1972) and named *P. silvaticum*. Both species may exist in the same

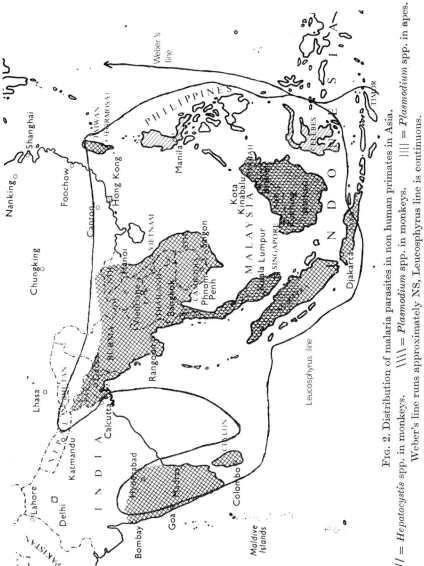

FIG. 2. Distribution of malaria parasites in non human primates in Asia.

//// = *Hepatocystis* spp. in monkeys. \\\\ = *Plasmodium* spp. in monkeys. |||| = *Plasmodium* spp. in apes.

Weber's line runs approximately NS, Leucosphyrus line is continuous.

animal. Neither species bears an exact resemblance to the parasites of the African apes (nor to those of man) and a human volunteer failed to develop malaria after being bitten by a mosquito heavily infected with *P. pitheci*.

Eight orangutan were studied for a month in the Sepilok reserve of 10 000 acres where confiscated animals are rehabilitated for life in the wild (de Silva, 1971). Six of the eight showed one or both of the two species of parasites, and it seems probable that conditions for transmission must be remarkably favourable, for at least two animals were uninfected on arrival in Sepilok and developed malaria some months later. This feature of the epizootiology is mentioned here, because it has a bearing on the possible phylogeny of the parasites. The forest contains a very small number of indigenous orangutan and a larger number of gibbons.

P. pitheci resembles *P. youngi* and *P. hylobati* of the gibbon (see below), but differs in a few details in the blood stages. *P. silvaticum* resembles *P. eylesi* of the gibbon but again there are minor differences in the erythrocytic, exoerythrocytic and sporogonic stages. Unfortunately it is not possible to carry out experimental work involving cross infections in orang-utan. The slight differences in the appearance of the parasites in the two apes might be the result of the well-known change in morphology of a species when transferred to another host, or perhaps a more likely explanation is that *P. pitheci* and *P. silvaticum* represent mutations of gibbon parasites.

Experimental work however has been able to show that a splenectomized chimpanzee and an *Aotus* monkey were resistant to *P. pitheci*, while a splenectomized chimpanzee proved to be highly susceptible to *P. silvaticum*. A splenectomized gibbon showed a very transient infection with *P. pitheci* but none with *P. silvaticum* (see Peters *et al.*, in press).

These observations on the malaria parasites of the orang-utan suggest that their affinities lie more with the species found in the gibbon than in the chimpanzee or man, and tend to support the idea that the hosts themselves are phylogenetically closer.

Four malaria parasites of gibbons have been described as follows: *P. hylobati* from *Hylobates moloch* from Java by Rodhain (in 1941), and from Sarawak by Collins *et al.* (in 1972). The species was also tentatively diagnosed by R. Desowitz (pers. comm.) from *H. concolor* from Thailand (in 1967) and has been seen in Indo-China.

P. youngi from *Hylobates lar* from Malaya by Eyles *et al.* (1964). This bears a close resemblance to the former of which it may be a subspecies.

P. eylesi from *Hylobates lar* from northern Malaya by Warren, Bennett, Sandosham & Coatney (1965). This species was apparently able to infect man.

P. jefferyi from *Hylobates lar* from northern Malaya by Warren, Coatney & Skinner (1966). The naturally infected gibbon was also found to harbour *P. youngi.*

Unnamed species of malaria parasites have also been encountered in the Siamang (*Symphalangus syndactylus*) over 60 years ago in Malaya and again in the same region by Rudolph Geigy in recent years. A malaria parasite, thought to be an unknown species of *Hepatocystis*, has been seen twice in *Hylobates concolor* by Shiroishi, Davis & Warren (1968). If these identifications are correct they will represent the first records of the presence of *Hepatocystis* in an ape, and they would confirm parasitologically the proximity of gibbons to monkeys (in which *Hepatocystis* is very common) and their distance from the higher apes (in which the genus is otherwise absent). It has been mentioned above (p. 384) that the two species of *Plasmodium*, occurring in the orang-utan, closely resemble *P. youngi* and *P. eylesi* of gibbons.

MALARIA PARASITES OF AFRICAN MONKEYS

(Figure 1.) The rarity of *Plasmodium* in monkeys in Africa presents an enigma which remains unsolved; it cannot be explained on the basis of inadequate surveys, which have been at least as extensive as those made in the Orient. Only one species has been found—*Plasmodium gonderi*. It is limited to a narrow belt along the West African Coast, from the mouth of the Congo River to the former British Cameroons and Liberia, though its presence in the intervening territories is unknown. *P. gonderi* is limited to two types of monkeys—mangabeys (*Cercocebus atys, C. galeritus* and possible other species) and drills (*Mandrillus leucophaeus*). Records of its presence in mangabeys in Central Africa and in the eastern Congo probably represent misidentifications of *Hepatocystis kochi*. The mangabey has a wide distribution in tropical Africa and the explanation of the absence of malaria in the hinterland may perhaps lie in the restricted distribution of the—as yet unidentified —anophelinevect or. *P. gonderi* is a stenoxenous parasite and, even experimentally, all observers have had great difficulty in finding a species of mosquito readily susceptible to infection.

A possible reason for the rarity of *Plasmodium* in African monkeys is that the niche is completely filled by *Hepatocystis* throughout the Ethiopian region. Experimentally however there is no evidence for such an explanation, as monkeys infected with *Heptocystis* spp. are susceptible to *Plasmodium* by inoculation. The ancient lineage of *Hepatocystis* (from which *Plasmodium* may have evolved) and the stability of the genus may have restricted the rate of mutation and only a single example (*P. gonderi*) has managed to survive under African conditions.

Hepatocystis kochi and related species were all placed originally in the genus *Plasmodium*, until the writer (Garnham, 1948) removed them to the family Haemoproteidae because of

1. the characteristic exoerythrocytic cycle in the parenchyma cells of the liver,
2. the presence in the blood of gametocyte stages only, and
3. transmission being effected by *Culicoides* instead of by mosquitoes.

Although there are a few curious gaps in the distribution of *Hepatocystis* (for instance in some places the baboons are heavily infected, while elsewhere they are entirely free) this parasite is otherwise found nearly everywhere in monkeys in tropical Africa. The most recent example is the record by Orihel & Garnham (in press) from Talapoins collected in Rio Muni.

The distribution of *H. kochi* is however affected by certain climatological conditions. Thus the parasite is not found in the Barbary ape of North Africa, nor at altitudes above 2000 m in tropical Africa, while it disappears south of Durban in South Africa. *Macaca fascicularis* was introduced two or three centuries ago into Mauritius, but the monkeys lost any malaria which they may have had originally and naturally they have acquired none since their arrival, for there are no indigenous monkeys on the island. Similarly, *Cercopithecus sabaeus* was imported two centuries ago and more recently from West Africa into the West Indies and recent surveys carried out on their descendants on the island of St. Kitts showed that no parasites have persisted. Probably the correct species of *Culicoides* was absent in these new localities and transmission of the parasite was unable to occur.

Some degree of speciation of *Hepatocystis* has taken place in African monkeys. The identification is dependent upon the morphological characters of the exoerythrocytic stages in the liver; for the blood stages all look alike as in other members of the family (Haemoproteidae). One of the distinctive features is the size of the "merocyst" in the liver and sometimes the size varies according to that of the host. Thus, the largest monkeys (*Papio*) exhibit the largest exoerythrocytic parasite (*Hepatocystis simiae*) and the smallest monkeys (the Talapoin) the smallest. Five species of *Hepatocystis* are now recognized in African monkeys: *H. kochi*, *H. simiae*, *H. bouillezi*, *H. cercopitheci* and *H.*, sp. nov. from the Talapoin. Most records are from various species of *Cerco-*

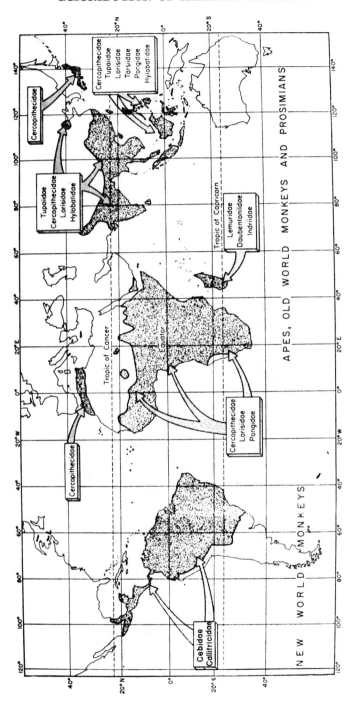

Fig. 3. Distribution of non human primates (Napier & Napier, 1967).

pithecus, but the identifications of the parasites are doubtful as few descriptions include that of the essential liver stages. Systematic work is handicapped also by lack of knowledge, except in the type species, of the invertebrate host (*Culicoides* spp.).

No species of *Hepatocystis* is pathogenic to the vertebrate host and if this parasite really inhibits the development of the lethal *Plasmodium*, this might partly explain the widespread distribution of the former in Africa; there is almost a vaccinating effect.

The absence of *Hepatocystis* from the Hominoidea is a striking feature in Africa where the apes (and man) are riddled with *Plasmodium*, yet the monkeys living in the same environment show a completely reverse picture: no or rare *Plasmodium* but gross infection with *Hepatocystis*.

MALARIA PARASITES OF ASIAN MONKEYS

(Figure 2.) In contrast to Africa. Asia is extensively infected with simian malaria (*sensu stricto*) though the distribution is not uniform. The existence of malaria is naturally governed by the presence or absence of the vertebrate and invertebrate hosts, and the eastern limit is provided by Weber's Line (Fig. 2) beyond which monkeys are missing (Fig. 3). Similarly, the northern and western limits are affected by the "leucosphyrus line", beyond which mosquitoes of this group are absent (Reid, 1968). *Anopheles leucosphyrus* and related species appear to be essential sylvatic vectors of simian malaria in Asia—see p. 401.

Warren (1973) has suggested that the actual species of *Plasmodium* may be divided zoogeographically by a line, running West to South through the Bay of Bengal separating S.E. Asia from India, Pakistan and Ceylon. Similar types of parasites are found in both but some degree of speciation has taken place. The following pairs have been identified from the two regions:

TABLE II

Pairs of Asian species

Eastern group	Western groups
P. inui inui	*P. inui shortti*
P. cynomolgi cynomolgi	*P. cynomolgi ceylonensis*
P. fieldi	*P. simiovale*
P. coatneyi	*P. fragile*
P. knowlesi	Absent

(From Warren, 1973)

The most widespread of these parasites is *P. inui* and besides the named subspecies (*P. i. shortti*) other subspecies undoubtedly exist, particularly in the two families of monkeys (macaques and langurs) and from different regions such as the highly sylvatic environment of Malaysia as compared with the drier country of Formosa.

The next commonest species is *P. cynomolgi* and here again a complex of parasites is found, which again can be primarily grouped according to the vertebrate host and the biogeography. It is difficult to know what taxonomic rank to accord these organisms, and they are variously termed species, subspecies or strains. The major differences relate to sporogony. It may require much adaption before a parasite from a langur monkey will develop readily in a macaque. On the other hand the parasites of the macaque easily infect all or most of the monkeys in the subfamily.

It seems likely that the Indo-Malaysian region provided, and has continued to provide ideal conditions for speciation of *Plasmodium*, and if one considers that lethality, quotidian periodicity of erythrocytic schizogony and relative rapidity of exoerythrocytic schizogony are indices of recent evolution, it should follow that *P. knowlesi*, in fulfilling these criteria, is the latest species to have evolved in the jungles of Malaysia.

At the periphery of the region, simian malaria becomes restricted to one or two species of parasites, i.e. on the Assam–Burmese border (Coatney *et al.*, 1971), Formosa and the Celebes. *Plasmodium* in Ceylon however exhibits almost as much speciation as the genus in Malaysia.

The incrimination of the natural vectors of the malaria parasites on numerous occasions in S.E. Asia is in striking contrast to the failure to find a single vector of *P. gonderi* or of the chimpanzee parasites in West Africa.

Man has been found naturally infected with *P. knowlesi* in Malaya on two occasions; while in the laboratory, either accidentally or deliberately, transmission of *P. cynomolgi bastianellii* and of other subspecies, and of *P. inui* and of its subspecies has occurred. (The susceptibility of man to primate malaria in general is usefully summarized in an appendix to a Technical Report of the World Health Organization, 1969).

The Russian workers, Sergiev & Tiburskaya (1965), suggested that the ancestor of the human parasite, *P. vivax*, was *P. cynomolgi bastianellii* present in the macaques of southern Asia, where transmission takes place throughout the year and the disease is characterized by short-term relapses. From this region man took the infection northwards and eventually into a colder climate, where transmission became limited to the summer, and where the infection was maintained from one season to the next by long-term relapses. This form of the parasite has thus undergone speciation and is now known as *P. vivax hibernans*.

Hepatocystis spp. are present in southern Asia, but infections are much less frequent than in Africa. *H. semnopitheci* has been reported from India (Himalayas, Poona, Assam), Ceylon, Thailand, East and West Malaysia, Indonesia and Formosa. Only from the last place has the parasite received a different name (*H. taiwanensis*). The exoerythrocytic stages of these Asian species of *Hepatocystis* present some curious features which led the late Professor H. Ray of Calcutta to suspect that a new genus was involved and the name, *Rayella*, was proposed primarily for the parasites of the Giant flying squirrel of Taiwan (see below), but thought by Das Gupta (1967) to be equally applicable to the parasite in the Himalayan langur. Until more is known about the life cycle of these parasites, their taxonomic position must remain uncertain.

The simian hosts of *H. semnopitheci* include a wide variety of macaques and leaf monkeys.

MALARIA PARASITES OF LATIN AMERICAN MONKEYS

(Figure 4.) Only two species of malaria parasites occur in American monkeys, *viz. P. brasilianum* and *P. simium*; the former is fairly common and the latter rare except in a few places. As in Asia, the distribution is probably dependent upon the presence of the appropriate species of *Anopheles* and Deane, who has made a profound study of all aspects of monkey malaria in Brazil (1972), concludes that certain acrodendrophilic species of mosquitoes (particularly, *Anopheles cruzi* and *A. neivai* and possibly *A. oswaldoi* and *Chagasia bonneae*) are the most important invertebrate hosts. *P. brasilianum* occurs in Brazil, Peru, Colombia, Venezuela and Panama. The record from the Yucatan probably refers to a piroplasm. In Brazil alone, 19 species of cebids have been found infected, but elsewhere the range is much less extensive. Some curious exceptions may be noted e.g. the absence of *P. brasilianum* in marmosets and *Aotus*. Slight differences in morphology and in the life cycle may occur in infections in different species of monkeys and subspecific differentiation of this quartan parasite is thought to have taken place.

P. simium has been found in the coastal forests of Rio Grande do Sul, Santa Catarina and Espirito Santo, and inland to São Paulo, where it was first identified in Howler monkeys (*Alouatta fusca*). Since then, the parasite has also been found in *Brachyteles arachnoides* and double infections with *P. brasilianum* are sometimes encountered.

Fɪɢ. 4. Distribution of malaria parasites in non human primates in New World.

\\\\ = *Plasmodium* spp. in monkeys.

Of great interest is the phylogeny of these two New World parasites of which there are two current views. The older one is that before the final separation of the continents, the original primate root grew in the land which was eventually to separate, and that this was infected with malaria parasites (Fig. 5). The quartan parasite (*P. brasilianum*) would presumably have been a stable form of the *P. inui* type, so widely

Malaria Parasites

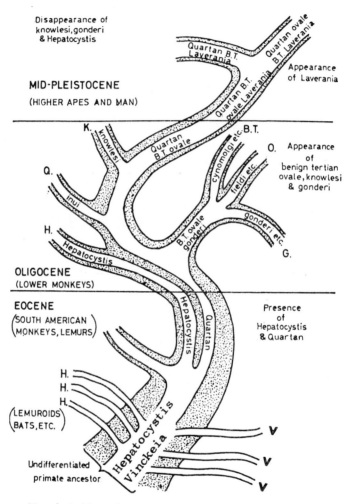

Hypothetical line of ascent of malaria parasites in evolution of primates.

Q = quartan type	O = *P. ovale*
H = *Hepatocystis*	G = *P. gonderi*
K = *P. knowlesi*	L = *Laverania*
BT = *P. cynomolgi* type	V = *Vinckeia*

Modified from Garnham (1963).

FIG. 5. Hypothetical line of ascent of malaria parasites in evolving primates (Garnham, 1966).

distributed in monkeys in the Old World today—and not the quartan parasite of the higher apes (which had not evolved at that time); the tertian parasite could have been one of several species, plentiful enough to-day, and represented perhaps by *P. cynomolgi*. Against this theory is the remarkable resemblance which the present day simian *P. brasilianum* of the New World bears to the human (or chimpanzee) *P. malariae* and its dissimilarity to *P. inui*. The more attractive view is based on the second group of facts; it is suggested that the malaria parasites of New World monkeys are of recent origin: they are the result of the importation, two to four centuries ago, of human species of *Plasmodium* in West African slaves or perhaps in the Conquistadores. Chronic quartan malaria infections persisted in the human population for years and during this time the carriers must have infected the local mosquitoes. Eventually, there would come a time when such mosquitoes bit monkeys instead of man, and after a little adaptation in the former, *P. brasilianum*, as later the parasite was to be called, spread throughout the simian population. The parasite still remained infective to man and today transmission easily takes place after blood or mosquito passage from monkeys to man. At all stages of the life cycle, the characteristic features of the human parasite are reproduced in the simian form.

Unfortunately for the second theory, *P. simium* has no obvious human counterpart and it would be necessary to assume that the human parasite had undergone a great degree of change in its new monkey host within a few centuries. However this species also is easily transmissible to man.

If the theory of a human origin of simian malaria in Latin America is correct, the absence of indigenous *Plasmodium* (and of *Hepatocystis*) could be ascribed to the early separation of the New World monkeys from the main primate line at a date prior to the evolution of malaria parasites in the latter, or in other words to the early date of colonization of the New World by monkeys before malaria had appeared in the Old World.

MALARIA PARASITES IN PROSIMIANS

If one accepts the theory that malaria did not appear in monkeys until *after* the branching off of the New World families, the absence of this infection should be even more apparent in those families which had separated still earlier. Such families are to be found in the prosimians which are widely distributed today in the Old World Tropics. The galagos, tarsiers, tree-shrews, Slow loris, lemurs and pottos have been extensively examined, and, with a single exception, they are all apparently free of both *Plasmodium* and *Hepatocystis*. As recently as a year ago, I examined further specimens of the Slow loris and tree-shrews in Borneo with a negative result. The habitat of these sylvatic animals is much the same throughout their range and it is difficult to understand their freedom from infection, except on the phylogenetic theory mentioned above.

However, lemurs are infected with two species of *Plasmodium* of the subgenus, *Vinckeia*. The infection in these animals is usually subpatent and can only be demonstrated after splenectomy. Garnham & Uilenberg (in press) have recently redescribed these species (*Plasmodium girardi* and *P. lemuris*) from *Lemur macaco macaco* and *L. m. fulvus*. The identity of a third blood parasite (*P. foleyi*) of *L. m. fulvus* Bück, Coudurier & Quesnel, 1952, remains doubtful, and it is possible that *P. lemuris* is a synonym of *P. foleyi*.

The presence of two species of malaria parasites of the subgenus *Vinckeia* in lemurs is interesting, because *Vinckeia* does not occur in other primates, but is essentially a parasite of the lower mammals. The isolation of Madagascar together with the great degree of speciation and the number of families of the lemurs ought to have been accompanied by a similar proliferation of the haemosporidiids. This has not occurred, however, and the rarity to-day of the hosts makes it unlikely that a full scale study of the phylogeny of the complex (including the vector) will ever be carried out.

MALARIA PARASITES IN OTHER MAMMALS INCLUDING BATS

Malaria parasites are as rare in the other arboreal denizens of the tropical rain forest as in the prosimians, but a few species have been reported usually in animals which are unobtrusive and uncommonly seen.

South East Asia

Malaria parasites (*sensu lato*) have been reported in a group of animals in S.E. Asia belonging to different Orders but with habits and habitats similar to those of the preceding groups. The animals all live in dense forest, feed on fruit or leaves, are nocturnal and fly to a greater or lesser extent. They include the following examples:

Colugo

The Colugo (*Cynocephalus variegatus*) possesses a highly distinctive species of *Plasmodium* (*P. sandoshami*) found in Malayan forests; it belongs to the subgenus *Vinckeia*. No details of its life cycle are known.

Flying squirrels

Petaurista petaurista and *Pet. elegans* harbour two species of malaria parasites of the subgenus—*P.* (*V.*) *booliati* in the former and an unnamed species in the latter. These examples were discovered in the forests of Selangor but the parasites probably occur elsewhere. Another species has been described from flying squirrels (*Petaurista p. grandis*) in Taiwan and was named *Plasmodium* (*V.*) *watteni*. On this island, the flying squirrel is also infected with a haemoproteid parasite which was placed in a new genus, *Rayella*, on account of the peculiar exoerythrocytic schizonts found in the liver. It was named *R. rayi*. The same parasite was first seen in the Indian flying squirrel (*Petaurista inornatus*) and later in the Himalayan *Pet. magnificus*. These animals are said to exhibit also exoerythrocytic schizonts of the genus *Hepatocystis*. It may be noted that the extensive surveys in northern India failed to reveal the presence of *Plasmodium* in these animals. The occurrence of haemosporidiids in the true squirrels is briefly discussed below (p. 347).

Fruit bats

Although *Plasmodium* is apparently absent from these animals, they are very commonly, almost universally, infected with *Hepatocystis pteropi* and over a wide area of tropical Asia, Australasia and in the Pacific islands as far east as Fiji. The chief host is *Pteropus* of which many species harbour *H. pteropi*: less often *Cynopterus* is infected, while a subspecies of the parasite was found in *Dobsonia moluccensis* in New Guinea. *Pteropus niger* in Mauritius is uninfected with *Hepatocystis*, just as the monkeys are uninfected with *H. kochi* (p. 386). It is interesting to note that the saturation of the African monkeys with *Hepatocystis* has apparently inhibited infection with *Plasmodium;* an analogous situation is now seen to prevail in fruit bats. A detailed analysis of the zoogeography of bat malaria (including the cosmopolitan *Polychromophilus* of insectivorous bats) is given by Garnham (1973).

Africa

Malaria parasites of the genus *Plasmodium* have been reported from scaly tails (Anomaluridae) and from fruit bats, but records are few and limited to a few foci.

Plasmodium anomaluri was found in the relict forest of Amani in East Africa in *Anomalurus fraseri*; its blood stages are typical of a *Vinckeia*. Two unnamed species have recently been found in West Africa (Côte d'Ivoire) by Killick-Kendrick & Bellier (1971), in *A. peli*.

The fruit bats of tropical Africa may be heavily infected with *Hepatocystis* but *Plasmodium* is rare. However, two species of the latter have been described. *Plasmodium (V.) roussetti* was found in *Roussettus leachi* in caves in Mont Hoyo in forested country in western Zaïre. *P. (V.) voltaicum* was discovered later in Ghana, in *Roussettus smithi* occupying caves in the banks of forested streams. Probably other examples (in *Lissonycteris*), not yet properly identified, have been seen in similar country near Brazzaville (Adam & Landau, 1970). It should be noted that roussettes have been examined frequently in other parts of tropical Africa and neither *Plasmodium* nor *Hepatocystis* has been encountered in the blood of these animals. However, other genera of fruit bats may be heavily infected with *Hepatocystis epomophori*. The type host is *Epomops franqueti*; several species of *Epomophorus* are also common hosts. *Hypsognathus monstrosus* in eastern Zaïre, and *Micropteropus pusillus* at the mouth of the Congo River and in the Southern Sudan harbour the parasites which, in all these localities and in different hosts, still retain a single name. Landau & Adam (in press) have recently found examples of *Hepatocystis perronae* in fruit bats (*Lissonycteris*) in caves near Brazzaville and in *Myonycteris* in the République Centrafricaine, but their identity remains uncertain. A full study of the life cycles of the chiropteran haemosporidiids of this region is planned for 1973.

MALARIA PARASITES IN INSECTIVORES AND MURID RODENTS

Insectivores

If the insectivores are thought to have arisen near the common ancestral source or source of the primates, the date must have been well before the branching of the Old World primate stock, because, like the New World monkeys and the prosimians, the insectivores are remarkably free of haemosporidian parasites of any sort, with a single exception. This is the elephant shrew; the two genera (*Elephantulus* and *Petrodromus*) are both commonly infected with *Plasmodium* (*incertae sedis*) *brodeni* over wide areas of tropical Africa from the Southern Sudan to Malawi and from Kenya to Zaïre. This parasite certainly does not belong to the Plasmodiidae and its position in the Haemoproteidae is still undefined in spite of several intensive studies. Unlike the parasites which so far have been described in this paper, the habitat of *P. brodeni* is not forest but savannah or semidesert.

Murid Rodents

The family Muridae as a whole is largely free of malaria with a few isolated exceptions as follows:

Plasmodium berghei and *P. vinckei* have now been found in semi-arboreal murids in four foci in tropical Africa: in the Katanga (Zaïre),

Republique Centrafricaine, Congo–Brazzaville and on the Nigerian coast. These parasites occur in pairs as eight subspecies (see Killick-Kendrick, in press). They are essentially linked with specialized anopheline hosts of which the first one to be discovered was *Anopheles dureni*. The host in the Katanga is *Grammomys surdaster* and in the other three foci, species of *Thamnomys*. No species of *Hepatocystis* exists in these animals.

Plasmodium atheruri has a restricted distribution near Lake Kivu in Zaïre and in Congo–Brazzaville, and is found in the brush-tailed porcupine, *Atherurus africanus*. Transmission is effected by the specialized vector, *A. smithi*.

In Asia the place of *Plasmodium* is usurped by *Hepatocystis* in squirrels. This is the third example of a niche being reserved for one genus but not available for the other. *Callosciurus* or other genera are very commonly infected with *Hepatocystis vassali* or its subspecies, from Ceylon and Southern India, through Assam and Burma, Thailand and Indo-China to Taiwan, and southwards into the Malaysian archipelago. Infection rates of 100% are often found.

DISCUSSION

Amongst several interesting papers which have dealt with phylogeny and parasitism in the Hominoidea, probably the most pertinent to the present paper was the article by Dunn (1966) who discussed the malaria parasites as an outstanding example of the use of this criterion for indicating phyletic relationships. Forty years ago Zuckerman (1933) suggested the inadvisability of drawing too specific conclusions from such data, but since that time many gaps in our knowledge have been filled, both of geographical and host distribution, and, more important, of the complete life cycles of numerous "key" species. The details of the latter (involving three cycles) have proved to be most significant in demonstrating that "ontogeny is a recapitulation of phylogeny".

Nevertheless, although striking features are now observable, their interpretation still entails much speculation. This discussion is confined particularly to some of the puzzles which have been uncovered but remain unsolved.

The general distribution of primate *Plasmodium* does not coincide with the distribution of the primates (Fig. 5), except in the case of man. In the non-human anthropoids, the coincidence of the parasite relates, with a few exceptions, to tropical rain forest. This is very strikingly seen in Africa, but is also plainly visible in the other continents. Probably in a former age when rain forest was more extensive, *Plasmodium* had a wider range; to-day its distribution represents a survival of a once much greater population. On the other hand *Hepatocystis* is well distributed in the savannah and riverine forest primates. The essential difference may be explained by the nature of the invertebrate host: *Plasmodium* requires sylvatic species of *Anopheles*; *Hepatocystis* requires the less specialized *Culicoides*.

The New World primates and other mammals exhibit no species of *Hepatocystis* and only two species of *Plasmodium*. This paper strengthens the hypothesis (Dunn, 1965) that the latter parasites were derived from a human source a few centuries ago. It is suggested that continental drift of the continents across the Atlantic Ocean occurred (see Tarling & Tarling, 1971) before the evolution of *Plasmodium* and *Hepatocystis* (Fig. 6); thus the New World monkeys were uninfected with these parasites when their hosts became isolated.

The distribution of malaria parasites in tropical Africa is puzzling. *Plasmodium* is common in the gorillas and chimpanzees in the western

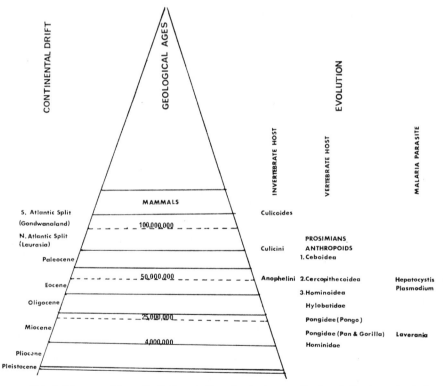

Fig. 6. Diagram of hypothetical evolution of malaria parasites in relation to geological ages, continental drift (Tarling & Tarling, 1971) and primate evolution (from Sarich, 1972; Lovejoy, Burstein & Heiple, 1972; and others).

half, and the species are almost identical with those of man. There is however only a single species of *Plasmodium* in the monkeys (*P. gonderi*) and this is confined to a limited region—incidentally to the region where dwells the true Negro, who is almost completely insusceptible to the human parasite, *P. vivax*, so common elsewhere and in all other races. Monkeys of nearly all genera and practically everywhere in tropical Africa are infected instead with species of *Hepatocystis*. A parallel anomaly is displayed by the fruit bats in which *Hepatocystis* is common and *Plasmodium* rare. The opposite effect is seen in Asian monkeys where the monkeys are heavily infected with *Plasmodium* but less often with *Hepatocystis*. These two genera may be grouped in pairs (see Table III) and this grouping probably denotes an evolutionary process, as *Hepatocystis* is thought to be the earlier form.

TABLE III

Species of Hepatocystis *and* Plasmodium *in mammals*
(*from Garnham, 1966*)

African monkeys	*H. kochi*	*P. gonderi*
Oriental monkeys	*H. semnopitheci*	*P. inui*
Lemurs	*H. foleyi*	*P. girardi*
Fruit bats	*H. epomophori*	*P. roussetti*
Rodents and shrews	"*P.*" *brodeni*	*P. berghei*
Antelope	"*P.*" *limnotragi*	*P. brucei*
Chevrotain	*H. fieldi*	*P. traguli*
Squirrels	*H. vassali*	*P. anomaluri*
Flying squirrels	*H. rayi*	*P. booliati*

The recent discovery of a second species of a malaria parasite (*P. silvaticum*) in an orang-utan and its resemblance to a parasite in the gibbons, confirms the idea that the orang-utan is more closely related to the gibbon than to the Hominoidea.

Figure 6 attempts to relate the phylogeny of these parasites to the evolution of their hosts and to other factors. There is a gap between the earlier cercopithecoids and the prosimians, and it is difficult to arrive at any hypothesis which explains the sudden appearance of *Hepatocystis* and *Plasmodium* (*Plasmodium*) in the former—for these forms are absent in the prosimians. Only the lemurs possess a *Plasmodium* of another subgenus (*Vinckeia*). This subgenus is fairly common in other lower mammals of the Old World, and other subgenera are very well represented in birds and lizards in both Old and New Worlds. All the malaria parasites are thought to have arisen from intestinal coccidia, yet the latter parasites are rare in the primates.

Mattingly (1965) has pointed out the possible significance of the respective dates of evolution of mosquitoes in relation to avian and primate malaria. The culicine mosquitoes may have evolved earlier than *Anopheles* and are the invertebrate hosts of the *Plasmodium* of birds, but they are unable to transmit the primate parasites which had to wait for the appearance of *Anopheles*. The vector of *Hepatocystis* is *Culicoides*, a genus which appeared earlier in evolution than the mosquitoes (Manwell, 1955). Another sandfly, *Phlebotomus*, is the invertebrate host of lizard malaria and this complex, as far as we know to-day, was the first of the malarias to have arisen.

Many conclusions have been drawn about the affinities of parasites, or of their hosts, as the result of various experimental manipulations, particularly splenectomy. If, for instance the spleen of a chimpanzee is removed, the animal at once becomes fully susceptible to the human species of *Plasmodium*: but if a rhesus monkey is splenectomized, this animal remains totally resistant. It was once thought that such experiments confirmed the closer relationship between man and ape than between man and monkey. But, when the full susceptibility of splenectomized *Aotus* and other South American monkeys to the human parasites was discovered (Geiman & Meagher, 1967; Porter & Young, 1967), the fallacy of such a conclusion became obvious.

This is not the place to speculate on the common factors present in various primates which may provide the clue to the curious range of their susceptibility to "foreign" species of *Plasmodium*. Professor N. A. Barnicot has drawn attention to the presence or absence of the Haemoglobin A_2 component in their respective erythrocytes (Barnicot, Wade & Cohen, 1970).

The susceptibility of a primate to the exoerythrocytic stages of a species of *Plasmodium*, accompanied by resistance to the erythrocytic stages, is thought to be a clear indication that the parasite is phylogenetically related to other species in which the *full* life cycle takes place. The liver stages are the primitive forms and they can probably develop in various primates which have arisen earlier in evolution. Unfortunately this relative susceptibility is not invariable and there are many exceptions.

The degree of susceptibility of various species of *Anopheles* to the primate parasites sometimes indicates the relative proximity of the latter, but more probably represents the effect of adaptation during the course of millennia. If there is restriction of contact between parasite, vertebrate and invertebrate hosts in a highly specialized habitat (e.g. *P.*

atheruri in the brush-tailed porcupine bitten by *A. vanthieli* in the burrows), transmission is limited to the single species of mosquito, and the geographical range of the infection is correspondingly narrowed. Such specialized conditions are however exceptional and more often the susceptible species is a free-ranging insect in the sylvatic environment. In these circumstances, many different species of *Anopheles* have the chance of eventually becoming adapted to the malaria parasites of the forest animals. Up to the present, no *natural* vector of the parasites of apes either in Africa or Asia, has been discovered, and attempts to transmit them with exotic species of *Anopheles* have rarely succeeded.

There is one group of *Anopheles* which is highly susceptible to the primate species of *Plasmodium*: this is the *leucosphyrus* complex, whose geographical distribution coincides (in Asia) with that of non-human primate malaria (Colless, 1956). Where the complex is absent, as in N.W. India, so is malaria. One member of the complex. *A. balabacensis*, is so susceptible, that it is almost a universal vector, capable of transmitting such foreign species of *Plasmodium* as the African *P. gonderi* and *P. schwetzi* and the South American *P. simium*, as well as most of the species occurring in the Asian macaques and apes.

Thus, although the type of invertebrate host may appear a useful clue to the distribution of the parasite, it appears to have little significance in regard to phylogeny. This is in agreement with the generally accepted view of the coccidial origin of malaria parasites in the intestine of vertebrates, rather than that they were primarily parasites of insects (as is the case in the haemoflagellates). In malaria, the role of mosquitoes was primarily accidental.

ACKNOWLEDGEMENT

I am grateful to Blackwell's Scientific Publications for Fig. 5 from Garnham, 1966.

REFERENCES

Adam, J. P. & Landau, I. (1970). *Plasmodium voltaicum* au Congo–Brazzaville. *J. Parasit.* **56**: 391–392.

Allison, A. C. (1961). Genetic factors in resistance to malaria. *Ann. N.Y. Acad. Sci.* **91**: 710–729.

Allison, A. C. & Clyde, D. F. (1961). Malaria in African children with deficient erythrocyte glucose-6-phosphate dehydrogenase. *Br. med. J.* **1961 1**: 1346–1349.

Barnicot, N. A., Wade, P. T. & Cohen, P. (1970). Evidence for a second Haemoglobin α-locus duplication in *Macaca irus. Nature, Lond.* **228**: 379–381.

S

Bray, R. S. (1963). The malaria parasites of primates and their importance to Man. *Ergeb. Mikrobiol. Immun-Forsch. exp. Ther.* **36**: 168–213.

Bray, R. S. (1964). A check-list of the parasitic protozoa of West Africa with some notes in their classification. *Bull. Inst, fr. Afr. noire* **26**: 238–315.

Bruce-Chwatt, L. J. (1965). Paleogenesis and paleoepidemiology of primate malaria. *Bull. Wld Hlth Org.* **32**: 363–387.

Bück, G., Coudurier, J. & Quesnel, J. J. (1952). Sur deux nouveaux plasmodium observés chez un lemurien de Madagascar splenectomisé. *Archs Inst. Pasteur Algér* **30**: 240–246.

Coatney, G. R. (1968). Simian malaria in man: facts, implications and predictions. *Am. J. trop. Med. Hyg..* **17**: 147–155.

Coatney, G. R., Collins, W. E., Warren, M. & Contacos, P. G. (1971). *The primate malarias.* Bethseda, Maryland: Government Printing Office.

Colless, D. H. (1956). The *Anopheles leucosphyrus* group. *Trans. R. ent. Soc. Lond.* **108**: 37–116.

Collins, W. E., Contacos, P. G., Garnham, P. C. C., Warren, W. & Skinner, J. C. (1972). *Plasmodium hylobati:* A malaria parasite of the gibbon. *J. Parasit.* **58**: 123–128.

Das Gupta, B. (1967). A new malarial parasite of the flying squirrel. *Parasitology* **57**: 467–474.

Deane, L. M. (1972). Malaria in Brazilian monkeys. *WHO/MAL/72* No. 774: 1–6.

Dunn, F. L. (1965). On the antiquity of malaria in the western hemisphere. *Human Biol.* **37**: 385–393.

Dunn, F. L. (1966). Patterns of parasitism in primates: Phylogenetic and ecological interpretations with particular reference to the Hominoidea. *Folia primatol.* **4**: 329–345.

Eyles, D. E., Fong, Y. L., Dunn, E. L., Guinn, E., Warren, M. & Sandosham, A. A. (1964). *Plasmodium youngi* n. sp. a malaria parasite of the Malayan gibbon, *Hylobates lar lar. Am. J. trop. Med. Hyg.* **13**: 248–255.

Garnham, P. C. C. (1948). The developmental cycle of *Hepatocystes (Plasmodium) kochi* in the monkey host. *Trans. R. Soc. trop. Med. Hyg.* **41**: 601–616.

Garnham, P. C. C. (1966). *Malaria parasites and other haemosporidia.* Oxford: Blackwell Scientific Publications.

Garnham, P. C. C. (1971). *Progress in parasitology.* London: Athlone Press.

Garnham, P. C. C. (1973). Bat malaria: zoogeography and possible course of evolution. *Commentar. pontif. Acad. Scient. (Civ. Vat.)* **11** (44): 1–15.

Garnham, P. C. C., Rajapaksa, N., Peters, W. & Killick-Kendrick, R. (1972). Malaria parasites of the orang-utan (*Pongo pygmaeus*). *Ann. trop. Med. Parasit.* **66**: 287–294.

Garnham, P. C. C. & Uilenberg, G. (in press). A revision of the malaria parasites of lemurs. *Parassitologia.*

Geiman, Q. M. & Meagher, M. J. (1967). Susceptibility of a New World monkey to *Plasmodium falciparum* from man. *Nature, Lond.* **215**: 437–439.

Halberstaedter, L. & Prowazek, S. von (1907). Untersuchungen über die malariaparasiten der Affen. *Arb. K. Gesundh. Amt.* **26**: 37–43.

Killick-Kendrick, R. (in press). Parasite protozoa of the blood of rodents 11: Haemogregarines, malaria parasites and piroplasms of the blood of rodents: an annotated check-list and host index. *Acta trop.*

Killick-Kendrick, R. & Bellier, L. (1971). Blood parasites of scaly-tailed squirrels in the Ivory Coast. *Trans. R. Soc. trop. Med. Hyg.* **65**: 430–431.

Landau, I. & Adam, J. D. (in press). Description de schizontes de rechute chez un nouvel hemoproteidae, *Hepatocystis perronae* n. sp. parasite de megachiroptères africains. *Cah. Off. Rech. Sci. Tech. Outre-Mer.*

Lovejoy, C. O., Burstein, A. H. & Heiple, K. G. (1972). Primate phylogeny and immunological distance. *Science, Wash.* **176**: 803–805.

Manwell, R. D. (1955). Some evolutionary possibilities in the history of the malaria parasites. *Indian J. Malar.* **2**: 247–253.

Mattingley, P. F. (1965). The evolution of parasite-arthropod vector systems. *Symp. Br. Soc. Parasit.* No. 3: 29–45.

Napier, J. R. & Napier, P. H. (1967). *A handbook of living primates.* London: Academic Press.

Orihel, T. C. & Garnham, P. C. C. (in press). A new species of *Hepatocystis* in the Talapoin. *J. Parasit.*

Peters, W., Garnham, P. C. C., Killick-Kendrick, R., Rajapaksa, N., Cheong, M. & Cadigan, F. C. (in press). Orang-utan malaria. *Phil. Trans. R. Soc.* (B).

Porter, J. A. Jr. & Young, M. D. (1967). The transfer of *Plasmodium falciparum* from man to the marmoset, *Saguinus geoffroyi. J. Parasit.* **53**: 845–846.

Reid, J. A. (1968). *Anopheline mosquitoes of Malaya and Borneo.* Government of Malaysia: Studies from the Institute for Medical Research, Malaysia.

Rodhain, J. (1941). Sur un Plasmodium du gibbon *Hylobates lensciscus* Geoff. *Acta biol. belg.* **1**: 118–123.

Sarich, V. M. (1972). Hominid origins revisited. In *Climbing man's family tree*: 450–460. McCown, T. D. & Kennedy, K. A. R. (eds). New York: Prentice Hall.

Schwetz, J. (1933). Sur une infection malarienne triple d'un chimpanzé. *Zentbl. Bakt. Parasitkde (Orig.)* **130**: 105–111.

Sergiev, P. G. & Tiburskaya, N. A. (1965). On the evolution of *Plasmodium vivax. Progress in Protozoology.* Internat. Congr. Series, **91**: 165–166. Excerp. Med. Foundation.

Shiroishi, T., Davis, J. & Warren, M. (1968). *Hepatocystis* in the white-cheeked gibbon, *Hylobates concolor. J. Parasit.* **54**: 168.

Silva, G. S. de (1971). Notes on the orang-utan rehabilitation project in Sabah. *Malay. Nat. J.* **24**: 50–77.

Tarling, D. H. & Tarling, M. P. (1971). *Continental drift.* London: Bell & Co.

Warren, M. (1973). Quoted in *"Report of a Meeting on Taxonomy of Malaria Parasites"* by P. C. C. Garnham. *J. Protozool.* **20**: 37–42.

Warren, M., Coatney, G. R. & Skinner, J. C. (1966). *Plasmodium jefferyi* sp. n. from *Hylobates lar* in Malaysia. *J. Parasit.* **52**: 9–13.

Warren, M., Bennett, G. F., Sandosham, A. A. & Coatney, G. R. (1965). *Plasmodium eylesi* sp. nov. a tertian malaria parasite from the white-handed gibbon, *Hylobates lar. Ann. trop. Med. Parasit.* **59**: 500–508.

Zuckerman, S. (1933). *Functional affinities of man, monkeys and apes.* New York: Harcourt and Brace.

DISCUSSION

BARNICOT (CHAIRMAN): Thank you, Professor Garnham, for presenting us with these very fascinating and puzzling data. We should like to hear the comments of primatologists and other evolutionists on your theory.

ZUCKERMAN: Could I ask a question? Is there any other creature whose life cycle is absolutely dependent on the primate in the way that the malaria parasite is tied to its host at a given stage of its life cycle? Although Cameron was able to show that some of the ectoparasites in African monkeys were the same as those found in man, and not found elsewhere, nevertheless the ectoparasites might have been transmitted by some Early Pleistocene animal keeper in an Early Pleistocene zoo to the animals, in the same way as one knows that tuberculosis can be transmitted to monkeys from an animal attendant (and *vice versa*). Are there any other such creatures? Viruses, I suppose, do not move away from their host at some given moment.

GARNHAM: I suppose that any of the protozoal parasites that have arthropod hosts—like, for instance, *Leishmania*, which is found in man, in rodents particularly as a reservoir, and the *Phlebotomus*, the sand fly—possess a very similar cycle. This must also occur in quite a number of other pathogens, including viruses, that have an invertebrate host and an animal reservoir.

ZUCKERMAN: I was wondering whether any had been studied.

GARNHAM: Trypanosomes of many species of monkeys, apes and man offer a parallel to malaria parasites in that their life cycle is dependent upon the vertebrate (and invertebrate) host; similarly various host specific intestinal protozoans (e.g. *Troglodytella* of the Chimpanzee). There are examples to be found in helminthic parasites, e.g. filarial worms, and in other pathogens including viruses in which part of the life cycle takes place in arthropods.

HEWETT-EMMETT: You said, I think, that in Africa only the drills and the mangabeys were affected of all the monkeys. Which of the species are affected in South East Asia? Are all macaques, or just some of them?

GARNHAM: Not *Macaca mulatta*, the Rhesus monkey, which as our Chairman has shown possesses a different type of haemoglobin. It has even been suggested that this might be the reason why the Rhesus monkey is not a natural host for *Plasmodium* (experimentally it is excellent). Practically all other species of *Macaca* and the langurs are hosts. It is very much the same as in South America where a tremendous variety of monkeys are hosts to *Plasmodium brasilianum*. Yet in Africa there are only two rare examples, which really present a conundrum.

GRAFTON ELLIOT SMITH: EGYPT AND DIFFUSIONISM

CHAIRMAN: G. DANIEL

The session opened with an address by the Chairman, Dr Glyn Daniel. This was followed by a formal discussion of his remarks, in which the participants were: Professor Daryll Forde, Professor Meyer Fortes, Professor E. R. Leach, Professor A. C. Renfrew, and Professor Stuart Piggott. Further contributions from the floor were made by Mr C. E. Joel and Mr G. Kraus, Mr R. A. Jairazbhoy and Professor J. V. S. Megaw.

It will be evident that some of the views expressed from the floor are totally at variance with those of the invited speakers, but the Editor feels that in fairness they should be included in the record of this session.

DARYLL FORDE

Daryll Forde died on May 3rd, 1973, six months after he had opened the discussion which followed Dr Glyn Daniel's paper. His death robs us of a scholar whose contributions enriched many fields of social anthropology. Daryll Forde's health had been failing for some time, and those who were present at the Symposium would have seen that he was in some physical distress as he spoke. But they would also have noticed that nonetheless he spoke without a note and without hesitation as he developed the ideas which are embodied in the transcript of the last contribution he was to make to a scientific debate. I know I speak for all who attended the Symposium when I express to his widow and children the thanks we owe to her late husband for the far-reaching researches he was able to make in a rich and full life.

24th August, 1973   Solly Zuckerman

Symp. zool. Soc. Lond. (1973) No. 33, 407–447

ELLIOT SMITH, EGYPT AND DIFFUSIONISM

GLYN DANIEL

St. John's College, Cambridge, England

So far in this Symposium we have been considering Elliot Smith as an anatomist and a physical anthropologist. Now we turn to his work in the fields of cultural anthropology and prehistoric archaeology. When Lord Zuckerman wrote to me in February 1972 he posed two questions for the Symposium's consideration concerning Elliot Smith: "How far-reaching has his influence been?" and "Has his interpretation been totally invalidated by recent developments, or are its adherents still justified in their support?"

Let us begin by considering the views of two of Elliot Smith's friends and colleagues, as set down in Dawson (1938). The first is by Professor H. A. Harris:

> "His periodical irruptions into the territory of the Egyptologists and anthropologists were so disturbing to the prevalent opinions of most academic minds that they provoked violent controversy. His vigour as a controversialist in these directions is apt to obscure his pioneer work in the unification of the fundamental sciences." (Harris, 1938: 180).

The second is from W. J. Perry who worked with Elliot Smith in Manchester and then came to fill the new Readership in Cultural Anthropology at University College, London.

> "Elliot Smith, by his anatomical researches and by his interpretation of the remains of primitive man, has left his mark upon physical anthro-pology, but, important as these contributions undoubtedly are, of themselves, it is highly probable that they will ultimately rank as inferior to his achievements in the domain of human behaviour, be it called cultural anthropology, ethnology or social anthropology." (Perry in Dawson, 1938: 205).

Elliot Smith advocated an extreme form of diffusionism which attempted to prove that almost all the sociocultural traits of interest to anthropologists were invented in Egypt, and nowhere else in the world: that they spread from Egypt to the rest of the world—although he and his disciples often had different views as to when this happened. Sometimes it was early on in dynastic Egypt: at other times the Heoliolithic Culture was diffused in the 10th century BC. This very extreme and rigid diffusionist view I have called the Egyptocentric hyperdiffusionist model or paradigm of the past (Daniel, 1962: 110). Its adumbration by Elliot Smith in the early years of this century and its persistent and often violent advocacy by him until his death in 1937 certainly earned him a place in the history of archaeology and anthropology. No one who concerns himself with the rise and fall of the models of the prehistoric past, can afford to neglect the English hyper-diffusionist school, just as they cannot neglect the German hyper-diffusionists, mainly members of the Roman Catholic clergy, who developed the Vienna-based Kulturkreis or Culture-Circle approach.

Professor Marvin Harris of Columbia writes:

> "Both of these movements were palpably bankrupt by mid-century: they require our attention today only as evidence of the international extent of the tide that was running against nomothetic principles." (Harris, 1968: 379).

This is too sweeping a statement, particularly for those of us here who are concerned with Elliot Smith as a scholar. We want to know how he came to hold his very extravagant views, how a scientist so distinguished in some fields, a man famous as a physical anthropologist, could become notorious as a cultural anthropologist. And to understand this we need to reflect on the state of thinking about cultural change at the end of the 19th century and the beginning of the present century. Only this can tell us how it was possible for Egyptocentric hyperdiffusionism to get such a hold on the minds of many people. Why were Elliot Smith's theories not "laughed out of court" at once? This is what happened to Lord Raglan (1939) who advocated Sumerocentric hyperdiffusionism: it was a parallel case to that put forward from 1911 onwards by Elliot Smith, and a more plausible one. But it was never taken seriously, just as no one takes seriously the extravagant views of those who believe that all civilization originated in lost continents like Atlantis or Mu, or in the highlands of Peru. Why, then, was the Elliot Smith Egyptian hypothesis so seriously considered? To answer this we must look at the development of his ideas in relation to the history of archaeological ideas, and in doing this we would have Elliot Smith with us. He quotes with approval (Elliot Smith, 1930: 120) Comte's remark that "No conception can be understood except through its history."

Before the beginning of systematic archaeology in the first half of the 19th century, theories of cultural change, and of the origins of nations and civilizations, were largely based on guesswork, on myths, legends and literary sources such as the Bible and Classical writers, and these hardly penetrated into the prehistoric past. Prehistoric archaeology

began between 1816 and 1859, during which period of 40 years the
great antiquity of man was demonstrated by the finds of Boucher de
Perthes in the Somme gravels, and Pengelly in south Devon,
and Danish scholars, such as C. J. Thomsen and J. J. A. Worsaae,
produced a relative time scale for the prehistoric past by proposing
the technological model of three successive ages of stone, bronze and
iron.

The literate civilizations of Egypt, Assyria, Greece and Rome were,
of course, well known then, and so, if less well known, were the pre-
Columbian civilizations of the Aztecs, Maya and Inca. A hundred
years of archaeology has added to this list of civilizations—the Sumer-
ians, the Hittites, the Minoans and Myceneans, the Harappans of the
Indus Valley, the people of Shang China, the Olmecs of coastal Mexico,
and more recently the early civilizations of Soviet Central Asia and
Iran. The question posed from the early 19th century onwards—and it is
still being posed—was this: how did man change from the Savage
hunter–fisher of Palaeolithic times to the Barbarian peasant-villager of
the Neolithic/Chalcolithic, and how did these barbarian agriculturists
achieve civilization, that is to say, how did they develop large, literate,
urban communities?

Two answers were suggested from the beginning. The first that man
had evolved his culture independently; that to biological evolution
there had succeeded cultural evolution, or social evolution. This was the
idea usually described as independent evolution. The other answer was
that new socio-cultural ideas came in from outside: this was the
idea labelled as diffusion. These two answers were very clearly
present in the writings of Thomsen and Worsaae, and were fully
discussed by Lubbock and Tylor. Several myths have grown up about
the diffusion versus independent evolution controversy of the 19th
century. The first is that if you argued for independent invention it
meant the independent invention of everything everywhere. It did not.
The second is that if you did not believe that everything was independently invented everywhere, you must believe that everything was
diffused from one centre. There grew up the notion in the second half of
the 19th century that there were only two positions for the cultural
anthropologist and archaeologist: you had to be a committed evolutionist or a committed diffusionist. But this is not so: what we are all
trying to do as archaeologists and anthropologists at present is to
study the cultural changes recorded by archaeology and history and
offer tentative explanations that are cogent and reasonable. There
is no necessary conflict between evolution and diffusion. As R. H.
Lowie said, looking back at the Anthropological Institute of the 19th

century, there "evolution lay down amicably beside diffusion" (Lowie, 1937: 87). So they could and should, but Elliot Smith would not let them.

He was particularly annoyed with E. B. Tylor whom he castigated wherever he could. Tylor was both an evolutionist and a diffusionist which was always confusing for those who were trying to force him into one or other of those moulds. Indeed it is very easy to make a list of quotations from Tylor's books to prove that he was on one or other side of the fence. He was on both sides and this is where we should be today. The position of evolution lying down amicably beside diffusion is where we are getting back to in the third quarter of this century. Now that the rapidly growing world archaeological record is being accurately and objectively dated by means of the natural sciences, we can see many well-attested examples of independent evolution as well as proven cases of diffusion. Only the most heavily blinkered diffusionists—and they still exist—would now deny that what we conceptualize as agriculture was independently developed in many parts of the world, such as the Near East, China, and Central America, and that what we group together mistakenly as megalithic monuments represent a wide variety of different structures separated in time, space and origins. But we should remember here in fairness that Elliot Smith was writing years before Libby had discovered Carbon 14 dating (Libby, 1963), and although he based his case extensively on the origins of agriculture and megaliths, he did not have at his disposal the detailed and exact knowledge that we now have.

Looking back at the history of archaeological and anthropological models in the last quarter of the 19th century it seems that perhaps it was inevitable that the evolutionists and diffusionists had to take their theories to extreme positions, often to violent extremes, before more moderate counsels prevailed. The extreme diffusionist position was first set out in England by a vigorous lady called Miss A. W. Buckland who, from 1878 onwards until 1893, wrote a series of papers all in the Journal of the Anthropological Institute. The papers had titles such as *Primitive agriculture, Prehistoric intercourse between East and West, Four as a sacred number, The serpent in connection with primitive metallurgy,* and so on. She believed that sun- and serpent-worship had spread agriculture, weaving, pottery and metals over the earth: she quoted ceremonial haircutting and sweating among the Navaho as evidence of intercourse with Japan. Miss Buckland declared that civilization was never, never, independently invented: it could not be, and it was not. What did her audiences think of these violent and uncompromising hyperdiffusionist views which so foreshadow Elliot Smith? In Lowie's words, "Whatever we may think of her evidence, she was certainly not lynched by her audience." (Lowie, 1937: 90).

But how did Elliot Smith get into this *galère*? Miss Buckland did not influence him at first, although later he wrote of her with approval, as he did also of Mrs Zelia Nuttall who had found strange things in her garden in Mexico (Nuttall, 1901). This was 10 years before the publication of *The ancient Egyptians* (Elliot Smith, 1911) and the delivery of Rivers's Presidential Address to the Section on Anthropology of the British Association (Rivers, 1912), an address which Elliot Smith declared "marked an epoch not only in Dr Rivers's own career, but also in the history of ethnology".

It must be confessed that my own University, and my own College in that University, were largely responsible for Elliot Smith's interests in cultural anthropology. It was a Johnian, J. T. Wilson, his Professor in Sydney, who sent him to England, where he met Alexander Macalister, then Professor of Anatomy at Cambridge and a Fellow of St. John's. Cambridge had just agreed to admit qualified graduates of other Universities. Rutherford was one of the first batch in 1895: Elliot Smith was admitted to St. John's in the following year, and was made a research fellow three years later. In 1900, on Macalister's recommendation, Elliot Smith became the first occupant of the chair of Anatomy at the Government Medical School in Cairo. It was in Egypt between 1900 and 1909 that he developed his ideas, but before he left St. John's he had been impressed by the admirable catholicity of Macalister's interests and had fallen under the influence of another Fellow, W. H. Rivers Rivers.

Rivers, a man of 29, joined St. John's in October 1893 as a Fellow Commoner, and in 1897 was appointed to the newly established Lectureship in Physiological and Experimental Psychology. Rivers and Elliot Smith knew each other at Cambridge in the years before Rivers went out with Haddon on the Torres Straits Expedition in 1898. They met again in Egypt and Rivers became a convinced diffusionist, although at first he was much influenced by the views of Theodore Waitz and Adolf Bastian. His conversion is recorded in his 1911 address to the British Association (Rivers, 1912), and in his preface to *The history of Melanesian society* (Rivers, 1914).

Elliot Smith edited Rivers's essays after the latter's untimely death in 1922 (Elliot Smith, 1926). In his introduction to this book Elliot Smith says that

"Rivers had been working at the subject for more than 12 years without seriously questioning the validity of the hypothesis of the independent development of culture. But when, in 1910, he began to examine the evidence he had collected in Melanesia, the conclusion was brought home to him that his data could not be forced into harmony with the doctrine then fashionable in ethnology.... The conviction was forced upon Rivers in 1911 that, if considerations of diffusion were eliminated from the comparative study of customs and beliefs, the result was confusion." (Elliot Smith, 1926: xvi.)

This in itself was a curious comment by Elliot Smith: he had somehow manoeuvred himself into the position of believing that all ethnologists and archaeologists in the late 19th century believed in nothing but evolution and independent development. I begin to doubt whether he ever read Montelius's book (*Der Orient und Europa* published in 1899), the year before he went to Cairo. The first words of that book are worth remembering now:

"At a time when the peoples of Europe were, so to speak, without any civilization whatsoever, the Orient and particularly the Euphrates region and the Nile, were already in enjoyment of a flourishing culture. The civilization which gradually dawned on our continent was for long only a pale reflection of oriental culture." (Montelius, 1899: 1.)

The problems raised by Rivers's first experience of anthropological field-work among the Torres Straits Islanders stimulated him to examine other groups of people. In the winter of 1900–1901 he went out to Egypt and, at the excavation camp of Dr Randall-MacIver and Anthony Wilkin at El Amrah, near Abydos, he investigated the colour vision of the workmen. He went to see Elliot Smith in Cairo and suggested to him that he should himself go to El Amrah to study what was, until then, the unknown phenomenon of the natural preservation of the brain in the crania of predynastic Egyptians. "So", wrote Elliot Smith a quarter of a century later, "incidentally and quite unwittingly, Dr Rivers was responsible for drawing me into anthropology . . . before long I had become definitely committed to the study of the anthropology of Egypt." (Elliot Smith, 1926: xiii–xiv.)

Elliot Smith had fallen in love with Cairo, "the gayest and most cosmopolitan city on the face of the earth, and intensely fascinating", he wrote. Part of the fascination was, naturally, the surface antiquities that were so obvious to any intelligent, alert and enquiring traveller like himself. He found the pyramids and the mastabas exciting and intriguing. In 1901 he writes, "I have not quite resisted the temptation to dabble in Egyptology." Next year he refers to "a sudden burst into anthropology", and in the following year he writes in a letter, "for the next six months I expect to be a slave to anthropology". He remained enslaved to Egyptology and cultural anthropology to the end of his life.

Slowly, as he worked and travelled in Egypt, Elliot Smith came to the view that practically the entire inventory of world culture originated in Egypt, and that this happened 6000 years ago. Prior to this the world was peopled by what he called Natural Man. Natural Man had no domesticated animals, agriculture, houses, clothing, religion, social organization, hereditary chiefs, formal laws, marriage customs and burial ceremonies. About 4000 BC all these cultural traits were developed in Egypt and constituted what he called the Original Archaic Civilization—a product of the Nile. The members of this Egyptian Archaic Civilization journeyed by land and sea over great distances in search of precious metals and other raw materials. They were "The Children of the Sun" journeying to collect "The Givers of Life". As they spread their ideas through diffusion and direct colonization, other centres of the Archaic Civilization were set up. Some of these prospered: others, like the Maya, died out or collapsed. According to Elliot Smith, many primitive cultures were decayed and impoverished remains of the Archaic Civilization; others were mixtures of Natural Man and degenerating high cultures; while others were mixtures of various degenerated cultures. But all was explained without remainder. Elliot Smith provided a total explanation of the history of human culture in terms of the Archaic Cultural Revolution in Egypt (Elliot Smith, 1930).

His views on the role of the Archaic Civilization of Egypt were first clearly set out in his *The ancient Egyptians* in 1911, and there then followed a spate of lectures, papers and books. Let us mention a few of the most important: *Migrations of early culture* (1915), *The evolution of the dragon* (1919), *Elephants and ethnologists* (1924), *In the beginning: the origin of civilisation* (1928), *Human history* (1930)—the clearest and best exposition of his views, and *The diffusion of culture* published in 1933 when he was already a sick man. He had a stroke in 1932 and wrote little between then and his death five years later. But from 1908 to 1932—for a quarter of a century—he poured out material on Egypt and diffusion. First we see him moving cautiously to his thesis, then it is stated firmly and categorically, and then he goes on championing it rigorously, vigorously, relentlessly— brooking and heeding no criticism.

If one had to be convinced of the monogenic origin of civilization and had to choose a place, Egypt was a very good choice, and it had of course been chosen many times before Elliot Smith fell in love with that country and became enslaved to Egyptology and cultural anthropology. Herodotus said that the Egyptians had borrowed nothing and had themselves originated all of their culture. There is a very good account of the development of ideas about the primacy of Egypt in cultural diffusion in Wortham (1971). He reminds us that Plato, Ammianus Marcellinus, and many other classical writers argued that Egypt was the land in which man first developed art, religion and science; and that many medieval and renaissance writers repeated the belief that Egypt was the original homeland. Sir Thomas Browne in 1658 said:

> "when the two ancient nations, Egyptians and Scythians, contended for antiquity, the Egyptians pleaded their antiquity from the fertility of their soil, inferring that men there first inhabited where they were with most facility sustained, and such a land did they conceive was Egypt." (Wortham, 1971.)

In the 18th century many saw Egypt as the cradle of the world. One such was James Burnett, Lord Monboddo, who was forced to conclude that most of the arts and sciences known throughout the world had been borrowed from Egypt, and who derived Sanskrit and Chinese from Egyptian. The anonymous author of *The origin and antiquity of our English weights and measures* (1706) derived them from Egypt. Charles Vallancey found Egyptian influences in the Celtic languages and in inscriptions at New Grange. In an article in *Archaeologia* for 1779 F. S. Schmidt is proving that the classical civilization of Greece came from Egypt and is identifying Egyptian colonies in Greece. J. T. Needham in his *Dissertation concerning the Egyptian inscription found at Turin* (1761) argues that the Egyptians had a colony in China. E. Daniel Clarke believed that certain scenes painted on the walls of Egyptian buildings proved that the Egyptians had civilized India and China. Perhaps the most extravagant and unusual theory was that set out by Robert Deverall in his *Andalusia* of 1805. Egyptian civilization, according to him, came from England, and the pyramids were built in the geographical image of England itself. Later, Deverall argued, the Egyptians colonized South America, Japan, and India.

Of course most of this speculation was pre-archaeological—although Daniel Clarke was making an archaeological observation from paintings —and it was all so mixed up with general theories of the origin of man, language, and culture, and so bedevilled by calculations of the measurement of the pyramids. Pyramidiocy and cultural diffusion theories were developing together for a while. The pyramidiots claimed to have uncovered secret information about the past, present and future locked up in the measurement of the Egyptian pyramids; and, before we dismiss this out of hand, let us recollect that it was his father's obsession with the theories of Charles Piazzi Smith that first got Flinders Petrie to Giza. And we should remember that it was not only cranks and crackpots who were advocating Egypt as the centre and origin of all things. In 1926 that distinguished and responsible American archaeologist James Breasted declared that, "It is now a finally established fact that civilization first rose in Egypt, followed a few centuries later in Babylonia."

Certainly Elliot Smith was not alone in his views: he had enthusiastic disciples and followers, and, as so often happens, they were more enthusiastic and often more dogmatic, than the master. Such a one was Wilfred Jackson who wrote *Shells as evidence of the migrations of early culture* in 1917, and Warren Dawson who wrote *Custom of couvade* in 1929. But the main disciple and follower was W. J. Perry who was Reader in Comparative Religion in the University of Manchester when Elliot Smith was Professor of Anatomy there, and then followed him when he moved to University College, London, where he was made Reader in Cultural Anthropology. Perry had also been inspired and encouraged by Rivers when he was an undergraduate in Cambridge: Rivers told him to learn Dutch and read the works of the Dutch missionaries in Melanesia as a key to early Melanesian society. Elliot Smith had a very high opinion of Perry and described his researches as "converting ethnology into a real science and shedding a brilliant light upon the early history of civilization". Perry wrote *The children of the sun* (1923), *The origin of magic and religion* (1923), *The megalithic culture of Indonesia* (1918), and a general summary designed for the ordinary reader called *The growth of civilisation*. This last book was published in 1924, and, as a schoolboy, beginning to be excited by archaeology and anthropology, I read it with avidity and found it convincing.

In this book Perry analyses the growth of the Egypt theory. He stresses two things that most affected Elliot Smith: first, embalming and mummification, and secondly, megalithic monuments. Elliot Smith had been so impressed by the pyramids and mastabas of ancient Egypt, and had studied the burial rituals of ancient Egypt with such care and fascination, that he believed that Egyptian burial practices must be the origin of all burial rites and practices, and that the mastabas were the prototypes of megalithic monuments, wherever found, from Ireland to Melanesia. Whatever the diversity of form and function of megaliths, he argued they were all versions of early Egyptian tombs. He found that he just could not believe that mummification and megalithic architecture could have been invented more than once, and he convinced himself that both practices had been diffused all over the world from Ancient Egypt.

But it was not quite so simple as Perry made out. Elliot Smith's ideas were grounded in many things. First in physical anthropology: he studied the anatomy of the ancient Egyptians and found the same physical type in many parts of the world: these constituted his Brown Race which he thought was the basic type of modern man, and, secondly, in sound archaeological reasoning. "I came to realize", he wrote in 1911, "the vast significance of the demonstration given by Professor George A. Reisner of Harvard in 1908 that it was the Egyptians who invented copper implements, and this inaugurated the Age of Metals." (Elliot Smith, 1911.) Then he was fascinated by the fertility of the Nile valley and argued for the origins of agriculture there. He said that in 4000 BC the inhabitants of the Nile Valley "appreciated the fortunate chance provided them" by what he called "a natural crop of barley"— and adopted a settled mode of life. "It was" he said, "the agricultural mode of life that furnished the favourable conditions of settled existence, conditions which brought with them the need for such things as represent the material foundations of civilization." (Elliot Smith, 1928.)

Yet, in the end, it was mummification and megaliths that forced his conversion to hyperdiffusionism. In July 1919 he was staying in his old Cambridge College, and after dinner went to Rivers's rooms for a discussion on many things. Of the company that evening was Siegfried Sassoon, who has recorded the occasion in his poem *Early chronology*. I quote three lines:

> And soon they floated
> Through desiccated forests; mangled myths;
> And argued easily round megaliths.

Since James Fergusson wrote the first book about megaliths a hundred years ago there have been many people who have argued easily round megaliths. But none of us has argued more easily and less convincingly than Elliot Smith. Certain as he was that the megalithic tombs of Europe were derived from the Egyptian mastabas, he had to deal with a difficult stage in his argument, namely, to find a type that appeared transitional between serdab and dolmen. This was admittedly not easy, as no such monument exists. Undismayed, Elliot Smith invented a transitional form of which he drew a section and plan, and Colin Renfrew has recently reminded us of this manoeuvre (Renfrew, 1967). This was a purely hypothetical monument but Elliot Smith remained unperturbed. "I do not think there can be any doubt that by a process such as I have roughly sketched, and in response to such dominating ideas as I have mentioned," he wrote, "the mastaba made for an Egyptian by alien hands in a foreign land would assume some such form as I have represented in section and plan." This was an all-time low in his methods of argument. Trait-chasing was one thing, making false analogies and parallels was another, but inventing transitional forms was not merely a negation of method but positively wicked.

The ancient Egyptians came out in 1911 (Elliot Smith, 1911) and its sub-title then was, "The ancient Egyptians and their influence upon the civilisation of Europe". When the second revised edition came out in 1923 it was sub-titled, "The ancient Egyptians and the origin of civilisation". The whole world and all its ancient civilizations were now in Egypt's debt and thrall. In the Library of the Faculty of Archaeology and Anthropology at Cambridge we have A. C. Haddon's copy of the 1923 edition, and he had copied into it notes taken from Flinders Petrie's copy. Petrie had peppered the pages with outspoken comments such as, "No such thing", "Nonsense", "What a romance", "No evidence", "No, No", "No evidence whatsoever", and then, "Here the author is cowrie-shell mad: models of steatopygous women are not copies of cowrie-shells". Above the preface Petrie had written, "The asserted facts are largely untrue and the vague statements unsupported", and in the early twenties Petrie had set out what, even then, was the main argument which Breasted had failed to appreciate, the main argument against the Egyptian origin of all things, namely that Sumerian civilization was earlier. These arguments were set out on a slip of paper stuck into the book; I have recently reproduced it (Daniel, 1972: 262). But, of course, Elliot Smith was writing before the modern evaluation of the role of Sumer in relation to Egypt (Frankfort, 1951; Kramer, 1958; Daniel, 1968).

In the preface to the 1923 edition of *The ancient Egyptians* Elliot Smith wrote:

> "Little did I realize when I was writing what was intended to be nothing more than a brief interim report . . . that this little book was destined to open up a new view—or rather to revise and extend an old and neglected method of the history of civilization."

The more he wrote and lectured, the more his *idée fixe* became to him unassailable. Everything started in Egypt—absolutely everything. When he was once asked what was taking place in the cultural development of the rest of the world when Egypt was allegedly laying the foundations of civilization he answered, without any hesitation,

"Nothing". W. J. Perry's comment on this statement is quite flabber-gasting:

> "The accumulation of fresh evidence during the twenty-odd years that have elapsed has tended to confirm the essential accuracy of what was then an astonishing generalization."

Those were Perry's words (Dawson, 1938: 214) and we can at least agree with three of them: it was "an astonishing generalization".

Elliot Smith once said, "The American problem remains the crux of the whole dispute regarding diffusion" (Elliot Smith, 1932: 109), and he soon crossed swords with anthropologists and archaeologists in America. His main critics were Boas, Dixon and Lowie. Roland B. Dixon's *The building of cultures* (New York, 1928) was written partly to demolish the Elliot Smith–Perry hypothesis, and this it did very skilfully.

> "If the heliolithic theory, as propounded by Elliot Smith, verges on the impossible, its later amplification by his disciple Perry passes even these bounds and enters into the realms of fantasy." (Dixon, 1928: 256).

R. H. Lowie was even more outspoken and referred to

> "the unfathomable ignorance of elementary ethnography" displayed by Elliot Smith and Perry, and said, "Here is no humble quest of the truth, no patient scrutiny of difficulties, no attempt to understand sincere criticism. Vehement reiteration takes the place of argument. . . . Every-thing is grist for his mill, everything is black or white. . . . In physical anthropology Elliot Smith controls the facts, hence, right or wrong, his judgements command respect; while in ethnography his crass ignorance darkens counsel (Lowie, 1937: 160, 167).

These are harsh, outspoken words, but fair comment. Perry's book (1924) was re-issued as a Penguin paperback in 1937 at a moment when many archaeologists and anthropologists were hoping that Egypto-centric hyperdiffusionism had died with its founder. In 1941 Penguin Books produced a new version of the prehistoric and protohistoric past: it was Gordon Childe's *What happened in history*.

Childe, like Elliot Smith an Australian, was the most important and influential synthesist of archaeology in the quarter century from 1925 to 1950. At first he was attracted to the Archaic Heliolithic Civilization but soon realized the impossibilities of the theory. Like Petrie and Frankfort he realized the primacy of Mesopotamia and that some, at least, of ancient Egyptian culture had come from outside. Childe (1925, 1929) and Daryll Forde (1930), did not deny diffusion. Indeed Childe preached it, and for the greater part of his life denied independent evolution. He preached and practised what I have called modified diffusion—perhaps modified hyperdiffusion would be a better phrase. Childe realized that so far as Europe was concerned most important inventions came from what he called the Most Ancient East, an area that included south-west Asia and Egypt. It was to him the nuclear area of the Old World. Inventions were spread by diffusion, invasion, and by all kinds of culture–contact from this area: but there was in this view no master–group engaged in this diffusion. He was not really very far away from the diffusionist position set out by Oscar Montelius.

Ten years after Elliot Smith's death and only a few years after the publication of Childe's *What happened in history*, Libby announced the techniques of radiocarbon dating. Now we are equipped with hundreds upon hundreds of C14 dates; we know the dates of the ancient civilizations of the world and of the peasant village communities out of which they evolved. We can now trace with certainty the course of some diffusion while at the same time observing multilinear evolutionary sequences. At this moment of the 20th century it seems that both the Egyptocentric hyperdiffusionism of Elliot Smith (and for that matter all monocentric systems), and the modified diffusionism of Childe are no longer suitable paradigms. Both were useful in their time. Elliot Smith's Egyptian paradigm was a useful one, and its extremes and extravagancies played a useful and important part in the development of theories of culture change. It was a challenge to those who were thinking loosely in terms of a vaguely ill-defined diffusion to Europe from somewhere between north Africa and the Hindu Kush. The challenge of Elliot Smith's definite and violent historicist position made people think: I suggest that the views of Childe and Forde were products of reaction to the Egyptian model.

Why did Elliot Smith himself not listen to criticism? Why did he pursue an idea that came to him in the first decade of this century? Why did he go on preaching it relentlessly through the next quarter-century and refusing to consider the possibility that it was wrong? First of all he had on his side men like Rivers and Perry—scholars he admired and whose views he respected. Secondly he was, at least in matters of cultural anthropology, remarkably obstinate. He gave the Huxley Memorial Lecture at the Imperial College of Science in 1928. It was called *Conversion in science*, and in it he said, "The set attitude of mind of a scholar may become almost indistinguishable from a delusion." He was not, quite naturally, thinking of himself when he said those words but this is what, unhappily, had happened to him. He had acquired a set attitude of mind with regard to the Egyptian origin of all culture and civilization, and this *idée fixe* had become, sadly, a delusion.

Some may see all this, impatiently, as a thing of the past, a curious episode in the history of archaeology and anthropology between 1911 and the mid-thirties. But the present still has its believers in Elliot Smith Egyptocentrism. Professor H. A. Harris, who worked with Elliot Smith in London before he came to the Chair of Anatomy in Cambridge, was a devoted believer in the heliolithic theory. Just before he died he said to me, "Mark my words, there will be a swing of the pendulum. The Egypt theory will be back in favour again, whatever you may say." We now know that a new magazine called *The New Diffusionist* has been founded to campaign for and develop Elliot Smith's ideas, and Mr Joel, the Editor, and Mr Jairazbhoy, who believes the Mexican pyramids are derived from Egypt, are contributing to this Symposium.

Then the success of the Ra expeditions, and the skilful writing of Thor Heyerdahl, have brought, in the last few years, a resurgence of thinking about Egypt as the cradle of civilization and a re-awakening of Elliot Smith's ideas. The successful crossing of the Atlantic in *Ra II* (the *Ra I* crossing was nearly a success) in 1970 in a reed boat, built in the shadow of the Pyramids, has obviously demonstrated that ancient Egyptians could have crossed the Atlantic in a reed boat. But what does this mean to students of prehistory and protohistory? It does not demonstrate that they did so, or, if they did, that they had any cultural influence on the development of pre-Columbian America. But to some people, who confuse the demonstration of a possibility in the present day with the establishment of a prehistorical fact, the *Ra* expeditions have done a great deal to rehabilitate the theories of Elliot Smith.

But the possibility of trans-Atlantic journeyings by Ancient Egyptians—and I do not deny their possibility—must be seen in the context of their historically known achievements in navigation and travel. Recently J. V. Luce of Trinity College, Dublin has summarized these achievements in these words:

> "There is no evidence that the ancient Egyptians ever traded further west than Crete . . . Crete for the Egyptians lay at the western limit of the world. There is no evidence that the ancient Egyptians ever traded, looked, much less went, any further west. Speculations about the Old Kingdom 'explorers' and 'colonists' diffusing Egyptian culture far and wide have no basis in fact, and are most implausible, given that the Egyptians never even explored their own river to its upper reaches." (Luce, 1971: 71–73).

I quote Luce because this is how we should approach the question of cultural origins at this moment of 20th-century archaeology and anthropology. We must seek out, and work out, and define in detail cultural origins and borrowings, and establish clearly independent origins here and diffusion there. Independent invention and diffusion do lie down together. As a basis for discussion let us remember the aim of this Symposium is to discuss the work and significance of Elliot Smith in cultural anthropology: it is not a general discussion of diffusion versus independent evolution. The following points might be concentrated on:

1. The 19th century had widely canvassed ideas of diffusion but had been vague in the choice of a precise area as a centre.

2. Elliot Smith chose Egypt as the central area, an area that had been chosen by many people before from Herodotus onwards. But his novelty was that he pursued the idea with exact archaeological parallels. His was the first clear statement of diffusion from one area in archaeological and anthropological terms.

3. His hyper-diffusionist model, by reaction, was largely the cause of the Forde-Childe diffusionist paradigm which dominated thinking in European prehistory in the thirties, forties and fifties of this century.

4. No one nowadays denies the reality of diffusion and the reality of independent invention. These are the two processes of cultural change that have gone on all the time, and C14 dating has been proving their reality.

5. No one denies the reality and achievement of the Ancient Egyptians, but most of us would deny that a case has been made out before, by, or after Elliot Smith that *all* civilization everywhere in the world was the product of Ancient Egypt. Such a doctrine is one of those *simpliste* comforts of unreason which bedevil archaeology and anthropology but of which we must take cognisance, particularly at this moment when we are discussing the work of a man, famous as an anatomist and physical anthropologist, who did so much from his own sincere conviction, to propagate the idea of Egyptocentric hyperdiffusion.

REFERENCES

The following list of references includes those cited in the discussion on pp. 424–447.

Aldred, C. (1961). *The Egyptians*. London: Thames and Hudson.
Aldred, C. (1965). *Egypt to the end of the Old Kingdom*. London: Thames and Hudson.
Allchin, Bridget. (1966). *The stone-tipped arrow: Late Stone-Age hunters of the tropical Old World*. London: Phoenix House.
Binford, L. R. (1968). Archaeological perspectives. In *New perspectives in archaeology*. Binford, L. R. & S. R. (eds). Chicago: Aldine.
Birmingham, Judy. (1969). "Fabricators" and "eloueras" in West Bengal. *Mankind* 7: 153.

British Museum, Department of Egyptian Antiquities. (1964). *A general guide to the Egyptian collections in the British Museum*. London: British Museum.

Childe, V. G. (1925). *The dawn of European civilization*. London: Kegan Paul.

Childe, V. G. (1929). *The Danube in prehistory*. Oxford: Clarendon Press.

Childe, V. G. (1941). *What happened in history*. Harmondsworth: Penguin Books.

Childe, V. G. (1958). Retrospect. *Antiquity* **32**: 70.

Clark, J. D. (1971). A re-examination of the evidence for agricultural origins in the Nile valley. *Proc. prehist. Soc.* **37** (2): 34 ff.

Coe, M. (1966). *The Maya*. London: Thames and Hudson.

Colbert de Beaulieu, J. B. (1958). Armorican coin hoards in the Channel Islands. *Proc. prehist. Soc.* **24**: 201–210.

Daniel, G. E. (1962). *The idea of prehistory*. London: Pitman.

Daniel, G. E. (1968). *The first civilizations: the archaeology of their origins*. London and New York: Thames and Hudson.

Daniel, G. E. (1971). From Worsaae to Childe: the models of prehistory. *Proc. prehist. Soc.* **37**: 140–153.

Daniel, G. E. (1972). Editorial. *Antiquity* **46**: 262.

Dawson, Warren R. (1938). *Sir Grafton Elliot Smith: a biographical record by his colleagues*. London: Jonathan Cape.

Dixon, Roland B. (1928). *The building of cultures*. New York: C. Scribner.

Elliot Smith, G. (1911). *The ancient Egyptians*. London: Harper Bros.

Elliot Smith, G. (1915). *Migrations of early culture*. Manchester: Univ. Press. London & New York: Longmans Green.

Elliot Smith, G. (1916). The influence of Ancient Egypt in the East and in America. *Bull. John Rylands Library*, Manchester, **1916** Jan–Mar.

Elliot Smith, G. (1917). Ancient mariners. *J. Manchr geogr. Soc.* **1917**.

Elliot Smith, G. (1919). *The evolution of the dragon*. Manchester: University Press. London & New York: Longmans Green.

Elliot Smith, G. (1923). *The ancient Egyptians*. (New and revised edition.) London: Harper Bros.

Elliot Smith, G. (1924). *Elephants and ethnologists*. London: Kegan Paul. New York: Dyson.

Elliot Smith, G. (ed.) (1926). Preface and Introduction. In *Psychology and ethnology*. Rivers, W. H. R. London: Kegan Paul.

Elliot Smith, G. (1927). *Human nature*. Conway Memorial Lecture. London: Watts & Co.

Elliot Smith, G. (1928). *In the beginning: the origin of civilization*. London: Howe.

Elliot Smith, G. (1930). *Human history*. London: Cape.

Elliot Smith, G. (1932). *In the beginning*. (2nd Ed.). London: Watts.

Elliot Smith, G. (1933). *The diffusion of culture*. London: Watts.

Evans-Pritchard, E. E. (1961). *Anthropology and history*. Manchester: University Press.

Firth, R. W. (ed.) (1957). *Man and culture*. An evaluation of the work of Bronislaw Malinowski. London: Routledge.

Forde, Daryll (1930). The early cultures of Atlantic Europe. *Am. Anthrop.* **32**: 19–100.

Frankfort, H. (1951). *The birth of civilization in the Near East*. London: Williams & Northgate.

Freud, S. (1910–13). *Totem and taboo*. (1950 Ed.). London: Routledge.

Harris, H. A. (1938). At University College, London. In *Sir Grafton Elliot Smith: a biographical record by his colleagues*. Dawson, Warren R. (ed.). London: Jonathan Cape.

Harris, Marvin (1968). *The rise of anthropological theory: a history of theories of culture*. New York: Thomas Y. Crowell.

Hocart, A. M. (1927). *Kingship*. Oxford: University Press.

Hocart, A. M. (1936). *Kings and councillors*. (1970 edition) edited by R. Needham. Chicago: University of Chicago Press.

Hooke, S. H. (ed.) (1958). *Myth, ritual and kingship*. Oxford: Clarendon Press.

Joel, C. E. (in press). Introduction to reprint of Perry, W. J. *The Children of the Sun*. Chicago: University Press.

Jones, Ernest (1951). *Essays in applied psycho-analysis* 2.

Kramer, S. (1958). *History begins at Sumer*. London: Thames and Hudson.

Kraus, G. (1972a). The Egyptian origin of the Olmecs: report of a lecture. *New Diffusionist* 2(7): 53–55.

Kraus, G. (1972b). The origin of cultures. Pt. V. Biological evolution and cultural origins. *New Diffusionist* 2(8): 88–117.

Kraus, G. (1973). *Homo sapiens in decline—a re-appraisal of natural selection*. Great Gransden: The New Diffusionist Press.

Lévi-Strauss, C. (1962). *La Pensée sauvage*. (English translation, *The savage mind*. (1962).) London: Weidenfeld & Nicholson.

Lewis, I. M. (ed.) (1968). *History and social anthropology*. ASA Monographs. London: Tavistock Publications.

Libby, Willard (1963). The accuracy of radiocarbon dates. *Science, Wash.* **140**: 278.

Lowie, R. H. (1937). *The history of ethnological theory*. New York: Farrar & Rheinhart.

Luce, J. V. (1971). Ancient explorers. In *The quest for America*: 53–95. Ashe, G. (ed.) London: Pall Mall Publ.

Mair, Lucy (1957). Malinowski and the study of social change. In *Man and culture*. Firth, R. W. (ed.) London: Routledge.

Mellaart, J. (1965). *Earliest civilizations of the Near East*. London: Thames and Hudson.

Montelius, O. G. O. (1899). *Der Orient und Europa*. Stockholm.

Morgan, Lewis H. (1877). *Ancient society*. (1964 edition) Edited by Leslie A. White. John Harvard Library, Harvard: University Press.

Morris, D. (1961). *The naked ape*. London: Jonathan Cape.

Nuttall, Z. (1901). *The fundamental principles of Old and New World civilizations: a comparative research based on the study of the ancient Mexican religious, sociological and calendrical systems*. Harvard.

Perry, W. J. (1918). *The megalithic culture of Indonesia*. Manchester: University Press. London: Longmans Green.

Perry, W. J. (1923a). *The children of the sun*. London: Methuen.

Perry, W. J. (1923b). *The origin of magic and religion*. London: Methuen.

Perry, W. J. (1924). *The growth of civilization*. London: Methuen.

Perry, W. J. (1935). *The primordial ocean*. London: Methuen.

Piddington, Ralph (1957). Malinowski's Theory of needs. In *Man and culture*. Firth, R. W. (ed.) London: Routledge.

Piggott, S. (1972). Dalladies. *Curr. Archaeol.* No. 34: 295–297.

Premack, Anne James & Premack, David (1972). Teaching language to an ape. *Scient. Am.* **227**: 92–99.

Pretty, Graeme L. (1969). The Macleay Museum mummy from Torres Straits: a postscript to Elliot Smith and the diffusion controversy. *Man* (N.S.) **4**: 24–43.

Radcliffe-Brown, A. R. (1922). *The Andaman Islanders*. Cambridge: Cambridge University Press.

Raglan, Lord (1939). *How came civilization*.

Renfrew, A. C. (1967). Colonialism and megalithismus. *Antiquity* **41**: 276–288.

Renfrew, A. C. (1970). New configurations in Old World archaeology. *Wld Archaeol.* **2**: 199–211.

Rice, T. T. (1957). *The Scythians*. London: Thames and Hudson.

Riley, C. L., Kelley, J. C., Pennington, C. W. & Rands, R. L. (eds.) (1971). *Man across the sea*. Problems of pre-Columbian Old World–New World contacts. Austin, Texas: University Press.

Rivers, W. H. R. (1912). The ethnological analysis of culture. *Rep. Br. Assoc. Advmt Sci.* **1911**: 490–499.

Rivers, W. H. Rivers (1914). *The history of Melanesian society*. Cambridge: Univ. Press.

Rivers, W. H. R. (1926). *Psychology and ethnology* (edited with preface and introduction by G. Elliot Smith). London: Kegan Paul.

Schmidt, Wilhelm (1926). *Die Ursprung der Gottesidee*. **1**. Munster: Aschendorff.

Seligmann, C. G. & B. Z. (1911). *The Veddas*. Cambridge: University Press.

Tiger, L. & Fox, R. (1972). *The imperial animal*. London: Secker & Warburg.

Turnbull, C. (1966). *Wayward servants*. London: Eyre & Spottiswoode.

Wittfögel, K. A. (1957). *Oriental despotism*. New Haven: Yale University Press.

Wortham, J. D. (1971). *The genesis of British Egyptology* 1549–1906. Oklahoma: Norman.

DISCUSSION

FORDE: I am not among those in this Symposium invited to outline the position of Elliot Smith's work in ethnology, but Dr Daniel has asked me to say a few words since I knew Elliot Smith and something of that side of his work from the early 'twenties onwards.

The Chairman himself has already given a thoroughly documented outline of the development of Elliot Smith's own writings, and the circumstances in which his ideas developed. There are just one or two small comments I should like to make. First of all, from conversations with him in the 'twenties and from the remarks of others who were round him, the central stimulus to Elliot Smith's thinking about diffusion from Egypt was the historical accident of his investigation of mummified or smoke-dried bodies in the Torres Straits which were preserved as ritual objects, and his previous study of Egyptian mummification. His work on Egyptian mummification had led him to write technical papers on the reasons why certain wrappings, incisions and preservations of tissues including fingernails and so forth, had been undertaken, and he saw—as I think a renewed discussion a few years ago emphasized—a number of remarkable and dysfunctional formal similarities between elements of wrappings on the Torres Straits mummies and Egyptian material. This, of course, was linked to his immense biological knowledge of ancient Egyptian material, and here I would again like to raise a question concerning one of Dr Daniel's remarks on his outlook on Egypt. Elliot Smith's earliest work in Egypt did stem more or less directly from his interest in physical anthropology in an attempt, by morphological examination of tomb remains, especially crania, to work out early sequences of populations in the Nile Valley. That is to say, he entered the broader anthropological field through his own métier.

From his study of mummification, the removal of organs, and the symbolic representation of organs in Egyptian tombs and wall paintings, he developed an interest in ritual. I think this was one of the important aspects of his work which was perhaps less influenced by Rivers than much of the rest.

I met Rivers only very briefly as a young man, and did not take much part in the conversation. This was in 1921. Elliot Smith's relation with Rivers again had come through the biological side. That is to say, his friendship with Head, the neurologist, who was a colleague of and fellow worker with Rivers was one of the links between them. In the 'twenties—which admittedly was fairly late on in the development of Elliot Smith's Egyptological thought—Rivers was much more concerned with kinship and social organizational studies than in heliolithic theories. And Rivers's *History of Melanesian society* (1914)—which was written as an attempt at historical reconstruction and has been too widely regarded as a complete waste of a good brain's time—attempted to assess the evidence in detail for a particular region not linked to any world-wide diffusionist theories.

A second point that I should like to make is that in his Manchester period Elliot Smith worked in relative isolation. I can remember his saying on several occasions that his work in cultural anthropology was, as it were, the lighter side of his life. I hesitate to say that this was his hobby, because it was not. As Dr Daniel has emphasized, he held his views firmly and with deep conviction. However, these were so to speak recreational activities when he was not having to teach anatomy in the university. At the John Rylands Library, and with one or two colleagues in the University of Manchester—such as Professor Canney—he began to explore other facets of Egyptian civilization, of Egyptian culture, which had bearings outside and in Europe particularly.

I think that Dr Daniel was quite right in his main point, that he shared initially a general interest in the influence of Egyptian civilization, on the transmission and traditional continuities of elements of Egyptian civilization in what might be called folk lore in Europe as well as in the building of megalithic monuments.

On the question of independent invention versus diffusion, Elliot Smith took an extraordinarily doctrinal position. As Dr Daniel has said, he

moved backward until he could assert that before Egypt there was
nothing. Part of the justification for this lay, of course, in the lack of
historical depth. I mean historical in the sense of reasonable evidence of
changes in material culture and basic economic organization of primitive
peoples as known in the first quarter of the 20th century. That is to say,
the non-literate populations, apart from the high civilizations of the Maya
and the Aztecs, were timeless snapshots, and these could be taken as stills,
so to speak, which had no historical depth behind them. The view of the
world that he took was that primitive peoples had no history, an outlook
which, of course, he shared with Lord Raglan.

Very little was going on in British ethnography in the period from the
end of the 19th century down to the Second World War to inject a time
perspective, because the growing point in British anthropology at that
time—as opposed to diffusionist views—was in the fields of kinship and
social organization. With Rivers earlier on, and between Elliot Smith and
Malinowski, there was very little possibility of discussion, not because
either of them was necessarily attacking the other's views, but because
they were working on entirely separate lines. By the 'twenties, Elliot
Smith was really very set in his ways, and he was really re-writing and
taking pieces out of earlier papers and out of contributions by Dawson and
others to write general books stressing his point of view.

In the London period he did not lack critics. Quietly sitting in the
Horniman Museum in London was H. S. Harrison, a very shrewd and
scholarly man with whom Elliot Smith had very good relations and fairly
frequent meetings. I listened to and took part in a number of debates on
this very question, but one could make no impact on Elliot Smith's point
of view at that stage in his career.

Then again in the 'twenties Lowie made a fairly lengthy visit to
London, he was in London for two or three weeks, and a number of
informal discussion meetings were arranged. Despite the fact that he, like
Elliot Smith, could have a very sharp pen, Lowie was an extremely
amiable man in conversation, and there was review after review of particular situations with regard to South East Asia, with regard to Central
America, and also to the question of trans-Pacific movements of populations, in which Lowie brought out the fact that the whole issue of independent invention versus diffusion was an empirical question requiring
close study of the data, and that in many cases the data did not yield any
clear-cut answer.

However, Elliot Smith himself, in the midst of a busy career in anatomy
and running a large department, did not have the time or, I think, the
disposition to go deeply into the ethnographic literature, and he remained
fixed in his ideas. The result, of course, so far as general anthropology is
concerned, must I think be accepted as negative. Little could he have
known the extent to which both further archaeological research and methods
of time dating would completely supersede the kind of point-blank
statements he sometimes pronounced.

In the field of prehistoric archaeology, Gordon Childe was working in London, and here again there was an interesting conflict at the time. Childe had a very strong animus against the German archaeologist, Kossina, who exerted considerable influence in Germany through his capacity to influence appointments. His doctrine was one of the Southern Baltic origin of European post-paleolithic cultures, linking them therefore to the Nordic peoples. Childe, in writing the first volume of *The dawn of history*, was very concerned with the need and importance of working out the evidence for cultural parallels and cultural transmissions from the ancient East to South Eastern Europe and Central Europe, and also through the Mediterranean. Childe was an empirical diffusionist, not of an extreme kind as you will all know, and he treated all these matters on the merits of the archaeological data at the time.

I came into the picture in a very small way through wanting to test Perry's sweeping hypothesis that the expansions of the prehistoric megalithic cultures of the Mediterranean and Atlantic Europe were the product of a metal prospecting people. I worked fairly intensively on the archaeological data as they existed at the time in Brittany and reached a non-proven conclusion which perhaps, with hindsight, was not unexpected.

I was working in association with Childe at the time and following him was attempting to work out typological distinctions between or within a given genus of monuments and various pottery styles. We were able to produce limited hypotheses concerning the spread of populations and of cultures in Europe. This was not something that greatly interested Elliot Smith. In fact, the details of European prehistory were not matters with which he concerned himself. As Dr Daniel has said, in the end he became so insistent on the heliolithic dogma that the work petered out into a dead end.

FORTES: When invited to take part in this discussion I found myself faced with a dilemma that has been accentuated by the Chairman's masterly review of the history of this controversy. I feared that what I could contribute, as a social anthropologist, might seem to be only marginally germane to your theme. There are however three aspects of Elliot Smith's position in the history of anthropology that particularly interest me as a social anthropologist, and I should like to say something about them. Firstly, there is the problem of how it came about that natural scientists of the calibre of Elliot Smith and Rivers could be so taken in by a theory of human society and human culture that was to others so obviously fallacious. This is not a question concerning merely these particular personalities, it is one of much more general implication with regard to the way in which the natural scientist approaches the facts of human social life. Next comes the question of what, in particular, it was in the doctrines and methods of Elliot Smith and his associates that aroused the objections, one might go further and say, even, the opposition and contempt, of the functionalists, using that term in the general sense of the kind of social anthropology that developed under the leadership of Malinowski and Radcliffe-Brown. Stimulated by the prospect of this Symposium I re-read some of the publications of Elliot Smith and his followers and I think I see a little more clearly than I did in the 1930's what might be the answer to this question.

The third question, a very important one, arises directly out of the letter which Lord Zuckerman addressed to intending participants in this Symposium. It is the question of what has been the residue, the surviving significance if any, for social anthropology today, of Elliot Smith's incursion into the field of ethnology. It is not only from the point of view of the uncritical extremism represented in such works as Morris (1967) and Tiger & Fox (1972) that Elliot Smith's most cherished beliefs about mankind now seem to be of doubtful validity. Serious anthropological research points in the same direction.

May I interpolate, since this is a commemorative occasion, that I did have a slight acquaintance with Elliot Smith. I attended his lectures at University College, on the evolution of the brain, and also had some casual contact with him later, on anthropological matters. I was better acquainted with Perry and Hocart, having been introduced to them by Evans Pritchard, who was kind and attentive to their views but remained, and still remains, sceptical of them (Pritchard, E. in Hocart, 1936). I found Perry rather uncritical, even naive, in his use of ethnography to formulate his general views and theories. Hocart, on the other hand, was a much more interesting man when he was talking about actual ethnographic and historical data. He was catalytic of ideas, and some of his arguments and theories were by no means incompatible with the kind of functionalism he purported to be opposing.

What, then, about the reasons why Elliot Smith, Rivers and their associates, adopted and so single-mindedly espoused their unicentric diffusionist doctrine? The basic reason, I would like to suggest, was their all-embracing commitment to 19th century, Darwinian evolutionary biology.* Of course, this was a way of thinking that dominated the ethnological thought of that period. According to this model, mankind was seen as the culmination of a ramifying, diverging evolutionary stream, a sequence of continuity broken by mutations, ultimately of relatively modern origin. The paradox to me, is that these were men who were very well accustomed to thinking analytically when it came to the investigations they were undertaking in the laboratory, whereas in their ethnological researches they were, from a modern anthropological point of view, credulously obsessed with a pseudo-historical dogma.

This Egypt-centered theory of diffusion, I would suggest, fits very well into a monogenetic Darwinian type of biological evolutionism—a concep-

---

* Summed up in the following of many such statements: "The principle of continuity which is the foundation of the theories of geology and biology, as enunciated respectively by Sir Charles Lyell and Charles Darwin, urgently needs to be rehabilitated and reapplied in the study of mankind, where it should play an even more extensive and significant purpose than in the Natural Sciences." (Elliot Smith, 1930: 18.)

tion of an evolutionary tree—and a habit of mind which prompts one always to seek explanations for contemporary states of affairs in terms not merely of origins, but of remote and ultimate origins. It seems to me that this is the way that the whole set of convictions established itself and got a grip on them; the alternative is to suppose that they suffered from delusion and megalomania, and that is not acceptable. The unilineal ethnological evolutionists started from the same theoretical premises, but applied them in ways which allowed for the manifest diversities of culture within the limits of the uniformity of mankind.

When we associate Elliot Smith primarily with his theory of diffusion, we think first of his often expressed hostility towards these unilineal ethnological evolutionists as reflected in his polemic with Tylor, but we must remember that his central interest was essentially like theirs, and that was to explain and account for what he referred to as "human nature" (Elliot Smith, 1927). What this meant for him was to seek out the unique origin, the *terminus a quo* of the whole of human development and man's cultural experience. Human nature to him was the state of affairs which had arisen in a long drawn out process of species evolution which came to a biological climax and, almost by definition, had to ante-date culture and civilization. It was a climax shared with the higher non-human primates until the development of the human brain and eye differentiated man from monkey and ape.

What is interesting is the intellectual process by which he thought up this original human nature. It was achieved by stripping off, in thought, by thinking away the facts of culture from primordial mankind. It was constructing a picture of mankind before mankind had a history, which meant that aside from speech, fire and a few elementary tools, human nature was a complete cultural blank in the beginning. It was much closer to the non-human primates than to ourselves or to earlier forms of higher civilization. This reductionist procedure practically compelled him to resort to a picture of what amounted to a Golden Age of humanity supposed to be reflected and recorded in the origin myths and stories of the simplest living peoples. This belief ran completely counter to the prevailing theories of the significance of such myths and stories, especially those of Malinowski and the functional school but also, up to a point, some of Hocart's ideas (Hocart, 1936: Ch. VIII). There is nothing that today sounds so pathetic as Elliot Smith's repeated assertion of man's innate peacefulness, before culture and, later, civilization, overtook and corrupted him.

However, what I want to emphasize here is not so much this dogma as the way he arrived at what he thought was evidence for it. It was by examining field research among very simple hunting and collecting peoples like the Bushmen, the Negritos of Malaya, the Veddas of Ceylon and others of that level of technology and social organization, and by subtracting, so to speak, from the ethnography the additions to their supposed pristine state which he attributed to influences that reached them by diffusion. Elliot Smith and his followers constantly asserted that the only "natural" form of social organization based on "biological foundations" (Elliot Smith, 1930: 256) was the Family. Tribes, villages, chiefs, ritual specialists, etc. were later elaborations due, in the last resort, to diffusion from Egypt (Elliot Smith, 1930: 7).

It is important to realize that this question of the fundamental, rock-bottom nature of human nature was a crucial one for Elliot Smith. But it is a question that every student of human social life, certainly every anthropologist and archaeologist, has to come to terms with at some time or another; and in every case there is bound to be, tacitly or overtly, a distinction drawn between what is attributable to man's organic nature and what society and culture add to it. Thus Tylor's theory of the independent development of beliefs about the soul and animism, though derided by Elliot Smith (Elliot Smith, 1933: Ch. 4), for assuming innate capacities common to all mankind is, in principle, a theory much on the same lines as his. Lewis Henry Morgan likewise postulated common mental capacities as the ultimate origin of human family systems and social institutions (White in Morgan, 1964). Malinowski's "basic needs" belong to the same class of assumptions (Piddington in Firth, 1957: 33–52) and so also, to come down to the present time, does Lévi-Strauss's postulate of the innate propensity of the human mind to classify by binary opposition (Lévi-Strauss, 1962 passim).

In this connexion, no view of human nature could have been proposed which was so antithetical to Elliot Smith's as the Freudian view. Elliot Smith's view of primordial man was, in a sense, a rationalistic and utilitarian one. This comes out in the notion that irrefutable proof of diffusion was to be found in "arbitrary" or "unnatural" or "useless" features of customs and institutions in a given area. The impossibility of objectively verifying or falsifying such attributions was one of the main criticisms made of the Egypt-centered diffusionist theory, especially on the part of the functionalists who argued that everything to be found in a given way of life could be shown to have meaning at some level or another be it only symbolically. Add to this Elliot Smith's belief in the innate "honesty", "decency" and peacefulness of primordial man (Elliot Smith, 1930: 73) and it is obvious that his views could not be reconciled with Freud's emphasis on sexuality and his image of human nature beginning with the first parricide due to incestuous wishes (Freud, 1910–1913). It is the more interesting, considering Elliot Smith's vehement rejection of Freudian theory (Elliot Smith in Rivers, 1926) to find his ideas about primitive mankind's preoccupation with questions of human origin and more especially his celebrated account of the diffusion of the cowrie shell as a symbol of female fecundity (Elliot Smith, 1930: 320–321) praised by Ernest Jones as affording anthropological confirmation of Freudian theory (Jones, 1951: 131). Curiously enough, poles apart as their theories of human nature were, Elliot Smith and Freud had one thing in common; that was an underlying pessimism about the influence of civilization in human development.

Implicit in the attention paid by Elliot Smith and his associates to simple hunting and collecting societies, was the belief that they were survivals of prehistoric communities. Analogous views were held by Tylor, Frazer and others of that period, but ethnological field research with strictly descriptive aims, e.g. Radcliffe-Brown among the Andamanes (Radcliffe-Brown, 1922), the Seligmanns among the Veddas (Seligmann & Seligmann, 1911), and even the field researches inspired by Father Wilhelm Schmidt of the Vienna School from 1908 onwards (Schmidt, 1926) was also undertaken in very simple societies, with the object of trying to establish what they conceived to be the essential characteristics of human society and culture. It is worth adding that no human group has as yet been discovered which is devoid of some form of extra-familial social organization, leadership. ritual practices, or aggressive tendencies.

I come to the conclusion that Elliot Smith's unbending diffusionist commitment with its dogmas and doctrines that have more the marks of a mythological system than of objective scientific analysis, was significantly conditioned, if not wholly determined, by his Darwinian evolutionist convictions. These gave him his basic explanatory model; but what is perhaps more to the point, these led him also to the image he formed for himself of fundamental human nature; and his unicentric diffusionist theory was a way of reconciling this image with the data of human culture and social organization and with movements in recorded history that struck his imagination.

I come finally to the question as to why Malinowski and the other functionalists of the 'twenties and 'thirties so uncompromisingly rejected Elliot Smith's views. It should be remembered that none of them denied the reality of the facts of social life generally referred to by the term diffusion. Indeed Malinowski was particularly interested in what was then called culture contact (Mair in Firth, 1957). One could go further and say that Malinowski and his colleagues regarded the selective acquisition of customs, institutions, artifacts, etc., by one people from another by diffusion as strong evidence supporting functionalist theory. Such borrowings might look bizarre to the outsider but must have meaning to the borrowers else they would never have been adopted.

I think the important point here is that Elliot Smith's diffusionist scheme was not fitted to formulate questions or suggest ways of answering questions that were relevant for the synchronic functionalist enquiries. Functionalist and post-functionalist social anthropology was and is concerned to find out how social systems work at a given time and how their customs and institutions hang together consistently with one another. They want to know what are the variables within such systems and how they are connected with one another. The parallels in biological science would be with anatomy and physiology rather than evolutionary theory. That is why the search for conjecturally remote origins of human cultural traits and their supposed transmission and distribution through the world by diffusion could have no bearing on functionalist research. It is a very interesting point for me that Elliot Smith and Rivers were medically trained biologists by profession originally, whereas Malinowski was a physicist and therefore trained, I suppose, to think in terms of the constancy of natural systems and not in terms of evolutionary chronology.

Must we then dismiss the whole diffusionist movement led by Elliot Smith and Rivers, as no longer of relevance for us today? I myself do not think so, though I agree with Dr Daniel's conclusion as to the general futility of the Egypt-centred bias in the study of culture. Evolutionist ethnology, even in its ultra-diffusionist form, had this continuing relevance, I suggest, that it did assume the possibility of developing a unified view of the sciences of man, or if not unified, a view of the sciences of man in their widest extent, which stresses their inter-connexion with one another and seeks to make use of their mutual complementarities, in a context combining historical perspective with synchronic analysis. This may be an aspiration which can never be fulfilled, but there is, I suggest, some virtue in holding on to it.

But there is another and all-important lesson to be learnt from the diffusionist movement. Studies of the social life of sub-human primates and of animal behaviour in general in the 40 years since Zuckerman's famous investigation, leave no doubt in our minds now that society is not a purely human condition. But diffusion, that is to say, the transmission and circulation of knowledge, skills, institutions, and artifacts, between generations within a human society and between societies is, I suggest, a very distinctive human characteristic. Monkeys, so far as I have been able to ascertain, cannot teach one another skills they may be taught by humans or pass them on from one monkey community to another. We now know that a chimpanzee can be taught to make sentences and thus communicate with its human teacher through the medium of manipulable pieces of plastic (see Premack & Premack, 1972). But there is no evidence that it can teach another chimpanzee this skill. In short, the facts and problems covered by the notion of diffusion lie at the very centre of human social and cultural life. They remain of critical importance for anthropological research.

LEACH: Coming at this rather late stage in this multiple Symposium I find that all the things I had planned to say have been said already one way or another, but I will try to comment on some points that have been made by Professor Fortes.

What I am interested in is how these two great scientists, Rivers and Elliot Smith—and I know more about Rivers than I know about Elliot Smith—could come to adopt such a bizarre ethnological theory when they were so scientific in their treatment of biological problems.

It seems to me that both these scholars tended to think of cultures as racial attributes. My charge of racialism is not meant to be tendentious; I am just trying to bring together the two halves of their interest—as scientists concerned with physical man, and as scientists concerned with cultural man. The failure to distinguish between race and culture was very general at the beginning of this century and, as has already been remarked, primitive culture was equated with the behaviour of animal species. The following passage from Elliot Smith (1932) illustrates the line of thought: "The habits of the gorilla and the chimpanzee by their analogies afford a startling corroboration of the truth of the sketch which I have given of the primitiveness of natural man". This natural man was an animal species, a special kind of primate. "Primitive man shows no more innate tendency than do the man-like apes to embark upon the invention of civilisation", and so on. So there is the idea that we start with this creature, natural man, a special primate, with no special cultural attributes at all.

With this baseline Elliot Smith then assumes that the invention of civilization was a kind of accidental mutation which occurred once only in one particular branch of the human family, namely the Mediterranean race as exemplified by the ancient Egyptians. The point here, which has not so far been made, is that Elliot Smith discussed the races of man as if they were quite separate sub-species of *Homo sapiens*. But in addition he assumed that races, like cultures, could be graded as high and low, primitive and advanced. Rivers argued in the same way. For example, he maintained that since Australian aborigines were technologically very primitive, but palpably all of the same race (in the sense that they all looked very much the same), it was paradoxical that their cultures were not uniform. Rivers declares categorically (Rivers, 1926: 130) that "The fundamental problem of Australian society is the coexistence of two forms of social organisation side by side". For reasons that are not very clear he had persuaded himself that what he calls "the dual organisation coupled with matrilineal descent" on the one hand, and what he calls "the totemic clans possessing patrilineal descent" on the other, were mutually inconsistent systems. That being so, it was not possible that both these features could both be equally indigenous attributes of the Australian aborigines considered as a race; one or other must therefore have been brought in from outside. This "proves" the diffusion of culture. You can see the line of reasoning here, even if you do not agree with the conclusion.

Rivers was not consistent in his attitudes, but if one studies the diffusionist side of his work, one gets the impression that he thought of elements of culture, whether linguistic or material or organizational, very much as if they were surface growths of the human body like fingernails or human hair. So long as these growths of the human body were attached to their natural parent, they could go on developing and growing, but once they were detached from the parent and used as fetishes by other people, then they simply became elements of diffusion and would either be preserved unchanged in a fossilized state or, more likely, would gradually decay away. Furthermore, cultural elements are detachable in differing degree.

Rivers takes it for granted that clothes, ornaments, utensils and weapons are very readily detached from their original creators; beliefs,

T

rituals and language only rather less so. But then there is a mysterious something which he refers to as "social structure" or "social organisation" which he says is "so deeply seated and so closely interwoven with the deepest instincts and sentiments of a people that it can only gradually suffer change" (Rivers, 1926: 134). The idea seems to be that when artifacts are shown to have moved around the world this is just an example of historical diffusion, but if you get the same social structure turning up in two different places this is evidence of the movement of people.

In this respect Rivers' scheme is rather more moderate than that of Elliot Smith. Elliot Smith clearly supposed that the ancient Egyptians had themselves wandered all over the world carrying their culture with them.

But to return to the general confusion between race and culture. Towards the end of his life, Elliot Smith was maintaining that the living races of mankind are precisely six in number, and that they had diverged from the main stock in the following order—Australian, Negro, Mongol, Alpine, Mediterranean and Nordic. He recognized of course that members of these "races" were inter-fertile, but he consistently showed a very marked reluctance to discuss race mixture. This is understandable because his whole analysis really depends on the proposition that these human sub-species can exist side by side as separate cultural entities without getting mixed up.

As we have heard, it was part of Elliot Smith's doctrine that a capacity for cultural innovation is a special characteristic of the Mediterranean race. This characteristic had first appeared as a kind of genetic mutation. This is why inventiveness occurred once only in one particular locality. Inventiveness is an attribute of a particular race, it is an attribute of a particular sub-species; therefore obviously it cannot belong to any of the other five sub-species.

Clearly this is a very strange way of looking at the evidence. How could Elliot Smith have come to look at things in this way? Surely the answer to this puzzle must lie in the fact that Elliot Smith was accustomed to dissect tissues rather than study the biochemistry of growth. His field was in the analysis of anatomy, and he evidently considered it an essential part of scientific procedure that the investigator should first reduce a complex object to its simplest possible components by taking the whole thing to pieces, and should then interpret the result by means of the principle of reductionism which Professor Fortes has already mentioned. Elliot Smith dissects human culture in this way; he then reduces all food-gatherers in the world into just one category which is ultra-primitive and ultra-peaceful, and finally by polarization boxes all other human cultures into another unity, "civilization", which is ultra-aggressive and ultra-creative.

The only other point I should like to make is that, because of Elliot Smith's assertion that all primitive peoples are examples of natural man and are therefore peaceful food-gatherers, and because he makes the unique polarization that the ancient Egyptians are the opposite of natural man, he is led to make this sort of statement: "The creation of the state involved not only the invention of a multitude of arts and crafts, but also a complicated social and political organization under the rulership of a king endowed with peculiar powers and authority over the lives of his subjects" (Elliot Smith, 1930: Ch. 5).

The context of this is irrigation in the Nile Valley, but you may find similar remarks in Wittfögel's study of oriental despotism, which is a proto-Marxist account of the beginnings of civilization (and irrigation) in ancient China (Wittfögel, 1957). There is this difference: for Elliot Smith the invention of the calendar and the art of mummification precedes the invention of agriculture and irrigation; with Wittfögel it is the other way round. For Elliot Smith religious development is a cause; technological organization is a consequence. For Wittfögel the creative moment is the leader's seizure of control over the means of production.

This ties in with what has been said before. Elliot Smith saw himself as a righteous apostle of Darwinian truth seeking to destroy the heresies of unilineal evolution. The latter doctrine, which traces back to Condorcet in the 18th century and even to Lucretius, assumes a uniform law of progress throughout mankind. Cultures and civilizations represent stages in the evolution of society; they are not the peculiarity of particular racial groups. At the beginning of this century the evolutionist views current among cultural anthropologists were of this kind; they had not derived from Darwin, but from Comte and Marx, McLennan and Morgan. They assumed that all human societies could be neatly slotted into such tidy exclusive categories as savagery, barbarism, feudalism, capitalism, oriental despotism. Wittfögel uses categories derived from this kind of theory.

As disciples of Darwin, Elliot Smith and Rivers rejected this unilineal scheme. In doing so, they were ranging themselves as opponents of Marxism, but not explicitly because they did not see the issue in those terms. But even so their treatment of hydraulic civilization manages to turn Marx (via Wittfögel) exactly upside down. However, their own scheme is just as simplistic as the unilineal theory which they were trying to upset. Everything is sharply black and white, primitive and civilized, passive and peaceful on the one hand, dominant and aggressive on the other. It is remarkable how much useful thinking they managed to extract from this highly exaggerated model.

In 1917 Elliot Smith published a world map which was supposed to show the lines of cultural diffusion which had resulted from the spread of Egyptian, Elamite and Sumerian prospectors travelling all round the world in pursuit of gold and copper between 3500 and 3000 BC. So far as I know, modern research would agree that the arrows shown on this map are, broadly speaking, the general lines of ancient trade and, if you simply reinterpret the map so that each arrow represents an ancient trade route, most people would now agree with it. But of course the dating as presented by Elliot Smith is crazy and the Egyptians had nothing to do with it.

I infer, then, that if Elliot Smith had presented his ethnological theories simply in the form of hypotheses, saying, "Here is an idea, let's see what we can do with it. Does the evidence stand up?", a good deal more of it would have stood up than one might suppose. And anyway one would have ended up with an enlightened view of the general pattern of trade routes throughout the world, even if one lost to view the heroic Egyptians. But of course this is not the way he thought, as we have been told again and again. He became fantastically dogmatic in the way he presented his arguments; he set up a total schematic system which had to be self-validating in the same sense as a schema by Freud or a schema by Lévi-Strauss; the reader is required to accept the whole packet without qualification as a matter of faith and religion. Any expression of doubt invited denunciation as an enemy. Personalities of this sort are by no means uncommon. Highly influential academics in all sorts of fields—of science, of sociology, of history—quite frequently present their theories as if they were dogma. Malinowski was one such. The academic disciples of such men are required to believe the whole truth, and not question the gospel even by one iota. Such dogma of course is always fallacious. In retrospect we can see that the gospel was not true. But we can learn a lot from enthusiasts who believe in the truth of their own gospel, and that is the sense in which I think we should view Elliot Smith's contribution.

What he taught us as regards ethnology was absolute rubbish; nevertheless he had great enthusiasm which generated a great deal of research. I would say just the same of Malinowski and Radcliffe Brown, who are Prof. Fortes' "patron saints". In exactly the same way nearly everything that they thought was false, but nevertheless they were very great men.

RENFREW: Sir Grafton Elliot Smith had an aspiration which we can still share today. His very first words, at the beginning of his book *Human history*, were:

> "By discovering a new world, Christopher Columbus compelled European statesmen and philosophers to think of mankind in terms of the world as a whole. Many attempts have been made during the last four centuries to give expression to this idea in a universal history of mankind. . . . The student of mankind, working in the frontier that separates—unfortunately the word is the appropriate one—Natural History from the Humanities is made to realize how the subject of his studies suffers from the conflicting allegiance. It would be a great gain if the benefits of the two disciplines could be merged in a Greater Humanity which might be called HUMAN HISTORY".

This was the aspiration and the aspiration was applied to a problem which is still perhaps the central problem of archaeology, and indeed of anthropology today—certainly in my view of archaeology. How do we explain the striking apparent similarities in the products of geographically remote communities—mummification in Egypt and Peru, to take a detailed case; metallurgy in Sumer and the Americas, to take a more fundamental one; the pyramids in Egypt and Mexico, to be specific again?

How can we establish or extract some broader pattern in human culture history, something more general than the mere narration of local and special historical sequences, which would be the only logical alternative?

Today we are still concerned to ask, as Elliot Smith did, not just how did *this* civilization arise, but how do civilizations arise?

What I should like to do very briefly is to comment personally on how this problem has fared. Clearly Elliot Smith did not solve it; he failed to solve it. The reason for this failure is, I think, a very simple one. It is not merely that diffusion did not take place to the extent that Elliot Smith believed, but essentially that "diffusion" is not in itself an explanation for culture change at all. Precisely the same may be said of the concept so often set up as an alternative explanation in antithesis to diffusion: "independent invention". For to assert independent invention in a particular case is merely to claim the absence of diffusion, and nothing more. Independent invention is not in itself an explanation of any kind. Our language is rich in these very general concepts which fraudulently purport to say something of culture change—"cultural evolution" is another. Yet in the sphere of human culture, to speak of evolution is no more than to assert continuity, not in itself very remarkable or rich in insight.

Dr Daniel rightly indicated in his discourse that diffusion is a fundamental feature of human culture; as Prof. Fortes stressed, one of its defining features. Diffusion, meaning the adoption by one human being of new customs and activity patterns learnt from another from beyond the residence unit, and hence the spacial dissemination of an innovation beyond its place of origin, certainly takes place. One difficult problem, to which Elliot Smith addressed himself, is to decide the extent to which it has in fact taken place in a particular case. This has been a problem in the study of European prehistory, to which I will refer in a moment. The statement that a particular innovation was transmitted by diffusion from place A to place B is indeed a difficult one to assess, since it is not strictly falsifiable once chronological priority has been established: you can never demonstrate positively that contact has *not* taken place, for this is a negative proposition.

This problem of establishing whether a specific innovation was learnt by a particular community through contact with its neighbours, or developed locally without such stimulus, interesting as it is, has only served to obscure more fundamental ones in the explanation of culture change. For the operation of diffusion, in any given case, the extent and nature of contact between communities, is an observation. As such it may be right or wrong: the contact may or may not have occurred. Yet of itself it explains nothing at all. In order to explain changes in culture, in the organization and life patterns of particular communities, the genesis of specific innovations is not the fundamental problem. More central is to explain how and why innovations of one particular kind came to be accepted at that time and at that place, where previously the community did not take to them. Such an explanation may require information about the entire functioning of the community—its subsistence, its demography, its social organization and so forth. The importance of its contacts with other communities, and their influence upon it may well be related to the bulk of trade between them, and lie in the changes in economic and social organization entailed, rather than in the flow of ideas accompanying this trade. Innovations do not generally arrive unannounced from afar, like messengers bearing glad tidings. Where Elliot Smith went wrong, I think, was to devote his energies to demonstrating that contact had taken place, without any but the most superficial analysis of what its effects in each case would have been upon the societies involved.

In the field of European prehistory, Gordon Childe was influenced by Elliot Smith, I think more than he realized, although he did admit that influence (Childe, 1958: 70) when he wrote his "Retrospect" in 1957. Referring to his book, *The dawn of European civilization*, he said:

> "Yet the sea-voyagers who diffused culture to Britain and Denmark in the first chapters in the first Dawn . . . though they do not hail from Egypt yet wear recognisably the emblems of the Children of the Sun".

He acknowledged that, as he set out to reconstruct European prehistory, his "sole unifying theme was the irradiation of European barbarism with oriental civilisation" (Childe, 1958).

European prehistorians over the past half century, setting out to reconstruct and explain what had happened, felt that they had done so once the "irradiation" or diffusion could be demonstrated. They rarely paused, as I understand it, to explain what had happened *within* these different societies (although Childe himself was a pioneer here). We would never have noticed the difference, I fear it is true to say, if it had not been for the impact of radiocarbon dating. The first radiocarbon dates had some puzzling consequences: they began to date some of the megaliths of Europe a little earlier than we expected. We began to have radiocarbon dates for the megaliths of Spain of about 2500, or maybe 2700 BC. But then came the calibration of radiocarbon dating, in which Dr Suess had so important a role, and that put all these radiocarbon dates very much earlier. Of course, this applies to Egypt as well, but the radiocarbon dates in Egypt had hitherto been too late, too recent. Few archaeologists had worried very much about this because the Egyptian historical chronology had seemed fairly well founded; the only people to worry were the physicists who suggested that the Egyptian historical chronology was wrong. Libby (1963) himself made this proposal, but for once it was not the physicists who were right, it was the archaeologists—not perhaps a very frequent occurrence—and on this occasion the physicists have had to modify their dates. So the corrected radiocarbon dates for Egypt now agree with Egyptian historical chronology. When you correct the European radiocarbon dates they are set right back several centuries earlier. So that we now have megaliths in Iberia long before there were pyramids in Egypt, and we have megaliths in Brittany about 2000 years before their supposed predecessors in the East Mediterranean. So where does this leave the traditional picture of European prehistory?

First of all, the diffusionist theory that the megaliths originated in Egypt or elsewhere in the East Mediterranean simply will not work. There was no such diffusion. More important, however, now we observe a great vacuum in our thinking, for how should we *explain* the construction of such monuments, with or without diffusion? How do we explain them now? We are forced now to do something of which I think Elliot Smith would have approved—I am glad to be able to say this since we are not, I fear, in this Symposium saying many things of which he would have approved. He would approve because we are now beginning to think again in more general terms. We are beginning to talk again about culture process, about what causes civilizations to emerge rather than this or that specific civilization. We are not seeking a unique explanation, nor are we taking a "monogenetic-biologist" explanation; we are trying to simply look at the regularities in the patterns of development of human societies.

It is not clear what form our explanation will take. Shall we have laws of culture process, as some of our American colleagues like Lewis Binford advocate? (Binford, 1968: 26). I do not know. Shall we speak in terms of systems theory? Shall we try to demonstrate, by being specific about culture systems, how they grow, how they develop, so that we might simulate them with the aid of computers? I hope so. But the exciting point is that suddenly, when we look at the megaliths of Brittany and see many of them built there before 4000 BC, we find it relevant to look again at Polynesia and say, "Well look, there are analogous monuments there, there is no connexion at all and absolutely no question of diffusion between the two, and yet there are processes going on in Brittany that may in some ways be similar to those that went on in Polynesia."

I think the short answer to this question is that Elliot Smith, with his energy and enthusiasm, distracted us all by insisting that diffusion was an explanation when, in fact, it was merely an observation. We are left now, after 40 or 50 years of inconclusive and largely irrelevant discussion, with the task of deciding just how culture change can properly be explained.

PIGGOTT: Coming as I do at the end of the discussion I am going to be very brief because, of course, when one comes at the end one always finds that everything that one thought one might say has already been said.

I think, though, that I might just make one general point. It is a general point which has lain behind what several other people have said. When one is turning to a situation such as was presented to archaeologists and anthropologists by Elliot Smith, whether one says that he was producing an explanation or a non-explanation, but when he was at least producing something which looked like a general theory, why was this attractive to him? We have seen some of the reasons—Prof. Fortes mentioned some of them—why he might think in that way because of his training as a biologist and anatomist. However, I think it is interesting to ask why this theory was acceptable to him and to his immediate circle, those working with him, and why it was also acceptable in the wider non-archaeological and non-anthropological circles.

I think that to archaeologists it appeared to be satisfactory because, like all theories of this kind, it seemed to fit the knowledge of archaeology, particularly of European pre-history, as it stood at that time in the 1920's. In the same way, when one produces a forgery such as the Piltdown forgery which has been referred to earlier in the Symposium, this is always conceived within the terms of the thinking, quite apart from the technical skills, of the time. In that case the evolutionary thinking was such that that very odd combination of jaw and skull did not seem so improbable as it would to us now. Similarly, with our ignorance rather than our knowledge of European pre-history, it could look possible in terms of explaining a number of phenomena by means of diffusionism, if not actually deriving it from Egypt. We were in an age of chronological lost innocence, lost with the initial radiocarbon dating and its subsequent calibration. So here was, at least in a modified form as Childe was using it, something acceptable in terms of our knowledge at that time.

One of the reasons why it is not acceptable now is that the knowledge has increased, and it has taken on a different form. To the non-archaeological and non-anthropological public it was satisfying because it was simplistic, and a simplistic theory is one that always appeals. There is one simple solution to all the problems. The problems are ignored, or they are reduced to such a simplified form that a simplified answer appears to fit, and this is always popular.

Behind that, I think, is what was the prevailing knowledge in educated men's minds, whether they were scientists or in the humanities, about what might have happened in pre-history. Remember that Childe wrote a book (1941) which he called, *What happened in history*; it was, of course, what he thought happened in history, as it would be if anyone else were to write. But what did people think? How were they guided in their thinking about the probability of certain happenings in the past? I think that Dr Daniel's reference to Herodotus here begins to point the way. Herodotus, and the antiquity of Egyptian culture, this was something that came into Greek thought and, therefore, was a very important component of Western European thought afterwards, just as the Chosen People of Israel was one of the other great components in Western thought—here was a unique people from whom, in fact, the one true religion did by a process of diffusion attain its spread as seen by the 19th and 20th centuries. Then, after all, the classical background was not so far behind people as it is now. People thought of the past in terms of the establishment of the Greek colonies in the West, and of the establishment of the Roman Empire, and then of the discovery of America by Europeans. Let us remember, too, that Elliot Smith was an Australian, and a British Empire was a demonstrable fact in his day. What a contrast between the scientific culture represented by him and his colleagues in Australia and the aboriginal inhabitants who had received the blessings of civilization brought in from outside!

So there was a whole series of apparent models of empires and the diffusion of higher technologies and what were thought to be other higher things as well, like religions, being transmitted in this way. I think that this was an inescapable thing in educated people's minds. Such a view therefore, though perhaps not in its most developed hyper-diffusionist form, was yet more acceptable to people than it would be if anyone tried to put forward such a view today.

Conversely, some of the reasons why we are now thinking in very contrary terms, thinking in terms of culture change and of independent invention, is because we are more aware of this. We are in a period of cultural change ourselves, we feel the necessity of recognizing this and, therefore, this becomes a more acceptable set of explanations than it would have been in the past.

Therefore, perhaps Elliot Smith's ideas were accepted not just because of congruent archaeological and anthropological thinking, but as part of a general tendency to view the ancient world in some sort of a diffusionist model based on a series of antecedent historical situations.

JAIRAZBHOY: I did not know exactly what my status would be here so I have not prepared a paper, but just to introduce myself let me tell you what I am not. I am not a hyper-diffusionist, I am not Egyptocentric, and I think I am not a fringe lunatic.

I should like to make some comments about Elliot Smith's ideas relating to pre-Columbian contact with America. I have nothing to say about the Maya reliefs that are purported to portray elephants, because that has not been settled one way or the other, and one person's view is as good as another's. One says it is a macaw, and it could be either.

Elliot Smith further claimed that the Mayas received the winged sun disk and the manikin sceptres from Egypt (Elliot Smith, 1916). After half a century of excavation there has been no confirmation, and these are not found in an earlier context, but this may only be because of the accident of discovery.

Take the case of the pyramid of Palenque. It has a corbel vaulted gallery leading to a tomb chamber with sarcophagus, very clearly with Egyptian affinities, and yet its date is only the seventh century AD. So we would not even consider this because there was no ancient Egypt at the time. This argument could have been advanced only two years ago. However, we now have the pyramid at Totimehuacan in Puebla which goes back to the sixth century BC, and there is just such a gallery leading to several corbel vaulted chambers inside. This opens up the situation altogether. In one stroke we go from the seventh century AD to the sixth century BC. In his time Elliot Smith did not know of the Olmecs—they have only been discovered in recent times—and had he known of them he would no doubt have turned his attention from the one or two vestiges among the Mayas to the hundred and more evidences of Egyptian contact which it has been my good fortune to discover among the Olmecs. My discovery must be a great blow to the isolationists who have been advocating the independent origins of American civilization. What I should like to know is, how long will they continue to turn their backs on the evidence?

In view of the wide shades of opinion represented at the Symposium let me distinguish between three types of Diffusionists. The first could be described as the Rampant. For them the door is wide open, and in their most extreme form they regard all ideas as having diffused everywhere at all times. In my view the validity of this position (even in a more moderate form) cannot be ascertained for many generations to come. The next type I would describe as the Reluctant. The door revolves and they go in and out again, for though overtly they allow for the possibility of diffusion, they do not in reality admit any. Hence to all intents and purposes they are Isolationists. And finally for the third type the door swings on hinges— it opens and closes according to the amount of evidence to prove the case.

Isolationists are non-starters, because to prove independent invention is an almost impossible task. So they shift the burden of proof on to the Diffusionists. But why when the proof is at hand do the former retreat? There is something very much amiss if scholarship spurns the evidence because of an entrenched bias.

KRAUS & JOEL: (*Observations submitted after the meeting*) Dr Daniel's address on Elliot Smith's work in culture history was bleakly negative and lacking in understanding. That Diffusionism has been through the years and still is referred to by writers on human history (Joel, in press) indicates that there is more to the theme of Egypt and Diffusionism than meets the prejudiced eye. In briefly considering Elliot Smith's theories in the light of discoveries and changing ideas in archaeology and anthropology since his death, we hope to put Daniel's view of Elliot Smith's achievements in these fields into better perspective. They still have a great potential for the interpretation of human history further along the lines of his best known book, *Human history*.

Twenty years after Darwin's death, the birth of genetics and the rediscovery of Mendel's experiments ruled out his Lamarckian explanation of inherited characters, but this was not held against him to discredit his theory of natural selection. Likewise Elliot Smith's great vision of the origin and dissemination of civilization cannot be written off because archaeological discoveries since his death have complicated the picture. They have yet to destroy the fundamental point which the documented history of culture overwhelmingly sustains, that diffusion was, and is, almost always at least one step ahead of self-generation of culture. That the reverse applied in prehistory is still to be demonstrated.

Although much remained to be revealed by archaeology when Elliot Smith was writing, the likenesses then known between widely separated cultural features in ancient times (Elliot Smith, 1915) were as open to one interpretation as another; they could equally well be explained in terms of spontaneous generation or of diffusion. For Elliot Smith the evidence clearly pointed to the latter, and he made his case for a scheme for culture history parallel to Darwin's for organic history, demonstrating the continuity and inter-relatedness of all its manifestations. As there was for Darwin no spontaneous generation of species of organisms, so for Elliot Smith there was no spontaneous generation of species of culture— all cultures were ultimately related.

On one point which influenced Elliot Smith in his adoption of the theory of the diffusion of culture, his views were completely in accord with recent developments. At an early stage of his investigations he realized that the evolution of the human brain could not be called upon to account for the origin of independent yet parallel cultures. The evolutionary factor had therefore to be excluded if the origin and spread of culture in its more complicated aspects was to be adequately explained. Consequently Elliot Smith suggested for the first appearance of culture the operation of historical circumstances, and for its secondary appearances the mechanism of cultural diffusion—even though he was necessarily unaware that there had prevailed, in effect, evolutionary stagnation in the human species. This has been revealed by a re-examination of Darwinian natural selection carried out over many years, which, in the light of modern genetics, shows that not only has there been human evolutionary stagnation for several hundreds of thousands of years, but that brain evolution has actually regressed since the times of Neanderthal and Cromagnon man. The full evidence for this claim cannot be presented here, but the reader is referred to Kraus (1972b) for an outline of the basic ideas, which are expounded in full in Kraus (1973). The theory implies the exclusion of the evolutionary factor when considering the origin and multi-presence of similar cultural traits during the post-Neanderthal period. The only reasonable alternative explanations of their existence are historical accident and diffusion, as Elliot Smith anticipated 50 years ago.

Two aspects of Elliot Smith's theme gained wide publicity: (a) that techniques of cultivation and stock-rearing originated in Egypt and had been disseminated thence to provide the basis of food-production wherever it occurred, and (b) that the civilization which developed from this foundation in Ancient Egypt provided the pattern for civilization elsewhere in the ancient world (Elliot Smith, 1911, 1923). Both these views have long been under fire and are now widely condemned as being out of touch with the archaeological realities revealed during the last 20 years. But the evidence adduced against his ideas is not sufficient to refute them altogether, and they have strong claims to be regarded as potentially valuable lines of approach to still unsolved problems.

Elliot Smith believed that civilization first arose in Egypt because the environmental conditions there were particularly favourable. The timing of the Nile flood favoured the growth cycle of cereals and facilitated the

development of basin irrigation; animals coming to the river to drink
would be tempted on to the cultivations where they could be corralled
and eventually controlled and domesticated; the dry climate desiccated
corpses so prompting the idea of preserving the body and hence the
development of mummification; the narrowness of the cultivated area
and the sharp definition of its boundary by the desert, together with the
ease of intercommunication between regions up and down the Nile, were
conducive to a closely integrated and coherent society; and the flood's
regularity provided a natural clock as a basis for year-reckoning and so
for a calendar. This "model" is basically that adopted by Egyptologists
now for the emergence of culture in the Nile valley. (Aldred, 1961, Chs. 3,
4, esp.: 65–66; British Museum, 1964: Ch. 1; Elliot Smith, 1930: 285–292).
But on the basis of recently developed physico-chemical methods of
determining the age of prehistoric materials, sites unearthed outside
Egypt which appear to have food-producing economies and cultures
have been dated millennia before the earliest demonstrable dates for the
Egyptian evidence, and Egypt's claim to priority is now almost totally
rejected (Renfrew, 1970; Mellaart, 1965; but see contra, Clark, 1971).
However, two factors could upset this conclusion: Egypt's earliest culti-
vators' remains may lie out of reach below the Nile mud; and the validity
of the new methods of dating is not yet unequivocally established. Nor
do these precocious essays in cultivation outside Egypt appear to develop
into distinguishable civilizations (Mellaart, 1965: 77). Thus Elliot Smith
may have been wrong about agriculture starting *only* in Egypt but not
necessarily about its setting the prehistoric pattern which prevailed over
much of the world. Food production may have begun earlier independently
of Egypt but it was there that cultivation and animal husbandry eventually
flourished, fostered the growth of the ancillary features of early civilization
and set the pattern for the rest of the world.

The second controversial aspect of Elliot Smith's theories follows on
the foregoing, namely that prehistoric cultures, early civilizations and
the "primitive" cultures of modern times were directly or ultimately
inspired by the civilization of ancient Egypt. That this civilization was,
overall, technologically superior to other early civilizations is difficult to
dispute. There is increasing recognition of Egypt's wide ranging influence
in the east Mediterranean and in the interior of Africa. (Aldred, 1965:
48). Similarities in the pattern of kingship in western Asia and in Africa
are best explained in terms of a historical relationship with Egyptian
kingship, the most developed expression of the institution in the ancient
world (Hocart, 1927; Hooke, 1958). In the same way, diffusion from Egypt
is the most feasible explanation of widespread and similar techniques of
mummification and of megalith construction.

Elliot Smith's work on mummification and on desiccated predynastic
bodies in Egypt was an outstanding and unquestioned contribution to
cultural anthropology. It was reasonable to compare the custom as it
occurred in Egypt and elsewhere to ascertain whether there was an

historical relationship. Although critics invariably contested Elliot Smith's conclusions on general grounds, they rarely did so by technological comparisons. During 1962–63 Pretty (South Australian Museum, Adelaide) examined the mummies of the Torres Straits islands and found little to contravene Elliot Smith's results (Pretty, 1969). There is also more evidence now of the widespread association of mummification with royalty, which supports the theory that both the practice and the institution had a common origin in Egypt (Perry, 1935: 217, 219, 241; Coe, 1966: 144; Rice, 1957: 88).

Megaliths were another critical piece of evidence in Elliot Smith's theory of diffusion to which little weight has been given. Despite 60 years of intensive study by experts since Elliot Smith wrote *The origin of the rock-cut tomb and dolmen*, the understanding of these monuments is little nearer general agreement. The use of mammoth stone blocks for building, with the "know-how" for handling, etc., reached its apogee in Egypt. The arbitrary distribution of the practice in prehistoric Europe, with its implications of sophisticated funerary customs among peoples otherwise culturally poorly developed raises questions seldom faced. Elliot Smith's treatment of the problem merits reconsideration despite recent claims for the inflated antiquity of some European megaliths (Piggott, 1972: 297; Renfrew, 1970: 206).

The principle that world prehistory was a unity is implied in this diffusionist theory. Before the advent of radiocarbon dating, this was widely accepted, for the Old World at least, by British prehistorians. This principle is now in the melting-pot, but is not yet invalidated. Elliot Smith realized that Old World influence in the formation of pre-Columbian civilization in the New World was logically inseparable from this principle. Although his views on this were consistently ridiculed and disparaged during his lifetime, it is now accepted that the possibility of such influence must be considered in any examination of the problem of American cultural origins (Riley et al., 1971). Elliot Smith did not maintain that the pre-Columbian civilizations of America were necessarily directly derived from Ancient Egypt, but that the pattern of civilization set in Egypt did eventually reach the New World through many and various channels. Evidence which has recently come to light suggests there were even direct links across the Atlantic between Ancient Egypt and America. Striking parallels between Egyptian mythical and symbolic features and those of the Olmec culture of the coast regions of the Gulf of Mexico can, it seems, only be accounted for by postulating direct transatlantic expeditions in the time of Rameses III. These links have not so far been contested by the experts, either in published work, or verbally at the recent International Congress of Americanists in Rome (Kraus, 1972a).

In other fields, too, Elliot Smith's ideas may be due for re-appraisal. His speculations on early religion, the concept of "life-givers" and the part played by symbolism were never systematized and have been totally ignored since in the historical and comparative study of religion. Some of his lectures which were printed (Elliot Smith, 1919) could still have much potential for the understanding of both ancient religion and the results of field studies of so-called primitive religion.

One rarely recognized consequence of Elliot Smith's historical approach to so-called primitive cultures, their political and social organization, customs, beliefs and myths, is the claim that they are derived from higher civilizations, rather than representing "primitive" stages in the development of civilization. Such cultures represent rather a devolution from civilization than an evolution towards it. Modern social anthropology is increasingly inclining towards a historical approach to the understanding of such primitive societies and this trend could eventually lead to a position that has much in common with Elliot Smith's in these matters (Evans-Pritchard, 1961; Lewis, 1968).

The character of pre-civilized man was a fundamental link in Elliot Smith's chain of reasoning about the nature of the original impulse to civilization, and he claimed to have found pre-civilized "natural" man in the "food-gatherers" who exist or existed until recently in various parts of the world, peaceful "children of nature". He was often derided for being naive in his estimation of the ethnological and ethological significance of these peoples, especially when used as a foil to the claims of the school who held man, as well as Nature, to be "red in tooth and claw", naturally aggressive and warlike, and his civilizations flawed by his nature. For Elliot Smith, civilization had educated mankind in violence and war; these evils were inherent in civilization, not in man. Recent field studies of such food-gathering peoples as survive, as Turnbull's (1966) on the Mbuti pygmies of the Ituri forest in Central Africa, completely bear out Elliot Smith's diagnosis from the records available in his time. A new evaluation of this evidence might well result in a fundamental reorientation of the enquiry into the beginnings of civilization.

In summary, Elliot Smith's diffusionism was the yeast which leavened the dough of antiquarian prehistory 50 to 60 years ago, and it is still far from being exhausted. If Dr Christopher Hill's recent verdict on another adventurer in culture history may be borrowed, Elliot Smith's "errors, if errors they be, are more fertile than most people's truths".

DANIEL (CHAIRMAN): Professor Megaw has just made the journey which Elliot Smith himself made, coming from Sydney to an academic post in England, and he would like to say a few words.

MEGAW: Thank you very much, Dr Daniel. I do feel, as a temporary New South Welshman, having worked at the University of Sydney which produced both Gordon Childe and Grafton Elliot Smith, that I might just say a few words about the sort of thoughts that were going through our minds when, I think as a result of independent evolution rather than hyper-diffusionism from Lord Zuckerman, we in Sydney were also designing our own celebrations commemorating Elliot Smith.

If I remember my historical archaeology correctly, at the time we started our discussions about six months ago I had just finished working on an Aboriginal site first observed in 1796 by Matthew Flinders, the circumnavigator and grandfather of William Flinders Petrie, and was just on my way to Grafton, the birthplace of Elliot Smith, to study a group of Celtic coins originally found in the Le Catillon hoard on the island of Jersey (Colbert de Beaulieu, 1958)—again no result of hyper-diffusionism.

But seriously, hyper-diffusionism or simply independent thought, the same ideas were going through our minds in Sydney as have been so ably put forward in discussion by Professors Leach and Renfrew. In other words, what went wrong with Elliot Smith, the brilliant anatomist whose work is still revered in the anatomical field at Sydney? I think it was Professor Leach who expressed the same feeling that we had, that Elliot Smith made an hypothesis his deduction. In other words, diffusionism became his be all and end all, instead of simply part of the process which has been called hypothetico-deductive.

The second point which I think my anthropological colleagues in Sydney are quite right in making, or rather those prehistorians who are working in the anthropological field, is that Elliot Smith paradoxically did *not* work in the field. The one result which has come out of recent work in Australian prehistory is, of course, to show that the Australian Aboriginal, past or present, was not just a fossilized primitive technologist.

As a final postscript it is nonetheless true, 100 years after Elliot Smith's birth, that his observations on Torres Strait mummification, based partially on the mummy which is still preserved within one of the museums in the University of Sydney (Pretty, 1969), are being followed in some degree by the fact that Torres Strait is still one of the key areas where archaeologists today are looking for some of the answers to the development, the evolution or the diffusion of prehistoric culture in Australasia.

Again, just as a specious historical footnote, it is also perhaps interesting to note that at the time I am speaking another of my ex-colleagues at Sydney is following, seriously and with good results, we hope, the microlithic trail from Australia back to India (where I think we are all agreed megaliths did *not* come from), and she is making observations and comparisons which were first made by Bridget Allchin of Cambridge University (Allchin, 1966; Birmingham, 1969).

If I can end on a suitably commemorative note, I think that there is indeed much of praise as well as of blame which we can direct to the man whom we commemorate today.

Symp. zool. Soc. Lond. (1973) No. 33, 449–453.

CLOSING REMARKS TO SYMPOSIUM

S. ZUCKERMAN

University of East Anglia, England

In his opening remarks to this section, Dr Glyn Daniel (the Chairman) reminded me that I had put two questions to him when I suggested that we devote part of this commemorative Symposium to a consideration of Elliot Smith's views about the diffusion of culture. What I first asked was how far-reaching had been Elliot Smith's influence in this particular area of his interests; my second question was whether the views which he propounded have been invalidated by recent developments, and whether those who still uphold them, are justified in their beliefs. I knew Elliot Smith well, but I was far less knowledgeable about his efforts to promote his particular theories of the diffusion of culture than was Daryll Forde, of whom I saw a lot in my early days in London.

The illuminating discussion of this section has fully answered both my questions. Equally, the fact that we have just learnt that another Symposium in commemoration of the centenary of Elliot Smith's birth has recently taken place in Sydney is an appropriate reminder that the issue which concerned him so greatly is certainly still very much alive. Obviously, however, Elliot Smith would not have generated as much heat as he did on the subject of culture if he had chosen to focus his views about its origin and diffusion on some period after the emergence of the civilizations of the Middle East. Our situation is totally different from what it was in Ancient Egypt. We operate in a world in which inventions are patented. Had there always been patent laws, this discussion might perhaps never have taken place. We would have known when and if technological ideas were unique, or when they were merely derivative.

In my opening remarks of yesterday, I said that Elliot Smith had transformed the intellectual climate of his time. That he certainly did. This Symposium makes it apparent that he also helped to transform the intellectual environment in which we, too, have to operate.

We have reviewed three of his main interests—the general field of primate evolution, the question of man's ancestry, and the process of cultural evolution. Our consideration of the first of these fields made us immediately aware of the enormous advances since Elliot Smith's time in the methods we use to analyse the evidence relating

U

to primate evolution and to primate relationships. Yesterday revealed the almost inconceivable rate at which biometric statistical techniques are now developing—at any rate, to me all but inconceivable when I look back to the early days when Yates and Healey first collaborated in my own biometric studies of primate skeletal material. The need for objectivity in the analysis of primate skeletal material is providing a spur for the advance of the very statistical methods which anatomists now have to use in their work. And let me add the thought that if today's anatomists do not really understand the statistical methods which they want to use, if they merely use them as routine devices, they will turn out to be far more culpable than I was when I made an inadvertent error of $\sqrt{2}$ in the calculation of certain standard deviations of the dimensions of primate teeth some years ago, an error on which Le Gros Clark seized—in spite of the fact that it applied to both sides of the equation with which I was concerned and was therefore irrelevant to the general conclusions I drew—when trying to counter criticisms I was making of statements which were then being published about certain unique anatomical characteristics of the australopithecines, and which he, unlike myself, accepted and defended.

We have been shown the fossil skull of a small brained creature, the shape and characteristics of which were far more human than any australopithecine skull that has ever been found. The difference between the two types is so striking that one would be justified in saying straight away that the australopithecine creatures, with their cranial crests and enormous jaws, were overwhelmingly ape-like in their general appearance, and that the new type might justifiably be termed hominoid—a much misused term, I should add. This conclusion would also follow from the preliminary indications which were given about the post-cranial skeleton of the new fossil creature. All this means that for a period far longer than we have evidence for the existence of the australopithecines, there lived a primate from which, without resorting to undue imagination, one could suppose the human line sprang. This possibility was of course mooted long ago, and the fact that it has now been justified means that for more than 25 years anatomists and anthropologists—I am talking about physical anthropologists now—have been turning themselves inside out, persuading themselves and others that the obviously simian characteristics of the australopithecine fossils could be reconciled with the model of some assumed proto-human type. Over these years I have been almost alone in challenging the conventional wisdom about the australopithecines—alone, that is to say, in conjunction with my colleagues in the school I built up in Birmingham—but I fear to little effect. The voice of higher authority had spoken, and its message in due course became incorporated in text books all over the world.

Someone said during the Symposium that the Piltdown story had cast doubts on the respectability of physical anthropology as a science. Some things were also revealed about defects of reconstruction of certain australopithecine remains which can also hardly add to the respectability of the subject. But in my view what above all has denied the study of the palaeontology of the higher Primates the right to be regarded as a serious science is the fact that over the years *ex cathedra* pronouncements about what constitutes a unique human characteristic in a bone have usually proved a nonsense. My belief is that they will always do so.

It could well be that some feature or group of features in a fossil bone—maybe those having some definable mechanical significance—proves to be more like the corresponding features in man than in the living apes. Almost invariably other features in the same region would be likely to turn out far more ape-like than human. In combination, we end up with something that differs from both men and apes, and which would thus be unique. What conclusion does one then draw, one might well ask. Are we to suppose that the fossils are ancestral to one group, or to the other, or to neither? This is the kind of question people try to answer, but we have to recognize that it is at the same time the sort of question which is not amenable to any answer which would be scientifically final. Alternatively, would our findings mean that these creatures used this part of their anatomy more in the way the living apes do, or more in the way we do? This is a problem of bio-mechanics and is much more susceptible to solution. But even so, the paper by Zuckerman *et al.* (pp. 71–165) on the pelvic bone shows how difficult it is to obtain an answer even to this more circumscribed question.

Unless we manage to eliminate the aura of theatre from our subject, it will never, in my view, become a respectable science. Arguments about the precise point in time when some pre-human line of Primates set off on their journey to manhood may have an esoteric interest, but they have no particular intellectual value. What does it matter scientifically how far back one has to go in tracking man's ancestry in order to define the branch of the main trunk of primate evolution, where *Homo erectus*, and the apes and australopithecines parted company? It is certainly more than a million years back, something like a hundred times longer than the origin of settled agriculture, and of village life. In this field of comparative study, we must avoid sensationalism if we are to achieve respectability—and I regret that there was a little sensationalism in the course of the discussion of the opening section (pp. 63–69). There is no need to turn debates on human ancestry into a new item in the Olympic Games, with competitors saying "We have a hominoid or hominid fossil from Africa (or Asia) and it is much older than your fossil." This may be grist to the mill of the daily Press, but it only brings the subject of anatomy into disrepute. This sort of thing is not science. It can only be eradicated by a proper respect for objectivity.

In this section we have enjoyed a review of Elliot Smith's interests in the cultural field. We are all much more critical today than he was. In my introductory paper to this Symposium (pp. 3–21), I quoted from the preface of one of his last books in which he insisted that he had never claimed that the spread of culture was always due to diffusionism; but that what he did insist was that Egypt was the cradle of civilization. Today we all agree that culture can diffuse. As Professor Fortes has said, one cannot conceive of human beings without accepting the idea of diffusion. Our very language, the language of this Symposium represents a vehicle for the diffusion of culture.

You, Mr Chairman, have said that through the forceful way in which Elliot Smith developed and advanced his views about Egypt, he brought about a clash, but a clash which inevitably stimulated the reaction which has yielded us the greater understanding we have today. Earlier in the Symposium I said that Elliot Smith never rejected new evidence when it was put before him. Had he survived into the days of radio-carbon dating, or any of our other modern dating methods, he would never have continued to insist that certain things which we now know to be false were true.

Professor Piggott also wisely reminded us of the fact that when one makes a new observation it is inevitably prejudiced not only by one's personal experience, but also by current pressures. Elliot Smith developed his ideas in the intellectual environment of his own day. Our ideas belong to the intellectual environment in which we live. But we are fortunate in having at our disposal a far greater body of fact than was available to him.

Professor Fortes also referred to the superficiality of writings like those of Tiger and Fox, and of the books of Morris and Ardrey. I could easily add to his small list the names of some other so-called ethologists. But even though Elliot Smith, as we were told today, was also always searching for some simple kind of generalization by which to describe the nature of primitive man, even though he accepted Atkinson's theory about the primitive horde, I am quite certain that, as a scientist, he, like Professor Fortes, would have set about the modern species of ethological popularizers, and that he too would have rejected their speculations as valueless in a scientific context.

In his opening remarks (pp. 407–421) our Chairman rightly reminded me that I had omitted one sentence from a passage I had quoted from Elliot Smith's writings. I was conscious of doing so, but I refrained, not because I was unaware that Elliot Smith was often looking into a mirror when he castigated others for the obstinacy with which they defended views contrary to his own, but because this Symposium had been organized to do honour to his memory, and because the views he frequently expressed about convention and conversion in thought were no more than a commonplace.

There is one last thing I wish to say. I have noticed with sorrow that there are few anatomists taking part in this section of the Symposium. If I am wrong, then I would say that there are not as many here as were present for the earlier section of the Symposium. If Elliot Smith had still been alive, I do not believe that this would have happened. One of his great achievements was to broaden the subject of anatomy. His Department in University College was unlike any other Department of Anatomy in the country. He was able to diffuse one culture into another, as this Symposium has done. Whatever else, he was not a man with a narrow view.

This Symposium has succeeded beyond my expectations, not only in reviving the memory of Elliot Smith, but also in showing the enormous advances that have been made and the further advances that lie ahead in the fields in which he was interested.

It only remains for me to thank you all for having participated in this Symposium. That I do most sincerely, not only on behalf of the Zoological Society, but also of the Anatomical Society.

AUTHOR INDEX

Numbers in italics refer to pages in the References at the end of each article.

A

Adam, J. D., 396, *402*
Adam, J. P., 396, *401*
Adamo, N. J., 188, 190, *195*
Adams, H. R., 358, *371*
Adams, L. M., 157, *162*
Adamson, L., 183, *196*
Adrian, E. D., 237, *248*
Adrian, H. O., 239, *249*
Air, G. M., 351, 353, *369, 373*
Albe-Fessard, D., 188, *195*
Aldred, C., *421*, 444
Allchin, Bridget, *421*, 447
Allison, A. C., 380, *401*
Anden, N.-E., 177, *195*
Andrews, D. F., 272, 273, 275, *293*
Arambourg, C., 56, *62*
Ariens Kappers, C. U., 171, 172, 173, 174, 176, 178, *196*, 219, *229*
Ashton, E. H., 30, 34, *46*, 76, 77, 78, 81, 84, 85, 108, 148, 157, *160, 161*, 265, *293, 299*
Attenborough, D., 284, 285, *293*

B

Bagley, C., 189, *196*
Bailey, P., 242, *248*
Bakardjieva, A., 353, *371*
Baker-Cohen, K. F., 191, 192, *196*
Bard, P., 237, *250*
Barham, W. W., 146, *161*
Barnabas, J., 302, *333*, 345, 346, 347, 348, 349, 350, 351, 355, 356, *369, 371, 372, 299*
Barnicot, N. A., 358, *369, 370, 373*, 400, *401*
Bauchot, R., 316, *333, 335*
Bean, R. B., 104, *161*
Beard, J. M., 351, *370*
Behrensmeyer, A. K., 56, 58, *63*
Bellier, L., 395, *402*
Bender, D. B., 248, *249*

Benetto, K., 185, 188, *196*
Bennett, G. F., 385, *403*
Berkowitz, E. C., 183, 188, *198*
Berquist, H., 190, *196*
Bertler, Å., 177, 191, *196*
Biederman-Thorson, M., 188, *196*
Biegert, J., 104, 146, *161*
Binford, L. R., *421*, 439
Birmingham, Judy, *421*, 447
Biserte, G., 353, *370, 371*
Boggon, R. H., 236, *250*
Bolton, E. T., 341, 362, *371*
Bonin, G. von, 242, *248*
Boulanger, Y., 353, *370*
Boyden, A., 341, *370*
Boyer, S. H., 342, 351, 358, *370*
Bradshaw, R. A., 353, *370*
Bray, R. S., 381, *402*
Bremer, F., 188, 189, *196*
Brimhall, B., 351, *371*
British Museum, *422*, 444
Brodmann, K., 235, 236, 242, *248*
Broek, A. J. P., van de, 75, *161*
Broom, R., 30, 34, *44*, 51, *51*, 78, 79, 85, 86, 156, 157, *161*
Bruce-Chwatt, L. J., 379, *402*
Bück, G., 394, *402*
Buettner-Janusch, J., 342, *370*
Buettner-Janusch, V., 342, *370*
Burchfiel, J. L., 245, 246, 248, *248*
Burkett, M. C., 12, *21*
Burstein, A. H., 398, *403*

C

Cadigan, F. C., 384, *403*
Cain, A. J., 305, 306, *333*
Cairney, J., 173, *196*
Campbell, A. W., 235, 236, *248*
Campbell, C. B. G., 188, 193, *196*, 306, 313, 314, 328, 329, *333, 334*
Carney, J., 55, *62*
Cartmill, M., 256, 288, *293*

SUBJECT INDEX

A

acetabulum
  of man 76, 114, 123
    Australopithecus 79–80
    fossil hominid 31–4, 38
    Primate 84, 95, 99, 101–3, 111,
      114, 122–7, 132–3, 135, 137,
      139–40, 142–3, 145, 148, 150,
      262
    mammal 72–3
acetylcholinesterase
  distribution in birds and reptiles
    176–7, 191–2
Acheulian Industry
  evidence at Olduvai 38
acromion
  of Primates 265
Adapidae 288
Adapinae 330
Adapis 317, 329, 332
  A. parisiensis 322–5
Aelolopithecus 288
Africa 379, 381, 385–6, 395–7, 401, 404,
    452
  Central 446, 380, 385–6
  East 29, 34, 42, 44, 53–4, 253, 395
  North 50, 419
  South 15–16, 29, 34, 37, 44, 48, 163,
    386
  West 380, 382, 385–6, 389, 393, 395
agriculture 15, 409–410, 413, 416, 435,
    443, 451
alligator 188–9
Alouatta 83, 112–13, 115–17, 119–22,
    156, 264, 266–7, 344
  A. fusca 390
Alouattinae 127
Alpine man 434
Amani forest 395
amblypods 318
amino acids
  in Primates 339–42, 344–9, 353–7,
    359–60, 366, 369, 374–5
Amphipithecus 288

Amphibia 195, 336
amygdaloid complex 171, 176–7
Andaman people 431
ankle, see talus
Anomaluridae 395
Anomaluris fraseri 395
  A. peli 395
anthropoid apes
  anatomy 17, 301, 330–1
  classification 9–10, 12, 306, 313–4,
    451
  distribution 387
  malaria 377–85, 400–1
  pelvic girdle 78, 93–4, 105, 111, 114,
    118, 127, 132–3, 135–43, 150–1,
    153, 156–7, 159, 165, 262
  Piltdown mandible 24
  protein data 339–42, 356, 364–6, 374
Anthropoidea
  malaria 377
  pelvic girdle 76, 81, 107, 123–31,
    144–8, 152, 154–5, 158
  protein data 339–40, 358, 362
  shoulder girdle 265
antibodies 340, 344
antigens, protein 338, 343–4
antiserum 343, 361
Aotes, Aotus 83, 112–13, 115–17, 119–22,
    262–3, 264–9, 332, 344, 362–3,
    384, 390, 400
Archaeoindris 288
archistriatum 171
  reptiles and birds 171–3, 175–7,
    190–1
Arctocebus 82, 112–4, 115–7, 119–22,
    145, 264, 266, 363–4
arginine
  in Primates 358
arm
  movements of in apes and *Australo-
    pithecus* 157
artefacts 433–4
  at Olduvai 55, 61, 65
  E. Rudolf 58, 65, 69